Ecotourism Programme Planning

Dedication

To Sam and Jessie:
'Long may you run'

Ecotourism Programme Planning

D.A. Fennell

Department of Recreation and Leisure Studies
Brock University
Ontario
Canada

CABI *Publishing*

CABI *Publishing* is a division of CAB *International*

CABI Publishing
CAB International
Wallingford
Oxon OX10 8DE
UK

Tel: +44 (0)1491 832111
Fax: +44 (0)1491 833508
E-mail: cabi@cabi.org
Web site: www.cabi-publishing.org

CABI Publishing
10 E 40th Street
Suite 3203
New York, NY 10016
USA

Tel: +1 212 481 7018
Fax: +1 212 686 7993
E-mail: cabi-nao@cabi.org

A catalogue record for this book is available from the British Library, London, UK.

Library of Congress Cataloging-in-Publication Data

Fennell, David A., 1963-
Ecotourism programme planning / D.A. Fennell.
p. cm.
Includes bibliographical references (p.).
ISBN 0-85199-610-8 (alk. paper)
1. Ecotourism. 2. Heritage tourism. I. Title.
G156.5.E26 F46 2002
338.4'791--dc21

2002006107

ISBN 0 85199 610 8

Typeset by Columns Design Ltd, Reading
Printed and bound in the UK by Cromwell Press, Trowbridge

Contents

Preface

The idea for this book came as a result of what was viewed as a critical lack of information in the literature on ecotourism regarding the link between the service provider and the participant. This missing link or puzzle piece (perhaps there are many more missing pieces!) is the concept of recreation programming, which is a tool that allows for the enjoyment of intrinsically rewarding leisure experiences. Programming focuses almost exclusively on the element of service, which thus positions or situates forms of recreation, such as ecotourism, as experiences. Programming is a comprehensive process which allows for the incorporation of almost all information that either directly or indirectly affects the service provider–participant relationship. For example, these days there is a great deal of discussion on important aspects of ecotourism such as leadership, skills, certification, accreditation and planning. All of these have been considered in the field of recreation and leisure studies for some time, and may be central aspects of the programme planning process. As such, this is a book about the relationships that exist between operators and tourists, and about how providers can effectively plan and implement the most and best of their ideas through informative and educational programmes that allow for an appreciation of natural history and the natural world in general. (Note: In this book service provider, operator and entrepreneur are used synonymously to refer to a person who offers ecotourism programmes to participants; while participant, client and ecotourist are meant to be synonymous in referring to someone who participates in these ecotourism opportunities.)

Initially it was the intent of the author to apply the recreation programming concept to tourism in a broad sense. This certainly could have been accomplished. However, it was decided to focus on one specific form of tourism – ecotourism – because of the background of the author, but also because one form of tourism is more suitably managed. This allowed for a more in-depth treatment of the programming concept, as well as enabling the programming model to be stretched in its application to ecotourism activities, social and community issues, and the natural environment, all of which are central to ecotourism.

The sheer number of topics in this book precludes the ability to focus on these in a comprehensive manner. The intent was not such at the start of this project. However, it was soon realized that ecotourism programme planning can be a significant undertaking. The decision that had to be made was either to eliminate many of these concepts, or to be as inclusive as possible. In hindsight the author is pleased about the decision to be inclusive because it provided the opportunity to: (i) cover many of the most salient concepts in recreation programming; (ii) include a broad discussion of the setting for ecotourism (e.g. community development and many ecological systems); and (iii) tie together the literature from many different disciplines, including marketing, recreation and leisure studies, environmental studies and ecology. This approach also meant that a number of references were required to cover this breadth of topics. The author views this as an advantage for students and academics who are interested in ecotourism and programming. The hope is for the book to act as a springboard in intensifying research efforts in this area so that academics and practitioners may benefit from ongoing efforts to improve our understanding of many of the issues found here within.

Finally, in writing this book there is recognition and respect for the volume of work that has come before, and which has been so instrumental in the development of the recreation and tourism fields. As acknowledged, the book attempts to reformulate a long established process. However, there is an integration or synthesis of ideas from many social and ecological disciplines that helps to frame programming in a different context; one which is thought to be most applicable to the nature of ecotourism.

Acknowledgements

This book would have been much more difficult to write had it not been for the assistance of a number of people. I wish to thank the following students for their help at various stages of the process: Kaven MacLeod, Agnes Nowaczek, Val Shephard and Terri Gregotski. Furthermore, a number of faculty at Brock University, Department of Recreation and Leisure Studies, aided in shaping this book. These individuals include Dave Telfer, Atsuko Hashimoto, Paula Johnson Tew, Ryan Plummer, Sue Arai and George Nogradi. I thank them for their interest and support. Most of this book was written while I acted as Chair of the Department of Recreation and Leisure Studies at Brock. In my attempts to balance research and administration, I am indebted to Sandy Notar, the Administration Assistant of the Department, who freed up a great deal of my time in allowing me to engage this book. I would also like to thank Betty Weiler for passing along some much needed information, and Erlet Cater for her wonderful insight. I wish to thank CAB *International* (particularly Rebecca Stubbs) for their willingness to invest in this project, and for viewing this book as an essential aspect of ecotourism. In addition, I would like to thank Barry Griffiths of Quest Nature Tours and Carla Carlson of Niagara Nature Tours for sharing some important information on their respective ecotour companies. My trip co-leader, Gregg Campbell, is acknowledged for his ideas and enthusiasm, as are my parents who, from the beginning, nurtured in me a love of the outdoors (thanks, Jake, for your helpful editorial comments!). Finally, I wish to thank my wife, Julie, and children – Sam and Jessie – for their ongoing support and understanding, and for how they have allowed me to discover the things in life that are truly most meaningful.

1

Recreation and Tourism

> Although the outward rationale for tourism has as many variations as there are tourists, the basic motivation seems to be the human need for recreation
>
> (Graburn, 1989, p. 36.)

Introduction

In the quote above, Graburn recognizes the importance of recreation as perhaps the key motivation behind pleasure travel. This sentiment has been echoed by other authors at various times, including Jansen-Verbeke and Dietvorst (1987), and Edginton *et al.* (1980), who suggest that in analysing the leisure discipline from the perspective of the individual, there is little difference between elements of leisure, recreation and tourism. A focus on the individual is adopted in this chapter, indeed in the book as a whole, in emphasizing what appears to be a critical absence of discourse on concepts and practices that are common to recreation and tourism. This chapter sets the stage for a more in-depth argument of tourism-as-recreation through the examination of a number of models which have a common recreation and tourism root. This is followed by a discussion of recreation programming as an integral component of recreation planning, which is contrasted against the tourism product development mindset. In particular the aspect of service is emphasized in illustrating how recreation programme planning provides the necessary focal point from which to implement tourism services that would greatly enhance the tourism industry, as has happened in the recreation industry.

Leisure, Recreation and Tourism

Despite the sense of familiarity of leisure in everyday life (i.e. people seem to sense what leisure is and when they are experiencing it) it is a concept that has proved difficult to define. Objectively, leisure can be compartmentalized on the basis of time spent engaging in activities that are not associated with work. Subjectively, leisure may occur when somebody feels they are at leisure, despite the fact that others may be experiencing feelings of non-leisure even though engaged in the same activity. As such, leisure has been defined in a number of ways, but most often in the context of time, activity and experience. As a function of *time*, leisure is compartmentalized along with any number of actions engaged in during the course of the day (e.g. time that is free from obligations such as paid employment, taking care of one's home, family and self). As an *activity*, leisure refers to endeavours which may be physical, intellectual, communicative, imaginative, contemplative, creative or emotional (Kelly and Godbey, 1992). Finally, as an *experience* (e.g. feelings of satisfaction and excitement), leisure is explained on the basis of personal perceptions, cognitions and evaluations (Csikszentmihalyi, 1975). One of the most persuasive arguments in favour of the experiential approach to defining leisure has been put forth by Neulinger (1974), who proposed an attitudinal framework to leisure involving two principal agents. The first, perceived freedom (and most important), relates to whether or not the activity is freely chosen or whether it is somehow constrained. The second, intrinsic motivation, relates to the rewards that are generated from participation. These may be internal, and come on the basis of participation for its own sake (contributing to a state of leisure), or external such as to impress peers or to earn money (taking away from a state of leisure). The classic example of this 'leisure as a state of mind' approach is the person who, during work, is able to maintain a high level of freedom and control in their efforts, while at the same time being motivated from within to complete work-related tasks (live to work). From this perspective, this person's work takes on more of a leisure-oriented state, than another person who is constrained, for example, by a boss and time, and who does not possess a level or type of motivation that is internally derived (work to live).

The concept of recreation has been equally difficult to interpret, especially given the common usage of both terms. Recreation is said to fall within the realm of leisure such that while all recreation can be considered leisure, not all leisure can be considered as recreation. Smith (1990a) defines recreation as 'a pleasurable activity, which may be relatively sedentary, largely pursued for intrinsic motivation during leisure' (p. 253). In recreation, there appears to be more emphasis placed on the activity component of the experience, which thus provides the recreational practitioner with the ability to plan and develop programmes at a number of different levels and for a number of different participants (i.e. the view that recreation is *organized* leisure). The same can be said for tourism, which also shares many of the same characteristics of recreation, and more broadly, leisure, as stated earlier. This point is quite effectively illustrated by Metelka (1981) who envisions tourism as 'Free spontaneous activity; synonymous with recreation. An activity done for its own sake, rather than for economic gain' (p. 90). Subsequent to this Metelka (1990) defined tourism in three ways: (i) the relationship and phenomena associated with the journeys and temporary visits of people travelling primarily for leisure and recreation; (ii) a subset of recreation; that form of recreation involving geographic mobility; and (iii) the industries and activities that provide and market the services needed for pleasure travel (p. 154).

Tourism differs from recreation and the leisure state when it is undertaken for purposes of work. More specifically, Mieczkowski (1981) has suggested that non-recreational travel includes travel for business, family reunions, and health and professional development (see Fig. 1.1). The dynamic that exists between tourism, recreation, and leisure is effectively illustrated in the following quote from Murphy (1985), who writes at length on the work of Mieczkowski:

> Recreation falls entirely within leisure since it is an experience during free or discretionary time which leads to some form of revitalization of the body and mind. Part of this recreational activity takes place outside of the local community and as a result travel becomes an important component, leading this form of recreation to be classified as tourism. Tourism's orb extends beyond recreation to become associated with business trips and family reunions; and beyond leisure itself into personal and business motives for travel, such as health and professional development.
>
> (p. 9)

The number of recreation activities that people engage in is exhaustive. In recreation these activities are typically grouped according to shared characteristics. Such groupings include dance, arts and crafts, social recreation, mental and literary activities, music, hobbies, and sports and games. Recreational activities grouped as 'outdoor recreation' have much in common with tourism, and particularly ecotourism, based on participation in activities that occur in a variety of outdoor settings, such as botanical gardens, zoos, parks and so on, and the experience derived from such participation. However, despite this critical link in experience and activity, many outdoor recreation texts discuss tourism only in the context of its economic worth at local, regional and national levels, if at all. For example, Jensen (1995) writes that every state and territory is involved in actively promoting and manag-

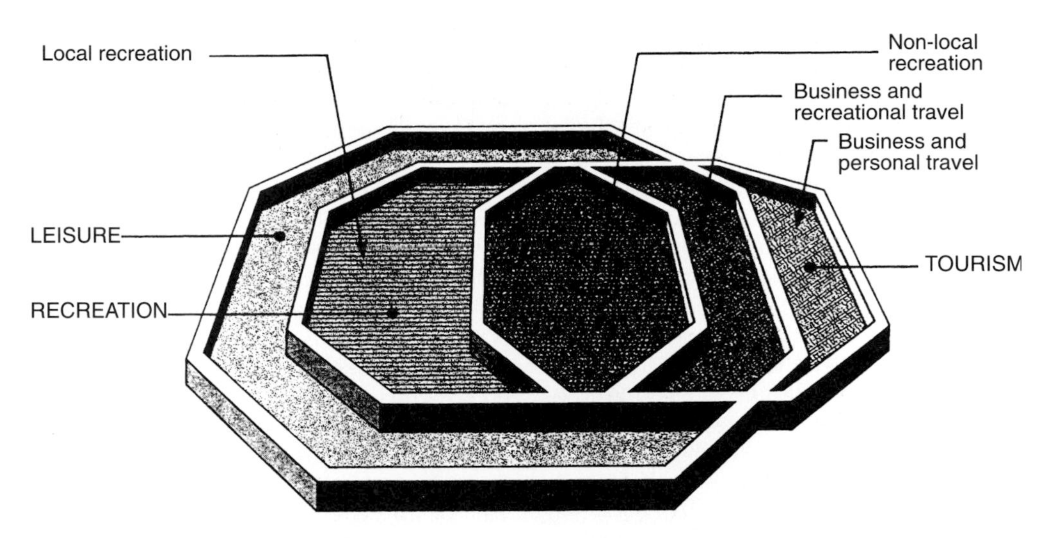

Fig. 1.1. Leisure, recreation and tourism. (Source: Mieczkowski, 1981, cited in Murphy, 1985.)

ing tourism. He suggests that the reasons for this are twofold: (i) increased revenue from taxes; and (ii) more jobs for residents. He goes on to discuss the receipts generated from tourism and a number of marketing case studies in the USA. Furthermore, in their discussion on economic impacts of outdoor recreation on local areas, Clawson and Knetsch (1978) and Knudson (1980) speak almost exclusively of tourism in the context of how money circulates within the community and who stands to benefit from the infusion of capital, despite the fact that very little effort was made to make a clear distinction between tourism and outdoor recreation, as in the following two quotes:

> Many towns and localities want more tourists or vacation businesses as a means of economic support for the community. And outdoor recreation has been suggested as a way of providing economic support to depressed rural areas.
>
> (Clawson and Knetsch, 1978, p. 230)

And in a section on 'expenditures on a typical vacation':

> Notions as to what happens to the recreationist's dollar are often sketchy. For a better understanding it is better to know: how much of the expenditure stays in the local area, however that is defined; who in the area receives it and who benefits most from recreation expenditures ...
>
> (Clawson and Knetsch, 1978, p. 232)

Crompton and Richardson (1986) support the notion that tourism has traditionally been thought of as a commercial activity as a result of its link to the private domain. In contrast, outdoor recreation has been viewed as a human/resource concern firmly rooted in the public domain. However, as Hall and Jenkins (1995) note, the polarization of public and private sectors has continually eroded in the western world since the 1950s. The resultant overlap between recreation and tourism is a topic in the work of Hall and Page (1999), who say that there is increasing convergence between the two concepts in terms of theory, activity and impacts. They further note that as society continues to change, especially in reference to government and government's role in tourism and recreation, the boundaries between public and private responsibilities in recreation and leisure change. Perceptually, this convergence is subject to debate as reported by McKercher (1996). He writes that the perceived differences between the tourists and outdoor recreationists occurs on a series of points along a continuum such that at one end tourism is said to be similar to all recreational uses of a park. At the other end, tourism is seen as an exploitive and conflicting park activity eliciting a significant amount of dislike by opinion leaders associated with national parks.

Disciplinary links

Tourism researchers have been preoccupied with developing ways in which to describe what is inherently a concept (tourism) that is intricately woven into the fabric of global life – economically, socially and environmentally – relying as it does on just about all levels of production and service. Definitions and models of tourism have thus ranged from those which are based on economics alone, to those which are inherently psychological, spatial or temporal. Some models incorporate many different variables. These 'whole systems' models typically deal with the movement of tourists from a generating region to a destination region, and back, and the associated industry that supports their activities in a variety of different environments.

Space – the movement of travellers beyond a predetermined spatial limit – seems to be the key criterion in distinguishing tourism from other forms of recreation. The importance of space as a defining characteristic of travel is reflected in definitions from countries such as Canada and the USA, who define a tourist on the basis of the distance travelled away from a traveller's place of residence or work (for example 80 km) (see Smith, 1990b). This text follows the guidelines established by the World Tourism Organization (WTO), which suggests that a tourist is a visitor who travels outside his usual environment for at least one night but not more than 6 months, and whose main purpose of visit is other than the exercise of an activity remunerated from within the country visited (see Var, 1992). Furthermore, tourism is 'the sum of the phenomena and relationships arising from the interaction of tourists, business suppliers, governments, and host communities in the process of attracting and hosting these tourists and other visitors' (McIntosh *et al.*, 1995, p.10).

Some of the earliest models describing outdoor recreation were conceived in the context of a broader spatial diffusion of people. Wolfe (1964), for example, was one of the first to examine the demand and supply aspects of recreation and tourism, as well as macrogeographic (national and international) and microgeographic (local and regional) travel patterns. He suggested that spatial imbalance was a key factor in determining the link between supply and demand, or rather the movement of people from places where they live to the places where recreational facilities are (Wolfe, 1964). (See also Plog, 1972; Hills and Lundgren, 1977; Britton, 1982; and Young, 1983 for examples of tourism models that are spatial in composition.) (As a point of interest, it is generally agreed (Mitchell, 1969; Pearce, 1979; Mitchell and Smith, 1985) that McMurry's article in the 1930 issue of the *Annals of the Association of American Geographers* on land use and recreation, was the first to recognize tourism as a distinct and significant form of land use.)

Although the movement of tourists from generating to destination countries on the surface may seem quite straightforward, in reality it can be rather complicated. With regard to adventure travel in Nepal, Zurick (1992) suggests that the tourist must enter and exit a complicated hierarchy of gateways. Tourists from the core generating regions move through an international gateway (semi-periphery), to a national gateway (periphery), and further to a regional gateway (periphery frontier), said to be the adventure region. Here, it is implied, traditional interests intersect with national development goals. This analysis, however, has other far-reaching implications. It is the international significance of tourism as an export agent that thrusts these frontier regions into the international marketplace. As frontier regions succumb to further intrusions, the uniqueness of these areas diminishes, as does the potential for travel to untouched areas in the future as these areas become fewer in number (see also Christaller, 1963; Turner and Ashe, 1975; Turner, 1976; Butler, 1980; Battisti, 1982; Plog, 1987; and Pearce, 1989 for a discussion of tourism and peripherality).

Another tourism-centred model derived from recreation is one developed by Klausner (1971), based on the premise that vacations, as a form of recreation, are fundamentally spatial, social and psychological (see Fig. 1.2). Spatially, Klausner suggests

Topographic relation	Direction of social movement	Psychological orientation		Illustration
Sedentary, open land	Centripetal	Active	1	Beach surfers
		Passive	2	Beach-front hotel guests
	Centrifugal	Active	3	Family camping on beach
		Passive	4	Stereotyped South Sea island paradise
Nomadic	Centripetal	Active	5	Teenage western summer tour
		Passive	6	Cruise ship
	Centrifugal	Active	7	Hiking
		Passive	8	Sightseeing
Sedentary, inner land	Centripetal	Active	9	Boy scout summer camp in mountains
		Passive	10	Country hotel guests
	Centrifugal	Active	11	Lone cabin in woods
		Passive	12	Private serviced lodge in mountains

Fig. 1.2. Typology of vacation forms. (Source: Klausner, 1971.)

that travel can take on one of three forms, or combinations of these types: (i) it may be 'Sedentary, Open Land' (deserts and beaches); (ii) it may be 'Sedentary, Inner Land' (forests); and (iii) it may be 'Nomadic', which appears to be an intermediary form between the two other poles, and where the process of travelling itself is the main motivation, rather than the destination. Krausner describes the social motivations of travellers on the basis of two distinct perspectives. The first, centrifugal, suggests that people want to move away from the centre and thus all other groups of persons (e.g. such as to visit a private serviced lodge in mountains). The other extreme, centripetal social relations, is a condition in which the individual actively seeks out others for a more gregarious experience. The third general dimension described by Klausner is the psychological energies of the vacation, which may be defined as active or passive. The active form is characterized by giving-out, expanding, and developing competence in relation to the environment; while the passive is typified by relaxing, taking-in and submitting or merging with the environment. Not surprisingly, Klausner suggests that his model is merely based on hypothesis. Nevertheless, it stands as an interesting perspective on how to classify a broad number of outdoor recreation and tourism activities.

The inclusion of a number of variables in describing tourism and tourists was also undertaken by Mathieson and Wall (1982), who emphasized the dynamic travel component and the implications of the stay within the destination. These elements are further elaborated upon through a consequential element, which underlines the various effects or impacts on the economic, physical and social subsystems with which the tourist is directly or indirectly in contact. This model quite nicely illustrates the characteristics of tourists and those of the destination that have the potential of creating the frequency and severity of impacts within a region. The framework also articulates the controls that may be employed to mitigate such effects, including finance, management strategies and policies, information carrying-capacity guidelines, and engineering controls (see Fig. 1.3).

One of the most oft quoted models of tourism and outdoor recreation research is

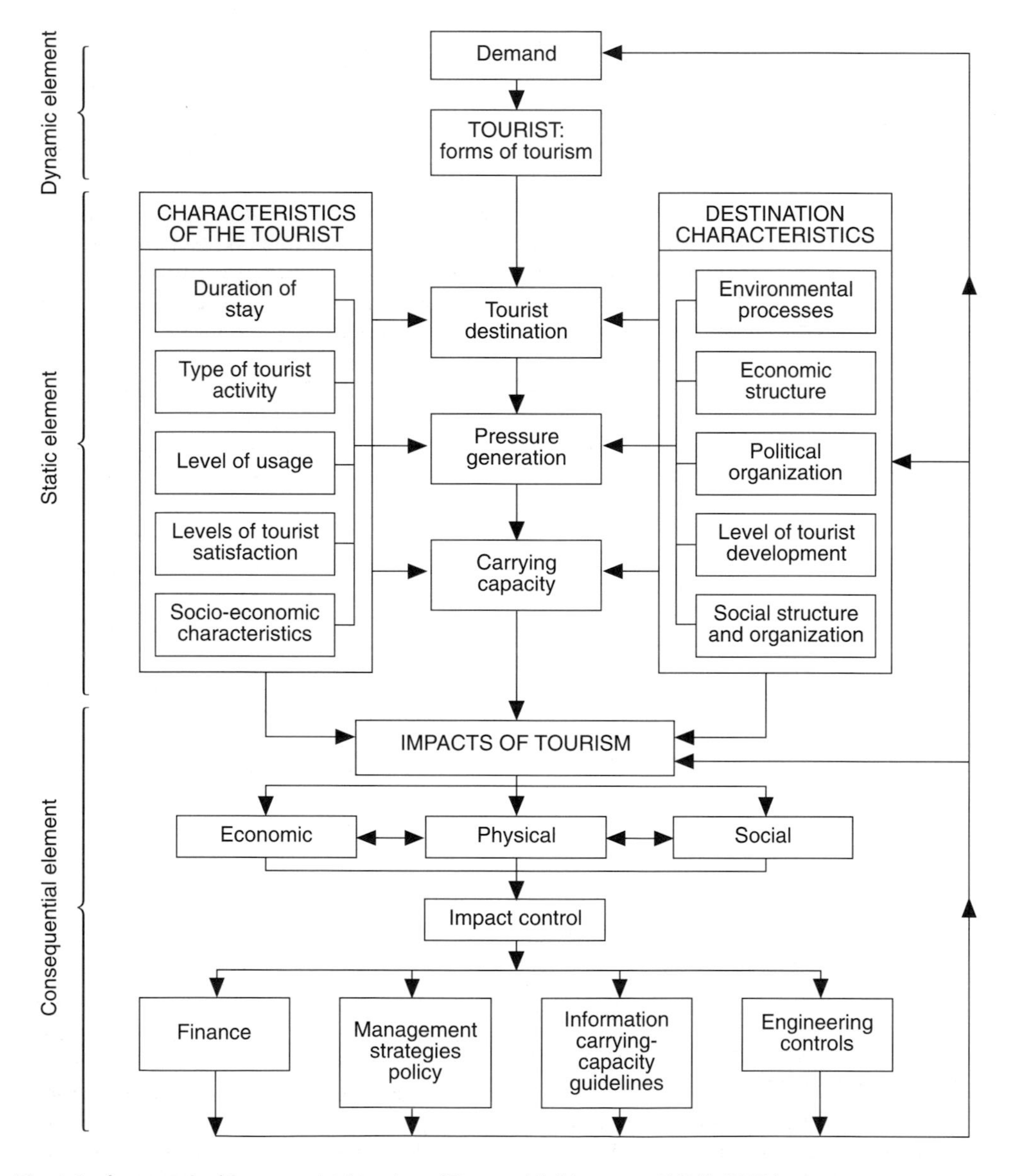

Fig. 1.3. A conceptual framework of tourism. (Source: Mathieson and Wall, 1982.)

one proposed by Clawson and Knetsch (1978), who examined the five major phases of the outdoor recreation experience. The uniqueness of the model rests in its movement away from a focus based solely on the tourist site itself, to one which is inclusive of a complete tourism experience. These authors argued that the experience begins with anticipation of the event and planning. If a positive decision is made based on anticipation and planning, the experience progresses to travel to the site itself. Depending on a number of factors, including the duration and timing of the trip, such travel may be a positive or negative experience. The third stage of the model, on-site experiences, makes reference to the combination of any number of recreational activities or events that occur within the destination region(s) itself. The fourth stage involves travel back to the generating region, which is reported to be a different experience by the authors on the basis of criteria which may include the route of return, the pace, and the frame of mind. The final stage of the experience involves recollection, which may be strong and lasting in the case of an enjoyable on-site experience, or dim and impressionless in reference to a dissatisfying experience.

The experiential–spatial aspects of tourist travel as developed by Clawson and Knetsch have opened the door for a number of different approaches to analysing the movement of tourists between regions. For example, Fennell (1996) suggested that travellers moving from core areas to peripheral ones experience a number of perceptual changes (following from the work of Zurick, as outlined above). The outward movement of tourists to the most peripheral areas sets in motion a different mindset which may be attributed to a number of different factors. And while extensive travel over great distances has influence on experience, it is this spatial aspect in combination with the penetration into peripheral and marginalized areas that may dramatically contribute to changing perceptions about space, place and time. These may include:

- *Distance from home.* This occurs in combination with other factors. That is, a vacation to Sydney, Australia, from rural Canada represents a large distance, but does not necessarily represent a periphery.
- *Familiarity with the destination.* A lack of knowledge about an environment or culture may add to the uniqueness or authenticity of travel and therefore represent a perceived periphery.
- *Unscheduled change.* Movement that occurs spontaneously, or that is not preplanned (e.g. from a regional periphery to the regional outer-periphery).
- *Psychological and/or physiological change.* Personal adaptations that have to be made in order to feel comfortable or maximize feelings of satisfaction or pleasure (culture, language, personal hygiene, food, shelter).
- *Distance from amenities.* This is a function of the previous factor, but reinforces the notion that a particular peripheral region does not, cannot, or will not offer goods and services that can be found at regions closer to the core.
- *Adaptation.* As tourists move out to the new periphery, their psychological state changes. Perturbations such as inadequate transport, fewer shops, poor facilities, etc., that are found at regions closer to the core, are accepted in regions of the perimeter. The state of the infrastructure plays less of a role in satisfying the tourist. With distance away from the core, comes a sacrifice of personal needs. Poor weather may also not pose a problem because a more 'frontier-like' attitude is adopted. (Apart from tourism where tourists bring along with them elements of home, such as might be found – the home environment – on a cruise liner.)
- *Population density.* Regions with a lower tourist and local population density may provide an indication that the periphery has, or is being reached.
- *Authenticity.* As this density decreases, the authenticity and/or incidence of natural and cultural attractions may increase in relation to the characteristics of the region. (The character of the region is important, as Costa Rica may have a higher species diversity in urban areas than Shetland has at the perimeter.)

- *Scale of attraction.* It is the region itself that may take on characteristics of 'attraction' in the outer-periphery or perimeter. For example, the capital city of Shetland, Lerwick, has many different types of small and large attractions. However, areas that are just as large spatially may have relatively few attractions and because of this, as well as the nature of its limited infrastructure and small size, may be an attraction in and of itself (e.g. such as the island of Foula off the coast of Shetland).
- *Symbolism.* The periphery or perimeter may be represented by experiential, environmental or cultural phenomena (symbols). Such phenomena include lochs, mountains, wind, solitude, barrenness, a rainforest or lack of banking machines. Consequently, the more these symbols are experienced, the greater might be the feeling of reaching the perimeter. Fear may also act as a symbol, as may the realization that there is a lack of knowledge of a region and its characteristics. Fear or anxiety of the unknown is an important feature of adventure tourism (characteristics that Csikszentmihalyi, 1975, has reported on regarding risk-taking in rock climbers).

Tourism Products

Kotler *et al.* (1996) write that a product is anything that may be offered to a market to satisfy a need or want. Usually a product is thought of as a physical object, such as a car or watch. Product however is seen as more than simply tangible items. For example, marketers use the expression goods and services to distinguish between physical products and intangible ones: goods are produced and services are performed. Tourism providers largely view their offerings as goods in the sense that they are able to stockpile them for use again and again (except in the case where they are offering one-off 'experiences' over the course of a day or weekend). Conversely, tourists view the product more as a service in the way in which experiences are immediately consumed on site, without the opportunity to take home any tangible product associated with such an experience. Consequently, the key issue between good and service seems to be the notion of 'stockpiling'. The tourism-as-service perspective is one advocated by Holloway (1994), who notes that tourists cannot inspect their package tour in the same way they can inspect a washing machine before the purchase. It is therefore a speculative investment involving a good deal of trust on the part of the purchaser. Furthermore, certain aspects of the tourism experience such as hotel rooms, meals and aeroplane seats, can be services in that they cannot be stockpiled. Once left empty, they cannot be used. Tourism products may be more easily understood through the following nine points, as outlined by Bécherel (1999):

- *Intangibility.* It is not actually a tangible 'thing' that is purchased, but rather some type of experience.
- *Perishability.* Things such as hotel nights and airline seats cannot be stockpiled for future use. If they are not used, they are lost forever.
- *Inelasticity of supply.* Tourism services are inelastic because they do not adapt well to changes in short- and long-term demand.
- *Elasticity of demand for tourism products.* On the other hand, demand for tourism services reacts very quickly to events and changes in the environment such as crime, recession and changing fashions.
- *Complementarity.* The tourism product is just not one single service, but rather a number of complementary sub-services, all interconnected and affecting each other both positively and negatively.
- *Inseparability.* Production and consumption take place at the same time. The tourist must be present when the service is performed to consume it.
- *Heterogeneity.* It is virtually impossible to create two identical tourism services. There will always be a difference in the service (hotel room, tour, etc.) at one level or another.
- *High fixed costs.* Initial costs of providing the basic elements of the tourism product such as accommodation, transport and so

on, is very high. As such, heavy investment is made without guarantee of return.

- *Labour intensity*. It involves the mobilization of employees who are well trained and prepared to work towards generating positive outcomes. In many cases, therefore, there is a high staff to client ratio to ensure positive experiences.

Interesting is the notion that products supplied by industry providers are not necessarily the same as those products perceived to be in demand by consumers. Whereas tourists are likely to have a more global view of the tourism product they are buying, individual suppliers, as outlined by Bull (1991), are more prone to segregating what they do away from what might be termed the overall 'industry'. To this end, there are a number of sectors that are generally accepted as comprising the tourism industry. These include: (i) carriers; (ii) accommodation; (iii) attractions; (iv) travel agencies; (v) food and beverages; (vi) tour operators; (vii) merchandisers; and (viii) public or private sector support services. However, because of this breadth between and among sectors, serious questions have been raised as to whether the tourism is an industry or rather a series of loosely attached sectors that accommodate the needs of tourists. Smith (S.L.J., 1998) asserts that tourism is not a conventional industry, but can be measured, from a supply-side perspective in a way that other industries are measured. In a long-standing debate, Leiper (in Smith, S.L.J., 1998) contends that tourism is best viewed as a mix of industries that vary in their involvement of the distribution of services to visitors. This is a perspective advocated by Tucker and Sundberg (1988) who write that 'tourism is not an industry in the conventional sense as there is no single production process, no homogenous product, and no locationally defined market' (p. 145).

Tourism Programmes?

Tourism and recreation may be differentiated, at least at one level, on the basis of the former's stronger tie to economic development. Philosophically, there is nothing wrong with such an emphasis. However, it is indeed unfortunate that too much of a focus on the financial aspects of tourism subsumes what is a very diverse human phenomenon, which should be defined at least equally on its behavioural and experiential characteristics. With this perspective in mind, it is the *purpose* of this book to provide a stronger link between the recreation literature (as well as related topics from marketing, ecology and resource management) and ecotourism through a focus on the interaction between participants, service providers and the environments in which they interact. The vehicle used in drawing this link is recreation programming, which acts as a common thread – a philosophy and methodology – enabling recreation professionals to deliver their programmes to a vast array of different populations in different settings. The book thus draws on a rich history of very good interpretations of the programming process in recreation, including the work of Russell (1982), Carpenter and Howe (1985), Edginton *et al.* (1998) and DeGraaf *et al.* (1999), and applies these to the concept of ecotourism.

Ecotourism programmes may greatly benefit from the recreation programming approach, which clearly recognizes the individual as the most central aspect of service provision. As inferred previously, it is unfortunate that this service orientation is at times neglected by those responsible for the planning, development and management of tourism. The literature supports a vast array of tourism development examples that discuss the value and benefit of tourism almost solely on the basis of economic impact. The same holds true for ecotourism, where there is an ever-expanding base of literature that questions the interests and values of those involved in the business side of ecotourism (see Munt and Higinio, 1993; Steele, 1993). The author has seen the focus on financial ends too frequently, especially at local and regional tourism planning meetings where certain groups are clear (perhaps too clear!) that the only reason for investing any resources (time, money or people) into tourism is to make money. To these groups there appears to be little interest in exploring tourism as a human experience that can be maximized to its fullest potential through appropriate planning, development and management.

The importance of a service orientation is evident in the perspective adopted by DeGraaf *et al.* (1999) in their servant leadership approach to recreation programming. These authors write that the service nature of the leisure organization, as well as the increasing demand for leisure services, cannot be ignored. They suggest that people are collecting experiences (services) where once they collected things (which would mean goods, based on our previous discussion). Programming, then, relates to the methods and philosophies used by service providers to help ensure that recreationists are able to exceed their expectations. To the author, the focus on programmes means a focus on participants; whereas the term products appears to have much more of a goods connotation. Indeed the programme concept has started to creep into the tourism literature. For example, the Department of Conservation (2000) in Western Australia has implied that programmes and products are essentially the same thing, as outlined in the following quote: 'Before developing the product (the ecotour or interpretive activities program), establish who the target audience is and what sort of experiences they might seek' (p. 19). It has been suggested time and time again in the literature that ecotourism is a form of tourism that is ethical, responsible, small scale and community-based. Consequently, this type of tourism may be useful as an instrument to underscore the human, service, values, elements which are central to the development of sound recreational programming (and thus a de-emphasis of the financial elements which are so frequently examined in this field). In addition, tourism employees coached in proper techniques of programming may be witness to the many tangible benefits which are generated for the programme and service provider, for other employees, and most of all, for the client. Finally, programmers, who may include programme planners, guides, interpreters and other service providers, are often required to 'wear many hats', through duties that typically range from administration to front-line interaction. Such a vantage point allows for a clear understanding of the importance of service to clients as well as to the broader goals of the organization.

Conclusion

At the outset of this chapter it was suggested that apart from the element of spatial movement, and associated perceptions connected to this movement, there are many commonalities between tourism and recreation, especially from the experiential standpoint. It is interesting, however, that this sentiment has not always been widely embraced. For example, a US Department of Commerce report on the state of the art in recreation and tourism suggested that while the 'recreationer frequently needs to acquire new skills, or new abilities in order to engage in a particular form of recreation ... Tourism ... does not normally involve a large measure of physical exertion, nor does it involve the acquisition of new skills' (cited in Edginton *et al.*, 1980). In the year 2002 it would be difficult to argue this point to the tourist on the Mayan Trail in Central America; the tourist visiting Antarctica; the tourist actively seeking to witness a rare insect in Borneo; or tourists making the trek to Cape Point in South Africa. More than ever, tourists want to live their travel experiences in the same way they live their recreation experiences. This means education, skill development, exercise and self actualization. It also means that travel needs to be programmed in the same way that recreation experiences have been for years. In order to maximize the experience of the tourist, a number of on- and off-site elements must be planned in a systematic way. These processes are considered in the pages that follow. It is also worth mentioning that although this book deals more specifically with ecotourism, the approach to programming used is such that it would apply to other forms of nature-based tourism, including adventure tourism and cultural tourism, but also more widely to include many other forms of tourism and outdoor recreation.

2

Foundational Aspects of Ecotourism

A crag is not just a rock, it is a living community of plants and animals. Who decides to climb it becomes part of it. He must judge how much of that community of which he is a part must die to satisfy his venture. None of it? Some of it? All of it?

(Wyatt, 1988)

Introduction

This chapter places ecotourism into the context of the broader literature on the land ethic and environmentalism. Examples of the growing level of research on ecotourism and sustainable tourism are included, along with a number of variables that have been used to define such terms. Ecotourism has emerged so quickly and with such intensity that it has done so without the development of widely accepted definitions and policies. The corollary of this is the evolution of an industry that has vastly different views in relation to ecological, social and economic spheres. The chapter also examines the ecotourist through a survey of past research, along with the concept of specialization, which may help to further scrutinize different types of tourists and ecotourists. Although decidedly ecologically based in nature, a number of philosophies, theories, and approaches are discussed in this chapter which may provide a basis from which to construct parameters and, in so doing, provide a framework to enable researchers and practitioners to better understand this elusive concept in attempts to conceptualize their own programmes and policies more effectively.

Ecotourism

What we view as ecotourism today is a relatively new phenomenon that has generated a significant amount of interest as well as controversy. Ecotourism has links to the ecodevelopment literature which began to appear in the 1970s through the work of Miller (1976, 1989) and others. Nelson (1994) writes that during the 1970s researchers were becoming increasingly dissatisfied with governments and industry who were too excessive and exploitative in their development policies. In reality, however, ecotourism has been with us for some time, in many regions around the world and represented through a number of different types of activities. For example, safari tourism in Africa has been around for decades, and Blangy and Nielson (1993) write that the American Museum of Natural History has conducted natural history tours since the 1950s. Both are examples of tourism activities and programmes which in all probability contain many of the practical and philosophical elements we attach to ecotourism in the 21st century.

While the genesis of the term itself is presently under debate, the literature points to Ceballos-Lascurain as the individual responsible for coining the term in 1983 (see van der Merwe, 1996). However, there is evidence that suggests that the term may have been coined as far back as 1965 in an article by Hetzer which appeared in *Links* magazine. This article was later reprinted in *Ecosphere*, a news bulletin of the International Ecology University. Part of the reprint is as follows (Hetzer, 1965, p. 1):

> We propose a program for the creation of competitive, UN-sponsored 'parallel organizations' whose role it would be to actively design, promote, and implement RESPONSIBLE ('alternative') TOURISM projects fulfilling our four main requirements:
>
> 1. minimum environmental impact
> 2. minimum impact on – and maximum respect for – host cultures
> 3. maximum economic benefits to host country 'grassroots'
> 4. maximum 're-creational' satisfaction to participating tourists.
>
> Tourism, if it fulfills AT LEAST the above requirements, can be a healthy and rewarding activity for the visiting tourist, an economically sound investment for the host area, and an environment-CONSERVING feature – an ecological tourism ('Eco-Tourism').

An early example of the use of the term in practice dates back to the 1970s, as suggested by Fennell (1998), who found evidence of Canadian government ecotours developed specifically for the Trans Canada Highway. These ecotours were initiated as a result of agreements between recreation and conservation branches and as part of a series of corridor studies emerging at that time. These joint efforts between government departments were designed to provide an enhanced sense of human–land relationships on the basis of a series of interpretive guides which reflected the range of different ecozones found along the course of the highway. Mathieson and Wall (1982) also discuss ecotours in the Canadian tundra as tourism programmes which were viewed by these authors as extending environmental appreciation through tourism. Ecotourism's emergence perhaps coincides or follows literature on the broader alternative tourism paradigm which underscores the dissatisfaction with conventional methods of tourism development and the social and cultural dependencies inherent in mass tourism (Dernoi, 1981).

Irrespective of ecotourism's etymology, the concept has generated a great deal of interest as reflected in the development of a very extensive base of literature (see Page and Dowling (2002) for a good overview of the growing base literature on ecotourism). The interest in ecotourism is not simply an academic phenomenon, but also a favourite in the popular media, as one can scarcely leaf through a magazine or major newspaper without encountering some type of ecotourism reference or advertisement. Some of these articles extol the virtues of ecotourism, while others seek to expose a vulnerable underbelly through case studies or situations which have been misrepresentative of the ecotourism 'ideal'. Although difficult to project, it appears as though the interest in ecotourism will continue at least for the near future, as new ecotourism industries, academic programmes, and associations of one form or another continue to surface in response to a level of demand that appears to be unprecedented.

Exploring the boundaries of ecotourism

McLaren (1998) writes that ecotourism is a multifaceted concept which includes a number of different forms of tourism such as nature travel, adventure travel, birding, camping, skiing, whale watching, archaeological digs, and so on, which occur in a number of environments. Hers is an example of a more comprehensive approach to ecotourism, and one quite frankly that is probably more realistic than many critics would like to admit. She feels that ecotour programmes cover a variety of experiences, 'from spartan, hard-core, bury-your-own-poop backpacking in special conservation zones to the purely hedonistic, luxury vacations at typical resorts' (McLaren, 1998, p. 97). Contrary to this standpoint, is the feeling that ecotourism is much more confined according to a number of clearly articulated tenets (see Burton, 1998; Fennell, 1999). In view of these two positions, researchers have tended to place ecotourism into the context of a broader nature-based tourism, in an effort to differentiate ecotourism, adventure tourism and cultural tourism (see Tisdell, 1996; Ewert and Shultis, 1997; Fennell, 2000b) from other types of tourism.

Such categorizations are not new, nor are they particular to tourism. For example, Jensen (1995) wrote that certain outdoor recreation activities can be viewed along a continuum from those which are resource-oriented to those which are user (activity)-oriented. The first end of the spectrum includes activities such as fishing, nature study, camping and wildlife viewing, which depend on natural resources and occur in natural settings. The opposite end of the continuum includes those unnatural or modified settings and the performance of activities or the witness of activities (e.g. sports, or a dramatic presentation).

However, even within the natural-resource based tourism sector (the resource-oriented sector as outlined by Jensen), there appear to be significant differences between various forms. One of the first to examine the relationships between various types of what might be considered nature tourism, broadly defined, was Graburn (1989). He identified two main types of tourism, nature tourism and cultural tourism, and compared and contrasted these through a number of other related sub-types. Cultural tourism was subdivided into historical tourism (museums and cathedrals of Europe) and ethnic tourism (visiting the Lapplanders of northern Scandinavia). Conversely, nature tourism included ecological tourism (e.g. wildlife viewing) and environmental tourism, which is further subdivided into recreational tourism (recreational activities in natural environments) and hunting and gathering tourism, which might include a hunting safari in Africa. Graburn felt that ethnic tourism forged a link between cultural tourism and nature tourism because of the link between culture and the natural world. In other related research, Ewert and Shultis (1997) distinguish between ecotourism, adventure tourism and indigenous tourism, as all falling under the heading of resource-based tourism. In addition, Reynolds and Braithwaite (2001) developed a conceptual framework of wildlife-based tourism, which is characterized as a form of tourism based principally on interactions with wildlife. It is a form of nature-based tourism that involves both non-consumptive activities (e.g. specialist animal watching and habitat specific tours), as well as those that are more consumptive in nature (e.g. hunting/fishing tours, and thrill-offering tours).

The relationship between adventure tourism, cultural tourism and ecotourism has been examined by Fennell (1999), who referred to these as ACE tourism (Fig. 2.1), as represented by the acronym involving each of the three tourism forms (all of which have been referred to as ecotourism in one form or another or at one time or another). He suggests that the overlap between these various forms has become much stronger in recent years to the point where operators, governments and other involved parties have treated them as being synonymous. Depending on the programme, its setting and other situational factors (e.g. accreditation, certification, regulations, and so on), ACE expands or contracts to represent different concentrations of adventure, culture and ecotourism. The conceptual framework illustrates that these three forms of tourism are often not mutually exclusive, but rather unique by virtue of the various concentrations of culture, adventure and nature. Consequently, ACE tourism evolves as a function of the setting in which it occurs, and other aspects including the market and competition from other sources. With the tremendous growth of ecotourism programmes

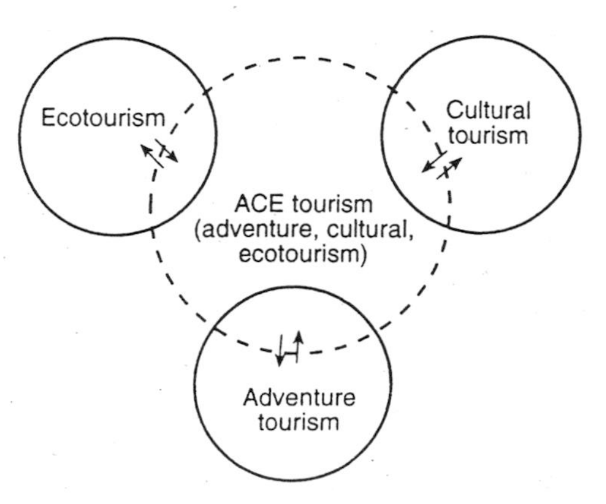

Fig. 2.1. ACE tourism. (Source: Fennell, 1999.)

around the world, this diversity in programming may be a function of many variables, which has necessitated the development of particular niches to satisfy the growing demand for culture, adventure and ecotourism experiences. The question remains, however: are these diverse programmes (i.e. those which include adventure, culture and ecotourism) ecotourism? (See also Blamey, 2001; Weaver, 2001.)

Important to this discussion is the feeling that not enough is being said, or written, to differentiate effectively between what can be very different forms of tourism (in reference to ecotourism and nature-based tourism). Burton (1998), for example, has argued that 'it is clear that strict ecotourism is a far cry from the everyday tourism activities that occur in a natural environment and that are known as nature-based tourism' (p. 756). Citing Griffiths (1993), she argues that ecotourists have distinct perceptions and beliefs that relate to the experience, and which are probably far removed from other forms of nature-based tourism. These may include some of the parameters established by Buckley (1994) who suggested that ecotourism was based on four main dimensions, including a nature base, support of conservation, sustainable management and environmental education; as well as those established by Fennell (2000b) who said that various forms of natural-resource based tourism differed on the basis of consumptiveness, natural resource values, technical skill orientations, and the basis of learning.

Consumptiveness was at the heart of a recent debate on billfish angling (fishing for marlin and sailfish) as ecotourism. Holland *et al.* (1998) argued that on the basis of the catch-and-release philosophy, economic support for resource conservation, and social responsibility to the local economy, among other factors, angling qualifies as a form of ecotourism – a status normally reserved for non-consumptive outdoor recreationists. In response to their work, Fennell (2000c) noted that fishing should *not* be viewed as a form of ecotourism because of the intention to entrap the animal, the pain which results from the entrapment, the consumptiveness of the activity (despite the catch-and-release philosophy discussed), and differing values and behaviour of anglers as compared to ecotourists. Fennell also suggested that although ecotourists at times place stress on an environment, through trampling of grasses, or trail impacts, that such effects were largely inadvertent and subject to management. Not so with fishing, where the main intention is to catch fish with the probable outcome of harming or killing them. It was noted that despite the angler's best intentions to eliminate stress on the animal, one can only do so to a point after which he or she must cease to pursue and capture the animal. As such, despite the fact that the fish were released back into the environment, there appears to be something intuitively wrong with fishing as ecotourism. The basic premise behind the assertions made by Fennell was to suggest that weak definitions of ecotourism are easily modified to fit a whole range of activities, which may be less than consistent with the ecological tenets (outlined below) of ecotourism – written or unwritten. While the jury is still out on this debate, it is nevertheless an interesting comparison and worthy of further scrutiny.

The issue involving consumptive vs. non-consumptive uses of wildlife/resources becomes more complicated in view of the value of the particular form of tourism to an economy. Theorists can argue at length over the philosophical importance of one form of use over another. However, to some tourism organizations and governmental decision makers the answer is a simple one. Whichever one pays the bills, is the one which will most likely continue to garner support. Such are the implications put forth in the work of Snepenger and Bowyer (1990), who examine differences between non-resident tourists making consumptive and non-consumptive use of Alaskan wildlife. In general, the authors write that due to the organization of the hunting and fishing industries in Alaska, tourists were able to contribute funds directly to the conservation of these species through licences and federal funds via multiple federal acts. No such mechanisms were in place for the non-consumptive tourists, contributing perhaps to less of an acceptance in the context of the social, economic and environmental 'culture' of Alaska.

Unfortunately with so many operators offering so many programmes in so many different settings, with little or no regulation, some academics have been severely critical of the industry and how it has evolved (see Wheeller, 1994). However, the critics do not just come from academic circles but also from the popular media. There is now a proliferation of articles that question the philosophy of ecotourism, and many of the tactics used by ecotourism operators (Brady, 2000):

> Its proponents portray eco-tourism as taking the high road to travel – as bringing inquisitive, sensitive, adventurous tourists into contact with unspoiled nature. Steady growth worldwide is forcing the business to confront its own troubling issues of sustainable development.
>
> (p. E3)

Many ecotourists and ecotourism theorists would suggest that there still exists a fundamental disjointedness in the ecotourism industry, which has allowed many to wrongfully use the 'eco' label. In many cases the tenets of ecotourism have been severely compromised by operators, with regional and national governments and international organizations and with little ability to employ corrective measures. In theory and in practice, the industry must get away from the belief that everything and anything can be ecotourism, as long as the programme has some tie (this is quite debatable in many cases) to the natural world, however defined.

Defining ecotourism

Like tourism, there is no shortage of ecotourism definitions. The infusion of these various interpretations has added to the mysticism surrounding the term, what it represents, where it occurs, who it applies to, and under what conditions. In a recent study involving 85 definitions of ecotourism, Fennell (2001a) isolated 20 keywords which have been used to define ecotourism in the past, along with the actors involved in developing the definitions, and the time period in which the definitions were developed. The words that were most frequently cited, of these 20, included: reference to where ecotourism occurs, e.g. natural areas (62.4% of the 85 definitions), conservation (61.2%), reference to culture (50.6%), benefits to locals (48.2%), education (41.2%), sustainability (25.9%) and impacts (25%). Most definitions were developed between the years of 1991 to 1993 (n = 25; 29.4%) followed closely by the period of time from 1994 to 1996 (n = 23; 27.1%). In general, conservation, education, ethics, sustainability, impacts and local benefits were better represented in the more recent definitions, showing the changing emphasis of ecotourism over time. Interesting was a stronger focus that governments placed on conservation, education and benefits to locals, as compared to the definitions written by individuals (e.g. researchers). Conversely, individual-based definitions more strongly emphasized ecology alone as a basis for definition (instead of in concert with socio-cultural and economic variables) as well as impacts. While these are the predominant trends, some definitions were unique by virtue of the move away from many of the most common variables included in such definitions, to emphasize terms such as stress, stewardship and spirituality.

In view of the results of the aforementioned study on definitions of ecotourism, it is defined in the context of this book as:

> An intrinsic, participatory and learning-based experience which is focused principally on the natural history of a region, along with other associated features of the man–land nexus. Its aim is to develop sustainably (conservation and human well being) through ethically based behaviour, programmes and models of tourism development which do not intentionally stress living and non-living elements of the environments in which it occurs.

This definition is a reflection of the fact that, at its root, ecotourism is based on the natural history of a region. The definition, however, leaves open the possibility of a cultural component, but only in as much as this cultural element is tied to the natural history or natural environment of the region. For example, delegates at an ecotourism workshop in 2000 in South Africa, representing a number of southern African nations, were in agreement that any definition of ecotourism had to be integrated with the cultural environment, but only in certain cases, because of the interrelations between environment and people. Although this tie is contestable, there is little question that a number of examples exist globally where the intricacies of the human–land relationship (i.e. the manner in which the natural world has influenced how people live) may help to interpret the extent, diversity, niche and so on, of a particular species or indeed, of an entire ecosystem.

Ecotourism is said to be intrinsic in the manner in which people value ecotourism, as an activity and as a source of income, *for its own sake*. This includes a respect for ecology but also for human agents who are involved in the experience (J.R. Butler, 1992, unpublished). It is participatory in the way that ecotourists should be involved in a first-hand (Utah Division of Travel Development, cited in Edwards *et al.*, 2000) and sensory (Fennell, 1999) experience, as opposed to some other level of involvement (e.g. helicopter tour over a city, which is more of an 'detached' experience). It is also education-based in the sense that there is the external push (from operators) and internal drive from the ecotourist to learn from his or her encounter with the natural history of the region visited. Although the level of involvement in the learning experience varies as a function of the programme and the willingness to learn, the focus is on education of the environment, as opposed to other motives like technical skill development, such as the skills needed to climb a mountain.

The definition views sustainable development in a slightly different context, from how it is normally viewed. While there has been strong debate as regards the emphasis on the aspect of development in sustainable development (a debate which is too extensive to include here), *develop(ing) sustainably* switches the focus from 'development', which is said to have too much of a tie to conventional models of development, to sustainability. In doing so, development perspectives in ecotourism may be more sustainably driven (or preservationist or ecocentric) and, as a result, act as a model for other forms of tourism and other sectors of the economy.

The concept of ethics is thought to be an important one, especially because the field of ethics has generated a great deal of interest in other disciplines such as environment, medicine and business. By virtue of its prime directives, stated here within, ethically based ecotourism initiatives could be an example to other types of tourism development, which may thus benefit not only tourists and operators, but also local people and the natural world. The concepts of stress and abuse are felt to be critical, especially these days when so many types of tourism programmes are actively using the ecotourism label. Some of the more consumptive activities, e.g. fishing, should not qualify as ecotourism because of the nature of the activity and the stress put on individuals of a species, despite the fact that whole populations of a species are, in resource conservation terms 'healthy'. Finally, the focus is not on particular environments in this definition, but rather on the aspect of natural history. This gets away from the belief that ecotourism can only occur in large natural areas, despite the fact that ecotourism occurs most often in these areas. For example, service providers may develop ecotourism programmes in urban areas, which are regions that have not typically been developed for ecotourism in the past.

The Ecotourist

Conceptually, we have known for some time that ecotourists are not a homogeneous group, but rather a group which differ across a number of different dimensions. One of the earliest papers to illustrate this was one written by Laarman and Durst (1987), and others from North Carolina State University, who examined ecotourists from hard- and soft-path perspectives, and along dimensions of nature-related interest and physical rigour. To these authors, hard-path ecotourists were those who were typically in a related discipline (e.g. botanists), and who spent significant amounts of time in the field in less than ideal conditions. On the other hand, the soft-path ecotourist was characterized as an individual looking for more of a general, casual experience, with less of a desire to stay in the field for extended periods of time. In other research, Lindberg (1991, p. 3) emphasized the importance of time, experience and travel, in the delineation of four basic ecotourist types:

- *Hard-core nature tourists.* Scientific researchers or members of tours specifically designed for education, removal of litter or similar purposes.
- *Dedicated nature tourists.* People who take trips specifically to see protected areas and who want to understand local natural and cultural history.
- *Mainstream nature tourists.* People who visit the Amazon, the Rwandan gorilla park, or other destinations primarily to take an unusual trip.
- *Casual nature tourists.* People who partake of nature incidentally as part of a broader trip.

In one of the earliest studies to examine ecotourists empirically, Fennell and Smale (1992), based on the work of Fennell (1990), used the Canadian Tourism Attitude and Motivation Study (CTAMS) (a database of the general attitudes, motivations and benefits of Canadian travellers) to compare a sample of Canadian ecotourists travelling to Costa Rica, against the average Canadian traveller. In general, ecotourists were found to be more active and adventuresome in their activity choices, while the average Canadian sought benefits that were more sedentary and family oriented. Furthermore, ecotourists were found to prefer outdoor attractions such as wilderness areas, parks and protected areas, and rural areas, while city- and resort-based attractions were more important to the average Canadian traveller. (See Kretchman and Eagles (1990), and Eagles (1992) for studies which employed the same research design in different regions, and found similar results.)

In revisiting this research 10 years later, Fennell (2002) wrote that it was unfortunate that studies on the motives and other behaviour patterns of ecotourists have either been one-off studies of a particular region, or have used the same methodology but in a number of different world regions. Missing from the list of studies undertaken under the aforementioned conditions are studies that employ a temporal or longitudinal analysis of ecotourists within one geographical region. With this in mind, Fennell examined Canadian ecotourists in Costa Rica using the CTAMS methodology as reported above, the same tour operator, as well as time as a key variable. The intent was to examine if ecotourists had changed in their benefits sought, motivations, and attitudes towards nature and natural resources in Costa Rica over a period of 10 years. In general, very few differences were found between the two groups regarding the importance of attractions and benefits, although 10 years apart. Differences were found in the average age of respondents, with the more recent group 10 years older on average; in the gender make-up, with females predominating; and, not surprisingly, in the price of a trip. Fennell speculated that the continuum which exists between hard- and soft-path ecotourists could be viewed as an inverted triangle (Fig. 2.2), as explained below:

> the specialised hard path sector is relatively small in comparison to the much larger and softer end of the market. The further up the inverted triangle, the greater the movement away from specialisation, expectations, and time spent with regard to the ecotourism experience … There is also an associated move away from activities and attractions that are strictly natural history based – the further one moves up the triangle – towards a range of activities which are much more diverse in nature, including general interest activities, adventure related activities, and cultural activities. Accordingly, the natural history realm represents a much smaller proportion of the overall mix of activities participated in for softer path ecotourists at the top of the triangle, than for hard path ecotourists at the base of the triangle.
>
> (p. x)

The figure also illustrates that, in order to accommodate larger numbers of tourists, there is an associated reliance on built or modified environments, such as hotels, transport, facilities or other required ameni-

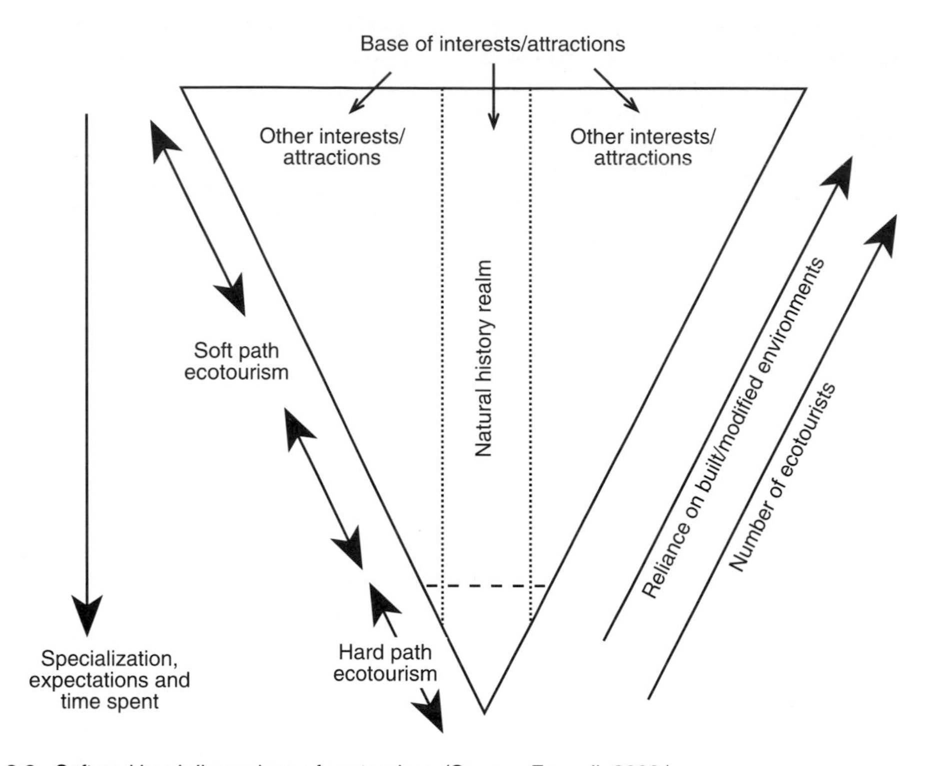

Fig. 2.2. Soft and hard dimensions of ecotourism. (Source: Fennell, 2002.)

ties. It further notes that the ability to achieve sustainability may become more of a challenge, in situations demanding increases in the tourism plant (i.e. the top of the inverted triangle).

Other recent, empirically based studies have also been helpful in allowing researchers to better understand the unique nature of different types of ecotourists. For example, in a cross-cultural study on the motives of Costa Ricans, North Americans and Europeans visiting five protected areas in Costa Rica, Wallace and Smith (1997) discovered that Costa Ricans felt that a number of trip-related motives were more important than did other sample groups. Such motives included seeing wildlife, enjoying nature, supporting nature conservation and studying nature. The authors clustered visitors into five groups on the basis of similar motivations. These included 'nature exclusive', 'social and nature oriented', 'autonomous and nature oriented', enthusiastic generalists' and 'less passionate'. The first group was characterized by their extreme focus on studying and enjoying nature, to the exclusion of other experience outcomes (see also the work of Hvenegaard and Dearden (1998) in Doi Inthanon National Park, Thailand, which identifies five different types of tourists on the basis of conservation interest and involvement, levels of interest in ecotourism, and other trip characteristics).

As stated earlier in this chapter, ecotourism programmes cover a variety of experiences, from Spartan, hard-core minimal impact camping, to the purely hedonistic luxury vacations at four-star resorts. The motivations for such travel obviously differ considerably between these two experiences, but not necessarily between sites. In their study of visitor characteristics and attitudes towards Kibale National Park, Uganda, Obua and Harding (1996) found that ecotourists were quite diffuse in their motivations within the park. Although wildlife viewing was participated in most often by participants, a whole range of other activities were participated in including nature walks, photography, 'the views', bird watching, camping, hiking, relaxing and picnicking (although these latter four were ranked far lower). Chadwick (1995) reports that even in Africa, there may be differences in the types of ecotourists and the things that they want or don't want. Citing the case of Botswana, Chadwick writes that there is a sub-set of ecotourists who may be termed 'game connoisseurs'. These individuals have already been to East Africa, with the hustle and bustle of getting a first glimpse of a lion, but would rather go to Botswana on subsequent trips, which offers these travellers an opportunity to get a bit deeper into the ecotourism experience. This ecotourism 'specialist' is examined in more detail, below.

Specialization

The concept of recreational specialization, as outlined by Bryan (1977), is based on the premise that just about all members of a particular form of outdoor recreation (e.g. anglers, cavers, canoeists) were traditionally viewed as one homogeneous group with little variation among individuals classified as such. For example, Bryan examined trout fishermen as an example of a 'recreational specialization', defined as a continuum of behaviour from the general to the specific. At one end of this continuum is the individual who has a general interest in an activity, while at the other end is someone who devotes significant amounts of time to an activity, especially to some special branch of the sport. Bryan (1977: 178) identified the following typology:

- *Occasional fishermen*. Those who fish infrequently because they are new to the activity and have not established it as a regular part of their leisure, or because it simply has not become a major interest.
- *Generalists*. Fishermen who have established the sport as a regular leisure activity and use a variety of techniques.
- *Technique specialists*. Anglers who specialize in a particular method, largely to the exclusion of other techniques.
- *Technique-setting specialists*. Highly committed anglers who specialize in method and have distinct preferences for specific water types on which to practise the activity.

Bryan subjected these various types to a series of questions related to equipment preference, species preference, water preference, management issues, angling history, social setting, distance travelled, vacation patterns and leisure priorities. In general the following conclusions were drawn from his research:

- Fishermen go through a predictable syndrome of angling experiences, usually moving into more specialized stages over time, with more commitment shown.
- The most specialized fishermen have in effect joined a leisure social world – a group of fellow sportsmen holding similar attitudes, beliefs and ideologies – engaging in similar behaviour, and having a sense of group identification.
- As level of angling specialization increases, attitudes and values about the sport change. Focus shifts from consumption of the fish to preservation and emphasis on the nature and setting of the activity.
- The values attendant to specialization are inextricably linked to the properties of the resource on which the sport is practised. As level of angling specialization increases, resource dependency increases. What appeals to the resource specialist is a resource setting allowing for predictability and manipulation, a degree of control so as to be able to determine the difference between luck and skill.

In a subsequent paper Bryan (1979) established that certain activities may be further sub-divided on the basis of specialization required. Water sports, for example, could be sub-divided into row boating, powerboating, canoeing, rafting and white-water canoeing. Each of these occur along a continuum on the basis of skill, commitment and involvement, such that row boating requires much less of these three factors than white-water rafting. As Ewert (1989) suggests, a change in specialization is coupled with a change in motivation. What motivates the rower is considerably different from what motivates the canoeist.

The concept of recreational specialization has been applied to other activities. For example, Hamilton-Smith (1981) used it to examine caving. Hamilton-Smith suggests that the boundary between caving as recreation and speleology as scientific study is a rather diffuse one. This research identified four levels of caving on the basis of level of skill and level of equipment. Category A includes tourists and outdoor education visitors. Guides, provision of equipment, marked trails and lighting are essential aspects for this experience. Category B includes casual visitors and low-skill cavers. Management resources are rejected (rails, lighting, etc.), and accessibility and ease of traversing caves is important. The third category, C, includes medium-skill cavers and speleologists. In this group, management resources are rejected, accessibility is somewhat important, and difficulty in traversing caves becomes more important. The final category, D, includes highly specialized cavers and speleologists. Setting preferences (remoteness) are very important, management resources are rejected (but not for researchers), and very specific characteristics are sought after (e.g. zoologists after bats). This need to classify cavers prompted the development of the following set of relationships between type of cave, the visitor, and environmental impact (Table 2.1).

Benefits of ecotourism

Driver *et al.* (1991a) write that the term 'benefit' has been used historically to prove that an activity, programme or other function has lead to some economic gain. As such, benefits and costs have been viewed as mechanisms by which to gauge the success of outdoor recreation ventures, for example, usually from an economic context. Archer and Cooper (1999) corroborate this usage of the word from the tourism perspective in suggesting that early work on the impact of tourism upon destination areas focused almost exclusively on the financial bottom line. This approach was adopted, say Archer and Cooper, not only because economic benefits or impacts were more easily quantifiable, but also because it was important to demonstrate to local communities that the huge transforming developments within their regions were actually benefiting host communities (and thus the implied assumption that they were worthwhile).

The meaning of the term benefit, however, has expanded considerably to consider the impacts or effects which recreation and travel have had on individuals and groups, beyond the focus on economic efficiency. The focus is on identifying the positive outcomes of participation, with the purpose of demonstrating to agencies (principally government) why recreation is such an important part of our lives. As funding cuts continued to constrain service providers in recreation, it became increasingly important to provide data to support the notion that people were in fact benefiting from their participation in such activities. This benefit-as-improved-condition, as outlined by Driver *et al.* (1991a),

Table 2.1. Relationship between a proposed classification of caves and specialization categories of cave visitors.

Proposed cave classification	Category of visitor	Assumed ecological impact
Limited access caves	Specialist speleologists and high-skill cavers	Minimal
Wild caves	Generalist speleologists and medium-skill cavers	Low
Adventure caves	Low-skill cavers, casual visitors, outdoor education visitors	High
Show caves	Tourists	High, but confined largely owing to the actions of management in providing necessary resources for visitors

Source: Hamilton-Smith (1981).

'refers to a *change* that is viewed to be advantageous – an improvement in condition, or a gain to an individual, a group, to society, or to another entity' (p. 4). The link between the policy maker and move away from the cost:benefit calculation is emphasized as follows (Driver and Peterson, 1986, p. 5):

> The noneconomic measures also help policymakers understand better those benefits not indexed by the economic measures, such as any spin-off (merit good or external) benefits to nonusers or to society at large, preservation/environmental benefits that might be realized in the future but are outside the sovereign consumer's economic decision calculus, and benefits of sustained yields management and attributable directly to flows of goods and services over a particular planning period.

In general, the benefits that are derived from recreation and leisure include those that are psychological (reducing depression and managing stress), social (empowering communities to develop more recreational facilities), economic (good physical fitness leads to higher job efficiency) and environmental (greenways which conserve plants reduce pollution) (for an extended list of benefits see Canadian Parks/Recreation Association, 1992). An example of the benefits-based approach to participation in outdoor adventure pursuits is outlined by Ewert (1989) in Table 2.2.

The benefits-based approach is one that could be very useful in ecotourism research, especially as a means by which to understand better how ecotourists compare against other types of tourists and outdoor enthusiasts. Given the nature of ecotourism, perhaps benefits would also include those related to education, in addition to ones outlined by Ewert. There might also be less of a focus on some of the sociological variables identified by Ewert, such as group cooperation and communication, for an activity which is less reliant on the group pulling together physically and emotionally.

Driver *et al.* (1991b) and his colleagues set out to identify a series of human needs that could be gratified by participation in leisure pursuits. Through a number of adjustments, these authors developed the Recreation Experience Preference (REP) scales. Since the early 1990s, over 40 REP scales have been developed to measure the benefits of leisure experiences. One of the key features of the REP, over other leisure scales, is its focus on outdoor recreation environments, and hence its applicability to ecotourism. This can be seen in the following examples of REP scales and the preference domains in which the scales are grouped.

1. *Enjoy nature.* Scenery; general nature experience.
2. *Share Similar values.* Be with friends; be with people having similar values.
3. *Outdoor learning.* General learning; exploration; learn about nature.
4. *Escape physical stressors.* Tranquillity/solitude; privacy; escape crowds; escape noise.
5. *Achievement.* Seeking excitement; social recognition; skill development.
6. *Risk Reduction.* Risk moderation; risk prevention.

Some of these domains have been discussed in the ecotourism literature, especially those related to the enjoyment of nature and outdoor learning, which correspond to the literature on ecotourist typolo-

Table 2.2. Benefits of outdoor adventure pursuits.

Psychological	Sociological	Educational	Physical
Self-concept	Compassion	Outdoor education	Fitness
Confidence	Group cooperation	Nature awareness	Skills
Self-efficacy	Respect for others	Conservation education	Strength
Sensation seeking	Communication	Problem solving	Coordination
Actualization	Behaviour feedback	Value clarification	Catharsis
Well-being	Friendship	Outdoor techniques	Exercise
Personal testing	Belonging	Improved academics	Balance

Source: Ewert (1989).

gies (see Laarman and Durst 1987; and Lindberg, 1991). Furthermore there is evidence to support the fact that some dedicated ecotourists and adventure tourists appear to prefer the company of like-minded or similarly skilled individuals in their experiences ('Share similar values'). For example, Ewert (1985) found that beginner climbers were more extrinsically motivated to participate in mountain climbing activity, whereas experienced climbers did so for intrinsic reasons.

Other domains of the REP, i.e. 'Achievement/stimulation' and 'Risk reduction', may also provide a theoretical ground from which to analyse the differences that exist between adventure tourists and ecotourists. Although there is a strong intuitive sense that ecotourists are in fact different from adventure tourists (see Dyess, 1997) along the lines of motivation and benefits sought, there is really only anecdotal evidence explaining how they differ. While the literature on adventure pursuits focuses on risk, skill and competence, Fennell (2000b) suggests that natural-resource based tourists, such as adventure tourists and ecotourists, may be differentiated on the basis of consumptive values, impact, focus of learning and reliance on technical skills, as suggested earlier.

Rolston III (1991) wrote that humans are the beneficiary of positive environmental benefits, but these are of a secondary nature (as opposed to other types of benefits which impact the individual directly). The primary benefactor is the environment itself whose members – plants, animals, geological formations, communities, ecosystems – are able to remain viable. An environmental benefit, according to Rolston III, maintains or improves or prevents degradation of the natural world. He suggests that there are a number of environmental benefits which may be derived from human actions, which are adapted, below:

- *Life support benefits*. Ecology is not subjective, but rather an objective entity. It must be preserved as the foundational support of all life – human and non-human.
- *Aesthetic benefits*. We preserve the world's landscapes in order to secure dynamic aesthetic encounters. We need pristine conditions, just like we need authentic encounters with wildlife, or those which are independent of *ex situ* gene banks (like zoos).
- *Scientific benefits*. We allow for discovery of scientific facts, and also allow tourists to travel with science as a basis for discovery.
- *Historical benefits*. Cultural and natural history of specific regions are appreciated. This allows us to better understand natural artefacts as they have evolved over time.
- *Endangered species/ecosystems benefits*. Although parks are for people, the deeper good being preserved is life itself. Wild creatures are of value in and of themselves, not simply for our use or amusement.
- *Religious/philosophical benefits*. Nature generates poetry, philosophy, religion (see the discussion on transcendentalism in this chapter), and at its deepest educational capacity we are awed and humbled by what nature has to offer.

Integrating the work of the aforementioned studies on benefits, a speculative list may be developed of benefits that may be derived from participation in ecotourism. In keeping with the trend on the recreational benefits approach, ecotourism benefits may be classified as psychological, sociological, educational and ecological. These are reported in Table 2.3.

Ecotourism's Philosophical Basis

The great scientific revolutions of the 1500s and 1600s radically shifted the relationship between people and nature, from one which was largely mystical and organic, to another that was static and mechanistic. The basis of change can be traced to the philosophy of Descartes (why anything can be said to be true), Newtonian physics (universe regulated by simple mathematical laws), as well as Bacon, who sought to control nature through observation by physical, rather than

Table 2.3. Benefits of ecotourism.

Psychological	Sociological	Educational	Ecological
Self-concept	Compassion	Outdoor education	Enjoy nature
Fitness	Share similar values	Nature awareness	Life support
Skills	Respect for others	Environmental education	Aesthetic
Self-discovery	Problem solving	Value clarification	Scientific
Actualization	Behaviour feedback	Ethics	Historical
Well-being	Friendship	Scientific	Ecosystem
Personal testing			Religious/philosophical

Fig. 2.3. One of the main aspects of ecotourism is provision of the opportunity to learn about the relationships that exist in natural settings.

mystical, means. Among others, these scholars formed the basis of reductionism, which sought to reduce complex phenomena into smaller more homogeneous component parts. The move away from a more holistic science was furthered through the empiricist philosophy of John Locke (1632–1704), who set the stage for our current development paradigm, where everything in nature is a waste until people transform it into usable things (see Short, 1991; Wearing and Neil, 1999). From the 1500s and 1600s onwards, history tells us that the motivations of humans in regards to the landscape were directed towards its transformation into living, usable spaces. Wilderness areas were loathed for their unproductivity and threatened the survival of mankind, because they not only feared these areas, but also that, for those living adjacent to such areas, people may revert to savagery themselves (Nash, 1982).

The framework for the appreciation of nature in North America and other parts of the world, stemmed from 16th- and 17th-century advances in science which revealed the world as part of a much broader harmonious universe. One interpretation was that only God could have created the world, the planets, the landscape, and those who would live upon it (Wellman, 1987). European Romanticism of the early 18th century was founded upon this interpretation of science, and human spirit emerged as a focal point which inspired people to transcend the materialistic condition of the time. Bowler

(1993) explained this as an analysis of not only the sensations that we collected from nature, but also the mind's efforts to interpret these stimuli. Some poets and theorists, who were advocates of this transcendentalism, including Ralph Waldo Emerson, Henry David Thoreau and John Muir, began to view the world as a harmonious system (a position well understood by Eastern philosophers) with nature as the focal point. The spiritual reference points of transcendentalism sought to scrutinize human land-use activities, especially those which defined the practices of emerging large corporations (by the late 1800s), as well as the associated 'frontier mentality' of the American psyche, which was a component part of the materialistic paradigm emerging at the time. The frontier mentality maintained that despite the physical boundaries of the American frontier, its resources were both bountiful and endless. As such, human perceptions of the inexhaustiblity of the landscape played a strong role in determining patterns of growth. The conservation movement of the late 1800s and early 1900s (in North America) did much to place issues of use and preservation firmly in the mind's eye of the public, and, especially in regard to wilderness preservation, the road was a rocky one.

This was quite apparent during the mid-1800s in attempts to establish Yellowstone as a national park. Following up on stories of Yellowstone's icy shores, smoking earth and multicoloured mud (Trefethen, 1975), correspondents from eastern newspapers generated enough public opinion to warrant a government sponsored exploration of the area. In their wisdom, politicians set aside the Yellowstone region to be held in trust as a national park in 1882 instead of allowing it to be exploited for the profit of a few. Although this was a significant step, policy over land use lagged far behind. Trefethen (1975) writes that poachers still shot buffalo and elk, vandals continued to smash archaeological artefacts, and tourists continued to remove items to take home. And despite the fact that regulations were being made, there was no authority for their enforcement. Twelve years later this was to change when the Lacey Bill for the protection of the Yellowstone Park was signed into law by Grover Cleveland. This legislation made it unlawful to kill wildlife, remove mineral deposits or destroy timber. Because of the intense recreational and consumptive pressures on the park, Yellowstone, early on, was managed to prevent the exploitation of wildlife, for the purpose of recreation, and for scientific study (Harroy, 1974). This was not so in Canada, where Banff (Canada's first national park, 1885) was created to finance the building of the Trans Continental railway from tourism dollars (Lothian, 1987). Even today, many of the same issues persist in Banff, including extreme recreational demand (fishing, hunting and so on), huge numbers of tourists in Banff town site, and resource extraction (Rollins, 1993).

Yellowstone and Banff were only two of many examples of natural regions subjected to conflicting views on land use. The penetration of Americans to the western, northern and southern boundaries of the continent, as mandated through an ideology of manifest destiny, prompted policy makers to consider strongly the importance of the emerging philosophies of conservation. Even a commitment to conservation however, created a great deal of controversy regarding how best to practise it. This was as a result of the interpretation of the concept, which according to Ortolano (1984) fell largely along three lines. The type most strongly advocated by government was a stance based on the *efficient use* of resources, meaning that resources such as fish and trees could be used indefinitely, but only if appropriately managed. This philosophy was cultivated by the US's first forester, Gifford Pinchot, who wrote that conservation should be the use of resources 'for the greatest good of the greatest number for the longest time' (see Nash, 1982). The second mode of conservationist thinking was based on *spirituality* and preservation. The transcendentalist John Muir was the most active supporter of this perspective, and was diametrically opposed to any sort of use which would disrupt the integrity of nature. The final view of conservation was based on *harmony*, which was based somewhere in between the preceding two. It recognized that technology, urbanization, industrialization, population growth, and transportation all contributed to a dissociation between man and nature. This was effectively articulated by George Perkins Marsh in his book *Man and Nature* (1877), which suggested that harmony between humans and the natural world might only be achieved on moral grounds.

This moral stance was adopted by Aldo Leopold (1991), in one of the most influential works on conservation to date. Leopold recognized, as did Marsh, that conservationism was only to work if both the spiritual and mechanistic sides could work together. In essence, Leopold was able to mesh science (the dispassionate and methodological realm) and humanism (the aesthetic and romantic views of Emerson, Thoreau and Muir) which meant the application of scientific reason to human values. His message was not only passionate from an idealistic standpoint, but also quite practical from social and economic levels. Animals, then, are more than just ecological resources, according to Leopold, they are social assets in that they have value just like a ticket to a symphony concert. One is a necessity, depending on who you ask, while the other may be less so. As part of a broader social and ecological community, however, the reliance on economics must go beyond self-interest:

> we asked the farmer to do what he conveniently could to save his soil, and he has done just that, and only that. The farmer who clears the woods off a 75 percent slope, turns his cows into the clearing, and dumps its rainfall, rocks, and soil into the community creek, is still (if otherwise decent) a respected member of society. If he puts lime on his fields and plants his crops on contour, he is still entitled to all the privileges and emoluments of his Soil Conservation District. The District is a beautiful piece of social machinery, but it is coughing along on two cylinders because we have been too timid, and too anxious for quick success, to tell the farmer the true magnitude of his obligations. Obligations have no meaning without conscience, and the problem we face is the extension of the social conscience from people to land.
>
> (Leopold, 1991, pp. 245–246)

Leopold believed that in making ethical economic decisions, people needed a firm image of the land and its biological functioning. Being ethical then may occur, he notes, only if people are able to see, feel, understand, love, or otherwise have faith in that which they do. Leopold's message unfortunately was ill timed. It emerged during the years following the Second World War; a time of great prosperity through technological advancement, urbanization, and population growth. If it was not recognized as a sign of the times, it was also not heeded as a warning of things to come. In general, any concerns over the environment were cast off in favour of growth, prosperity and change. Until the 1960s.

Environmentalism

Many cite Rachel Carson as providing the necessary catalyst for fuelling the most recent major era of environmentalism, from the early 1960s onward. Her book *Silent Spring* (1962) brought to light the effects of the indiscriminate use of chemicals on the natural world. Carson's thesis hit squarely on the notion that humanity's influence over nature was based on arrogance and shortsightedness. Arrogance because of the need to control nature, and shortsightedness because of the inability to understand the implications of the cumulative impacts of these poisions. In a similarly resourceful manner, White (1967) furthered the cause of environmentalism in a terse view of Western civilization, democracy and religion. Indeed, the seeds of our current environmental woes are traced, by White, to the democratic culture which viewed humans and nature as two different things, with the former as master of the latter.

The consequence of such forceful indictments against humanity was the realization that the human race was vulnerable. This accompanied a further realization that humanity was just one part of a broader ecological community in which each part had a central role. Additionally, it was no longer acceptable to view the environment as 'sound' simply because it appeared to be clean or aesthetically pleasing. Environmentalism thus became less about 'window dressing' and more about, in the words of O'Riordan (1976), a philosophy of human conduct. This set the stage for the development of diametrically opposed camps (ecocentrism versus technocentrism) that continue to influence the policies that guide our actions. The *ecocentric* line (McConnell, cited in O'Riordan) espouses a perspective based on natural laws, harmony, humility, responsibility, and on ends and the proper kind of means. Conversely, *technocentrism* is described as being more values-free, scientific, control oriented, arrogant, manipulative, and with a focus more on the means instead of ends (see Hays, 1959). Such clearly disparate positions continue to prevail in contemporary environmentalism underscoring the paradoxes, possibilities and challenges. Yet an environmentalism based on such extremes is clearly rudimentary in this day and age of growing anxiety over what must be done, when, and by whom, through often forceful means.

Mowforth and Munt (1998), for example, write that environmentalism embraces a broad range of different ideals, with particular implications for tourism. These authors provide a useful matrix of the spectrum of environmentalism as it relates to tourism. It includes resource conservation, at one end, followed by human welfare ecology, preservationism, and ecocentrism (see also Hunter, 1997). The matrix also includes an overview of the organizations associated with each view, its place of origin, views on resources and sustainability, application to tourism, and policies and political programmes. The more moderate environmentalist views, such as sustainable development, continue to be contrasted against those which are more radical and which seek to attack and reinvent society at its root. Deep ecologies (antihumanistic anarchism and primitivism), the classical left (humanistic eco-anarchism and eco-Marxism) and eco-feminism (see Lewis, 1992 for a description of these types) have rallied significant followings and articulated many alternatives to the technocentric Western paradigm. Unfortunately, although these have been quite useful in identifying the shortcomings of the technocentric position, these more radical approaches fall short as a consequence of their counterproductivity. The forcefulness and inflexibility of some of these perspectives is unpalatable to policy makers who must juggle social, economic and ecological issues. As such, the failings of these more radical approaches become the rallying cry of more moderate paradigms which allow policy makers and conservationists to find common ground, and avoid the wholesale deconstruction of society.

Sustainable development and tourism

As implied above, sustainable development (SD) is a paradigm that provides a much needed impetus for political, economic and structural change within society. The technocentric focus, which is so heavily emphasized in the Western approach to development, is tempered with a view of development which 'meets the needs of the present generation without compromising the ability of future generations to meet their own needs' (World Commission on Environment and Development, 1987, p. 43). Although such definitions of SD make perfect sense in theory, the concept has been difficult to put into practice at a socio-economic level because of too many different social systems, each with their own normative values and expectations, at too many different scales. While this seems a formidable task to overcome, the essential parameters of the concept are such that SD may lend itself to various situations and conditions at different levels, depending on the nature and intricacies of the systems involved. This means that despite the rather voluminous literature that exists to deflate the term, it may simply be too early to work out how the term is used. As Robinson *et al.* (1990) suggest, perhaps it is a process and not a final state; or that it should best be viewed as a guide for better planning and management of global systems as a result of the various complexities which are inherent in sustainability (Nelson, 1991). To this we may add that sustainability is perhaps an ethic which must underlie all of the decisions that we make which, either knowingly or unknowingly, have ripple effects across social, economic and ecological systems.

One of the challenges facing *sustainable tourism* has been the disassociation between it and sustainable development, and the debate on the meaning of the latter, which has only served to develop an overly simplistic approach to the former. This has lead to the belief that the current views of sustainable tourism development fail to operate within the general aims and requirements of sustainable development (Hunter, 1995), as well as that there are several meanings associated with SD, as outlined above. For example, and in general, the business community has adopted a development-oriented perspective, while those in the conservation realm have adopted the more ecocentric view (McKercher, 1993). Not surprisingly, the two camps are often diametrically opposed to one another with little evidence of significant conciliation.

Sustainable tourism is a specific term used to denote the application of the principles of sustainable development to the particular context of tourism. The differentiation between (i) sustainable development in the context of tourism and (ii) sustainable tourism, have been outlined by R.W. Butler (1993) through the following definitions:

1. 'Tourism which is developed and maintained in an area (community, environment) in such a manner and at such a scale that it remains viable over an indefinite period and does not degrade or alter the environment (human and physical) in which it exists to such a degree that it prohibits the successful development and well being of other activities and processes' (p. 29).
2. And that sustainable tourism is … 'tourism which is in a form which can maintain its viability in an area for an indefinite period of time' (p. 29).

Beyond the definitional perspective, a number of principles of sustainable tourism have been suggested to aid in our understanding of the issues confronting the industry. Examples of these include 'using resources sustainably', 'maintaining diversity', 'integrating tourism into planning', 'involving local communities' and 'marketing tourism responsibly' (see Eber, 1992, for an example of a paper on principles of sustainable tourism). While such principles have been important in addressing some of the key areas or concerns that sustainable tourism ought to address, they fail by not enabling decision makers effectively to use or operationalize their meaning.

Partially in response to some of these criticisms, the literature on sustainable tourism has pushed towards the development of a series of indicators for sustainability that may be applied at both cosmopolitan and local levels (Farrell, 1999). Foremost among the actors involved in this process is the World Tourism Organization, which has developed a set of indicators through the work of Manning and colleagues. Indicators measure information with which decision makers may reduce the chances of unknowingly making poor decisions (Manning, 1996). The WTO approach, according to Manning (1999), has been to use indicators in the following way.

1. Identify a small core of indicators that may be useful in almost any situation;
2. Supplement these with additional indicators known to be useful in particular ecosystems or types of destinations; and
3. Require a scanning process for risks not covered by the aforementioned indicator sets, which produces further indicators critical to the management of the particular site/destination.

Others have proposed the use of environmental economics as a manner with which to implement the principles of sustainable tourism. For example, Garrod and Fyall (1998) applied the constant-capital rule which provides a means by which to define the concept through the examination of macro- and micro-economic perspectives. The first of these involves the use of environmental balance sheets, through conventional accounting practices; while the latter involves the application of social cost:benefit analysis which would apply at the site-specific level of tourism development.

As researchers continue to search for the most effective models to operationalize sustainable tourism, there is general consensus that the human condition is especially important, as clearly outlined in the definition proposed by the Bruntland report of 1987. Increasingly, however, development that meets the needs of people, both now and in the future, must be accompanied by a willingness to do so through careful consideration of the natural world, which is a variable of considerable importance. Ecotourism, as one of our most ecologically principled forms of tourism, at least in theory, may be a model from which to test the ideals of sustainable development. In doing so, the application and the lessons learned may provide guidance for the industry and beyond.

The Environment as a System

Despite the magnitude of our planet, all life exists within a thin film of air, water, soil and rocks which is only approximately 15 km in thickness. This region of living and non-living things is complex and bound together in an intricate set of relationships (Miller and Armstrong, 1982). Living things or biotic factors may be loosely organized into three main categories, including producers (such as all green plants which are self-feeding), consumers (all animals) and reducers (including those protists, fungi, and bacteria that break down organic compounds in dead plants and animals) (Wallace *et al.*, 1981). Non-living or abiotic factors including temperature, moisture, light, texture and composition of soils, gasses and chemicals, gravity, pressure and sound have an effect on the living ones (Brewer, 1979).

These biotic and abiotic factors, along with their interactions with each other and the physical world, comprise an ecosystem. In fact, the foregoing description provides us with the basic parts of an operational definition of the term. An ecosystem is thus: 'a community of organisms and their physical environment interacting as an ecological unit' (Lincoln and Boxshall, 1990). Here, the sun is the principal agent of life, enabling producers to flourish, which in turn are food for macroconsumers (animals) and microconsumers (reducers or decomposers). The nutrients that are derived from this relationship are drawn back into the cycle of life, with a corresponding flow of energy in the form of heat (see Fig. 2.4).

The study of this very complex system is ecology, which may be differentiated from other forms of biology by being less reductionist and more holistic in its focus. Ecology evolved from the field of natural history, which slowly became antiquated as scientists pursued other more 'legitimate' areas of study. It was first defined by Ernst Haeckel over 100 years ago, and gets its name from the Greek word *oikos*, meaning home. Loosely translated, ecology is the study of the environment, or more formally, the study of the interaction between organisms and their environment. This field of research may be studied chiefly from three main perspectives: (i) the ecology of individual organisms (e.g. behaviour and habitat selection); (ii) population ecology (e.g. predator–prey relations and evolution); and (iii) community and ecosystem ecology, which examines interactions, ecological diversity and organization within and between communities and ecosystems (Brewer, 1979).

At the broadest level, the distribution of life on the planet may be understood on a grand scale. Biomes include, for example, grasslands, tropical rainforests, tundra and deserts, and are homogeneous by virtue of the types of plants that are endemic to the region. If one is able to understand the type and distribution of plants in a biome, with good confidence we may make some inferences about the macroclimate, which is the principal shaper of soils, surface topography and the distribution of life (Bailey, 1998). These major biomes in turn may be subdivided further into ecodivisions to reflect more accurately the specific conditions that occur in these regions. An example of the tundra ecodivision in Canada, as well as the implications for ecotourism, is offered in the following description (Fennell, 2001, pp. 116–117):

> *Tundra*
> Lying beyond the alpine tree line, this region is marked by slow-growing, low-formation and mainly closed vegetation of dwarf-shrubs, moss and lichens. Normal January temperatures range from about −20 to −30°C, while normal July temperatures range from 10

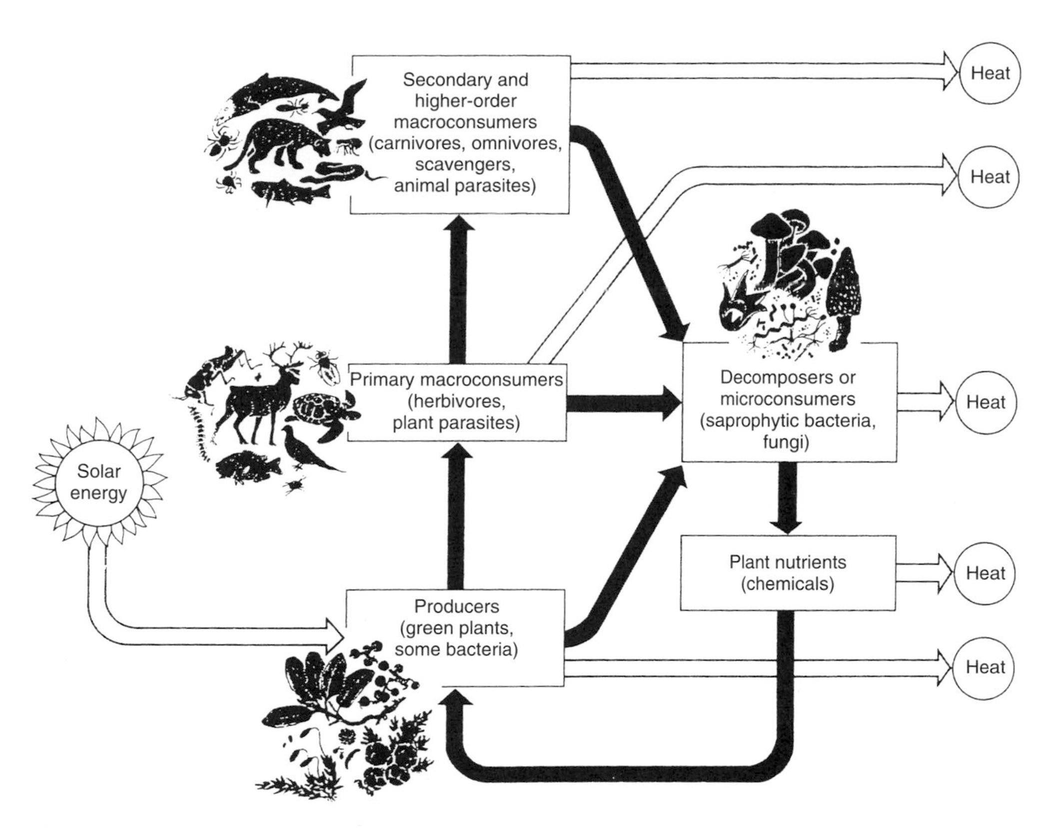

Fig. 2.4. Basic components of an ecosystem. (Source: Miller and Armstrong, 1982.)

to 15°C. In Churchill, Manitoba, a thriving tourist industry has developed around polar bears, which are forced to spend three or four months ashore due to the sea-ice melts. In Canada, there are 13 populations totalling some 15,000 bears (Churchill Northern Studies Centre, 1999), which feed typically on seals, walruses, beluga whales, and narwhals. Late October is an excellent time to view polar bears in Churchill, as the bears congregate along the shores of Hudson Bay in anticipation of the formation of ice. Tundra buggies are used to help tourists view the bears, in addition to arctic foxes, ptarmigan, and caribou. In the US, Katmai National Park, 300 miles from Anchorage, Alaska, is one of world's most accessible locations from which to view brown bears. In July the Brook's River in the heart of the park attracts an abundance of bears which feed on sockeye salmon. Although bears are the principal attraction, the surrounding lakes, forests, mountains, and marshland are habitat to a diversity of birds, flowering plants, and whales (US National Parks Net, 1999).

Figure 2.5 illustrates the various ecodivisions found within Anglo-America and some examples of attractions which fall within these various regions. Although the link between ecoregions and ecotourism is a natural one, there is little in the way of information which documents successes for biodiversity conservation. For example, how can ecotourism in the Prairie ecodivision be planned and developed such that it preserves species like the black-footed ferret, the whooping crane or the burrowing owl? By forging a stronger link between ecosystems management and ecotourism, the latter will be a more effective agent of conservation. Acknowledging the potential for ecotourism to contribute to biodiversity and ecosystem functioning, Gössling (1999) writes that this may only happen through recognition of the values of biodiversity in economic models, governmental incentives programmes for conservation, better land-use policies, and

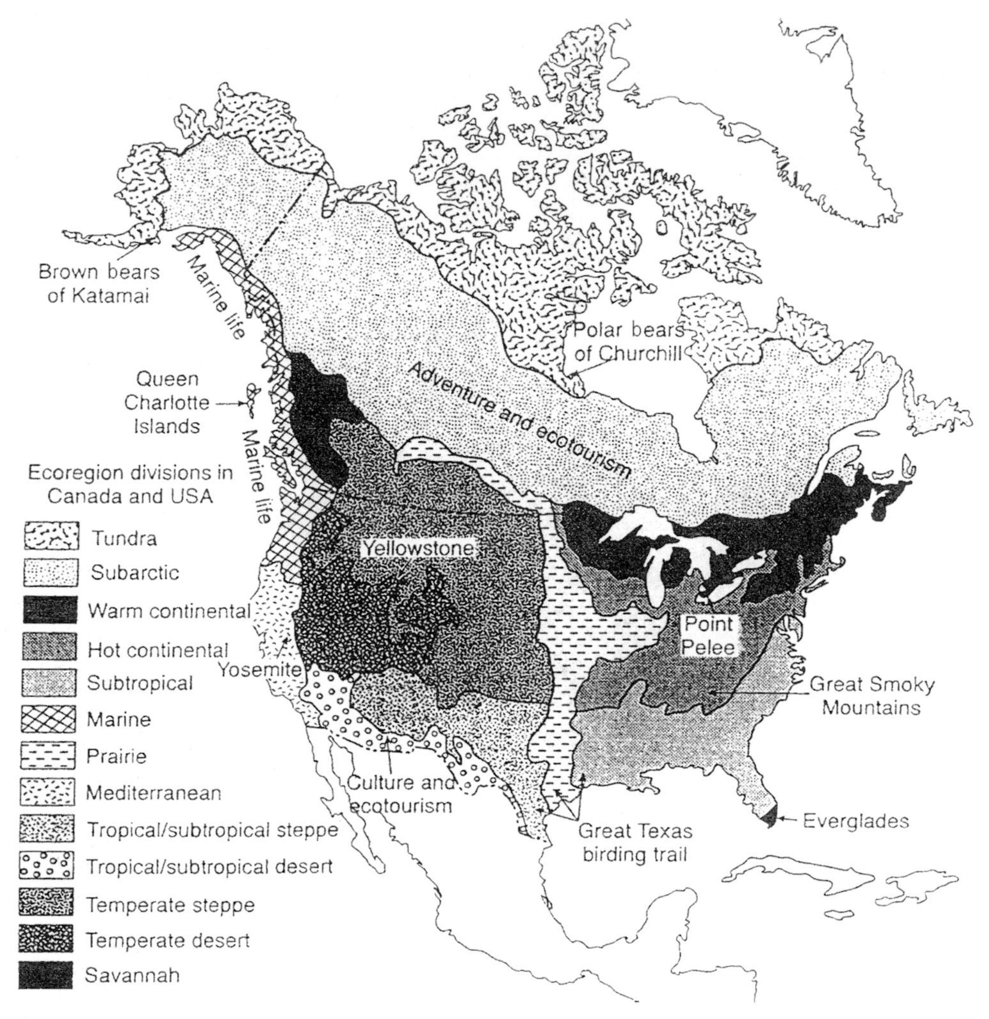

Fig. 2.5. Ecoregion divisions in Anglo-America. (Source: Adapted from Bailey, 1998, cited in Fennell, 2000.)

direct revenues channelled to protected areas. Stated more strongly, it is the failure of economic systems, and anti-environment policies which undermine the ability to protect biodiversity (Costanza and Folke, 1997).

Systems analysis

As alluded to above, the systems approach is essentially a perspective that allows for the understanding of the various components and interrelationships between components of an entity. The entity can be the environment, a community, an economic system or a manufactured 'thing'. While it is important to be able to understand the component parts of systems (more reductionist in focus), it is becoming increasingly important to examine these entities as contributing to and influencing the broader whole. For example, resources are no longer treated as discrete entities which may be isolated from the ecosystem, but rather as central to the dynamism of the broader region (Berkes and Folke, 1998). In addition to interrelationships between parts, systems are marked by feedbacks which affect the future behaviour of the system. These may be negative or positive in nature, as well as economic (ebbs and flows of the economy), social (e.g. organizational policies which create greater freedom in the work place), and/or ecological (e.g. poor water quality from outdated sewerage practices). Systems are also subject to perturbations on the basis of scale. Social systems, for example, at the local scale are said to be better able to respond to environmental feedbacks than those with a larger centralized focus. As such we cannot make the assumption, as suggested by Young (1995), that systems are inherently the same, just larger or smaller. Each has characteristics that make them both unique and necessary, and thus able to contribute at different levels. This cross-scale approach is especially applicable to natural resource issues which tend to be have implications at a number of scales (Folke *et al.*, 1998).

The flexibility of the systems analysis approach makes it especially adaptable to systems that are holistic and interrelated (e.g. social and natural systems combined). Consequently, what is required to understand these interrelated phenomena is open-mindedness in relation to space and time, and various interdisciplinary and integrated methods of enquiry (Holling *et al.*, 1998). The systems idea has become fully ingrained in the manner in which we view just about everything. This is especially true of how we define and manage organizations. One of the benefits of a systems approach is that the organization is viewed as an overall conglomeration of subsystems (e.g. technology and management) each interacting with one other, all of which are sensitive to change, which is felt to make the organization more robust. Such an organic approach to the social environment has meant that 'by exploring the parallels between organisms and organizations in terms of organic functioning, relations with the environment, relations between species, and the wider ecology, it has been possible to produce different explanations that have very practical implications for organization and management' (Morgan, 1986, cited in Hall, 2000). The systems approach has also been applied in a tourism context. One of the most oft-quoted system analogies is one by Mill and Morrison (1985), who define tourism as a system of interrelated parts. They note that the tourism system is like a spider's web, and by touching one part of the web, reverberations will be felt throughout the tourism industry, including the market, transportation, activities within the destination, and how products are marketed. The following two sections, human ecology and complexity, are representative of the integration of ideas and methods as regards human–environment relationships, which thus allow us to continue to modify our perspectives on systems and related philosophies.

Human ecology

Developed in the early part of the 20th century by geographers such as H.H. Barrows and Robert Park, human ecology is a theoretical approach that attempts to combine culture and ecology. As suggested by Johnston (1986), human ecology is the application of ecological concepts to the study of the relations between people and their physical and social environment (see also Entrikin, 1980). The task of human ecologists, therefore, is to 'describe systems and to explain the interconnections among people and environments which sustain the system and keep it equilibrated' (Porter, 1978, p. 19). Perhaps one of the strongest advocates of human ecology was C.O. Sauer who recognized the power that human beings have in transforming the natural world, and thus realized that humans needed to be placed in the total ecology of the organic world (Leighly, 1987).

More recently, Peterson (1996) has identified four corners of human ecology that span the range of values that human beings define and evaluate their relationship with the earth. The first of these, 'domination', underscores the notion that humans rule over nature. Humans can be bad rulers, by trying to exploit and dominate for short-term personal gain at any expense, or good rulers by protecting and serving for the well-being of all. The second corner, 'stewardship' puts humanity in the position of a caretaker of earth in general, the human race, all living things or a particular species. The third corner, 'participation' implies a symbiotic relationship by constructing complementary and cooperative relationships between people and other beings. Human beings are seen as one of many life forms considered to be equal partners. The final corner, 'abdication', according to Peterson, suggests that humans must renounce all claim to the right to prosper when that right conflicts with the functional values of other species. Humans maintain a number of relationships with other species as competitors, as predators and as prey. Humans must play their part in this grand scheme, even if it means extinction, or reduced quantity or quality of life (although it is difficult to image human beings submitting to malaria-carrying mosquitoes, or other parasites).

Human ecology has been studied only tangentially in the context of tourism research. Examples of such work include the efforts of Berkes (1984), who analysed the resource competition between local and non-local (tourists) anglers on Lake Erie. In additon, Castro and Begossi (1996) studied niche and competition between local and non-local anglers of the Rio Grande, Brazil. However, one of the most dramatic links between tourism and human ecology was drawn by Murphy (1983) through his work connecting biological communities and tourism communities. Murphy said that just as plants, animals, predators and prey have to coexist, so too do natural tourist attractions, local resident reactions, the industry's investment (and return), and visitor satisfaction. Specifically, Murphy equated the tourist industry with predators, and the visitor as the industry's prey. The rationale followed that 'visitors are prey because the community, especially the tourist industry, feeds on them and the revenue they bring' (Murphy, 1983, p. 187). These papers correspond well to the thoughts of Zimmerer (1994) in his views of the 'new ecology'. He suggests that ecology, like other forms of knowledge, is socially constructed. As such, he feels it is imperative in studying human–environment relations that we need to consider different epistemologies in the examination of human behaviour and social organization, especially from the perspectives of time, space and subjectivity (unequal capacities of organisms as they respond to environmental heterogeneity).

Based partly on the work of Murphy, as stated above, Fennell and Butler (2002) constructed a model of the organization and development of a tourism industry through the use of human ecological principles. Tourism habitat was discussed as the places tourists or tourism groups visit (their range of spatial movement), and tourism niche, as the function of a tourism group within the destination. Fennell and Butler also incorporated the law of competitive exclusion (Gause's law), in examining tourism stakeholder interactions. Gause's law implies that two species cannot indefinitely live together and interact with the environment in the same way (Wallace *et al.*, 1981). Under laboratory conditions, Gause found that two species of *Paramecium* could not coexist. The most efficient feeder always out-competed the other when dining on the same bacterial source. On the other hand, it has been argued that competition in the wild rarely results in extinction, as Gause thought, but rather in a subdivision of habitat (MacArthur, 1957), where each species adapts and comes to live where it does best.

These concepts provided the basis for the following discussion on four different relationships that exist between different tourism stakeholder groups, including different types of tourists, local people, tourism industry brokers, and other land users. The relationships examined included: (i) predatory, meaning a stakeholder has a dominating, high level of impact and influence on other stakeholders; (ii) competitory, where there is open competition for resources among stakeholders; (iii) neutral, with very little impact on other stakeholders or on the resource base; or (iv) symbiotic, implying a shared or beneficial coexistence among stakeholders and the environment, or stakeholders and each other. These relationships were operationalized through a number of measures of pressure including: (i) visitor type and numbers relative to locals; (ii) displacement/interference with other tourism activities, locals, or other resource activities;

(iii) patterns of dispersion in space and time; (iv) nature of impact (direct); (v) nature of impact (indirect); (vi) knowledge/sensitivity to ecological features; (vii) degree of infrastructural and human support required; (viii) local acceptance of type (xenophobic); and (x) nature of activities.

Complexity

Earlier in the chapter, a brief discussion on sustainable tourism sought to identify some of the concerns raised within the sustainable tourism debate, and principally two methods (sustainable indicators and economics) as a means by which to understand better how the principles of sustainable tourism may work for tourism. One of the basic conclusions which may be drawn from the state of sustainable development at the start of the 21st century, is that there are no simple, straightforward solutions emerging from the application of sustainable development, nor is there universal acceptance about how it can actually help in our attempts to be effective global stewards. Sustainable development is thus an example of an inherently complex phenomenon – as is tourism as identified by Crick (1989) – which needs to address issues from many different perspectives, and according to the vastly different environments in which these issues take place. At the same time however, sustainability is inclusive and adaptable enough (in conserving social, economic and ecological systems) to provide guidance to some of our most pressing resource management issues.

Complexity underscores the fact that systems, whether they be social, ecological, or both, are non-linear and uncertain (Scoones, 1999). In fact, at its heart, complexity recognizes that social systems cannot be analysed independently of natural systems, and that these need to be studied as a linked whole (Berkes and Folke, 1998), as suggested above. Natural and social worlds are seen to be organized hierarchically by scale, and an understanding of this gives policy makers the ability to implement governance according to the conditions of the setting. For example, community and resource decisions within an aboriginal community might best be made by the community itself, whereas issues related to the direct and indirect effects of air and water pollution might best be made through international institutions. Complexity is thus viewed as the antithesis of reductionism and of the linear, equilibrium theory of nature which suggests that there is a predictable pattern of events that determine cause-and-effect relationships (see Perrings *et al.*, 1995).

The inherent adaptability of complexity theory makes it particularly relevant to SD, as suggested by Roe (1998), who has identified a number of different interpretations of SD through complexity. From a Girardian economics perspective, Roe illustrates that SD is a 'social convention which for a time underwrites and stabilizes decision-making under high uncertainty in a way that its subscribers believe keeps options open for the future' (p. 42). From a cultural theory standpoint, SD must be defined differently in different cultures, because their human and resource needs are so different. From a critical theory perspective, SD is a new class version of resource managerialism that serves to globalize the techno-managerial elite's control over everyday life. And finally, from the local justice framework, it is a cycle of justice and injustice that consistently comes back to the notion that we ought to manage resources today in a manner that allows us to keep open certain options for their management in the future. Roe suggests that, in cases where there are issues which are uncertain and complex, triangulation is warranted. Consequently, there may be many theoretical and applied options available to decision makers for the purpose of solving the problem, not simply one 'correct' choice (or more likely one convenient choice), with the result of eliminating bias and building confidence.

Conclusion

This chapter has sought to provide a foundation for an understanding of ecotourism and ecotourists, through an evaluation of past research on ecotourism and related research on specialization and benefits. A definition of ecotourism was offered on the basis of an analysis of over 85 different definitions of the concept. While the various stakeholders involved in tourism will continue to interpret the concept in different ways, it is important for these stakeholders to make attempts to isolate what may be the most salient principles of the concept in their policies and practices. A great deal of effort was placed into linking ecotourism with the literature on conservation and environmentalism, as well as the literature on the environment and systems. Although subject to debate, the philosophical basis of ecotourism must come from those areas that recognize the importance of systems approach to human endeavours, such as tourism.

In the quote at the start of the chapter, the author aptly recognizes that the environments that we choose to recreate in, and on, are indeed living things. In sharing these settings, the various stakeholders associated with the ecotourism industry have a responsibility to modify their behaviours subject to the viability and health of the system. Perhaps the most important part of the quote is the concept of judgement: judgement on behalf of the service provider, ecotourists, government and so on, about what is and is not acceptable behaviour. This takes us to issues related to supply, impacts, norms, carrying capacity and ethics, which may be found in the following chapter.

3

The Programme Setting: Going 'In'

I only went out for a walk, and finally concluded to stay out till sundown, for going out, I found, was really going in.

(John Muir)

Introduction

Recreation programmes are often conducted within fairly defined boundaries, including, for example, gymnasiums, urban parks, school yards, pools. On the other hand, ecotourism programmes rely extensively on natural areas (broadly defined) which are often more diverse spatially, ecologically and socially. This chapter involves a discussion of ecotourism supply, including a focus on biodiversity, island biogeography and nature reserve design, as well as a brief look at public and private reserve systems. The direction of the chapter then switches to an overview of the extensive resource management issues inherent within ecotourism. Topics include the carrying-capacity debate, which is linked to the concept of ecological footprints, the precautionary principle, discourse analysis, adaptive management, encounter norms, and codes of ethics. A discussion of community development is included as a means by which to underscore the importance of the community as host, as well as the many economic and resource issues that surface from the various ecotourism–community interactions.

Ecotourism Supply

Biodiversity

The planet is composed of a vast array of genetic information, species and ecosystems, the totality of which is representative of the world's biodiversity. Approximately 1.4 million species have been identified by scientists, with estimates ranging from 2 million to 100 million (see May, 1992, and Erwin, 1991). The link between ecosystems and species is a critical one as the destruction of the former contributes directly to the disappearance of the latter. In fact, habitat loss is found to be the greatest contributor to biodiversity loss. Biological researchers suggest that we are currently immersed in a period of mass extinction of species (Myers, 1990), with estimates of species losses numbering at least 27,000 per year, and a loss of 50% or more by the year 2000 (Myers, 1993).

An understanding of the biodiversity of a region, or indeed the planet, is important not only for the genetic information that may someday cure various illnesses, but to act as benchmarks in understanding the effects that humans have on the world. More specifically, Ehrlich and Ehrlich (1992) argue that biodiversity should be valued on the basis of the following:

- *Ethically*. Humans have an ethical, stewardship responsibility towards humanity's only known living companions in the universe.
- *Aesthetically*. Each organism is an irreplaceable treasure that is beautiful in design, which supports many economies, through, for example, ecotourism.
- *Direct economic*. These plants and animals contribute food, medicine, and a huge variety of industrial products.
- *Indirect economic*. These plants and animals supply people with an array of ecosystem services, such as generating and maintaining soils and controlling pests.

One must only think about wildlife, oils, plants extracts, fibres, agriculture and aquaculture to appreciate how thoroughly we are reliant upon the natural world. The conservation of biodiversity is thus critical for the long-term functioning of the biosphere, as well as for our own comfort and survival. The Global Biodiversity Strategy (World Resources Institute, 1992) endorsed by the World Resources Institute (WRI), The World Conservation Union (IUCN) and the United Nations Environment Programme (UNEP), has recognized the imminent responsibility of humanity to strive for an expanded capacity to conserve biodiversity. This, they suggest, can be accomplished by: (i) acknowledging that people use resources for a number of different ends; (ii) understanding that conservation is the key to ensuring the survival of species; and (iii) long-term study of ecosystems, species and human behaviours.

Island biogeography and nature reserve design

Our efforts to safeguard terrestrial and marine habitats have spanned well over 100 years. Over this time, parks management has evolved to a point where habitat and species protection have become increasingly more important. This is a result of specific management techniques employed by parks management specialists, but also a result of a better understanding of the shape and dynamics of the park itself. This latter element was not widely considered until 1963, when MacArthur and Wilson published their theory of island biogeography. In general, the theory considers why fewer animals exist on islands that are further from large land masses (i.e. islands) than those closer to a major land mass, such as a continent. The theory is based on the following aspects:

1. Immigration varies with distance from the mainland.
2. Extinction of species varies with area and distance.
3. An island with a stepping-stone island between it and the source area had higher immigration rates.
4. Island clusters raise immigration rates.
5. A low pool of immigrants reduces the number of species on islands.
6. A species is more likely to die out on an island with a high extinction rate.
7. The turnover rate of species was greater on near islands.
8. Islands further from the source of colonization have fewer species.
9. Smaller islands have fewer species.
10. Species numbers grow with increasing area.

While the specifics of the theory will be omitted for the purposes of this discussion, as well as some of its criticisms (e.g. the theory is based principally on the colonization of avifauna, and could not describe the heterogeneity of the real world), its implications for the design of parks and protected areas were monumental. Especially, that small, insular, fragmented habitats did not make effective protected areas.

Figure 3.1 demonstrates the pros and cons of various reserve designs on the basis of the theory of island biogeography (from Diamond, 1975). It is based on three general rules for the design of reserves which have been proposed by various authors (see Blouin and Connor, 1985) to ensure for the maximum number of species possible, including: (i) reserves should be as large as

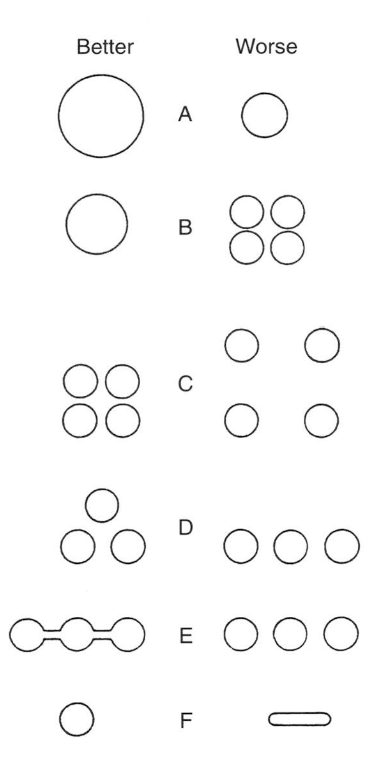

Fig. 3.1. Principles for design of faunal reserves. (Source: Diamond, 1975, cited in Shafer, 1990.)

possible; (ii) they should be placed as close together as possible to ensure for inter-reserve movement; and (iii) they should be as circular as possible. The debate on SLOSS (Single Large or Several Small reserves) has been an active one, but also an inconclusive one, with several authors siding on one perspective or the other. In an excellent overview of the debate and of reserve design in general, Shafer (1990) writes that due to the mixed results on these two perspectives, policy makers need to be cautious on how they develop reserves. This is especially true given the unique characteristics of the region and the flora and fauna of the 'island' in question. Game (1980) for example, writes that soil properties might be better in long, linear reserves, while Diamond and May (1976) suggest that the edge effect is a critical determinant of biodiversity. Others (Simberloff and Abele, 1976; Burkey, 1988) suggest that autecological (the study of individual species) considerations must be detailed for the management of reserves. Some of these are summarized as follows (see Shafer, 1990).

- *Home-range requirements.* Estimates of space needed for survival. Studies suggest that eight wolves require 600 km^2, eight cougars require 760 km^2, and eight grizzly bears require 600 km^2.
- *Large carnivores as indicators.* Keystone species or those with the highest area requirements should be used to establish reserve design, with associated protection for most other species.
- *Minimum viable population* (MVP). Reserves should be determined on the basis of the minimum number of individuals of a species to ensure its survival. This is done by determining the area requirements for one individual, and multiplying by the MVP, for the overall area requirement for the species. This figure is difficult to ascertain because of demographic, environmental, genetic and catastrophic stochastic (random) events.
- *Effective population size.* Reserve size based on the genetic fitness of a population, and the 50/500 rule of 500 individuals for the genetic fitness of the population, each requiring 50 km^2. These two figures multiplied suggest a reserve size of 25,000 km^2.

Biodiversity conservation can occur on many levels and through a number of different management philosophies. The huge network of parks and protected areas that exist throughout the world is an example of *in situ* (in the field) gene banks, which are critical for maintaining populations of animals in their natural state, and relatively free from human interference. Ecotourism thrives in these protected areas, providing much needed capital for conservation and infrastructure in both developed and developing countries. In addition to the protected areas network, there are a large number of *ex situ* gene banks which contribute to the conservation of genetic material, and which complement *in situ* gene banks through the transfer of genes to areas around the world for the purpose, for example, of avoiding inbreeding which results from small, unhealthy populations. These *ex situ* centres include microbial networks, plant networks, botanical gardens, zoological parks and aquaria. There is little doubt that these conservation mechanisms provide less of an authentic experience to the ecotourist, and the extent to which these qualify as authentic ecotourism experiences is still open to debate.

Public reserves

Woodley (1993) writes that parks are at one end of a continuum of human land use which ranges from cities to pristine wilderness areas. Even so, there are significant differences in how parks are managed, especially in regards to the degree to which they emphasize development (use) and conservation (conservation itself falls along a continuum ranging from extractive use, at the most development-oriented pole, through to multiple use, protection and preservation). At one end of the scale, parks are seen as humanity's most progressive thinking relative to ecosystem management and biodiversity conservation. At the other end, they can be nothing more than symbols which hold little value and even less influence.

Owing to the internationalization of the national parks concept, interest was shown in developing a *lingua franca*, or common language of conservation (Miller, 1989). He illustrates that in 1901 the nations of Europe and colonial Africa prepared a document on the nomenclature of concepts related to parks, and later revised and accepted the document in 1933 in London. The London convention defined a national park as:

> an area (a) placed under public control, the boundaries of which shall not be altered or any portion of capable of alienation except by the competent legislative authority, (b) set aside for the propogation, protection and preservation of wild animal life and wild vegetation, and for the preservation of objects of aesthetic, geologic, prehistoric, historical or archaeological, or other scientific interest for the benefit, advantage, and enjoyment of the general public, (c) in which the hunting, killing or capturing of fauna and the destruction or collection of flora is prohibited except by or under the direction or control of the park authorities.
>
> (Miller, 1989, p. 30)

Other conventions for similar purposes followed, according to Miller, including the Convention on Nature Protection and Wildlife Preservation in the Western Hemisphere, 1940, which defined national parks, national reserves, nature monuments, and strict wilderness reserves. Delegates to The Tenth General Assembly of the IUCN in India in 1969 further articulated differences between different natural areas, with particular attention to how national parks differed from 'scientific reserves', 'natural reserves', 'special reserves' and other such areas.

This international scrutiny of the park concept from a policy standpoint was concurrent to escalating demand, especially during the era after the Second World War, as a result of increased disposable incomes, more of a focus on the vacation especially in outdoor settings, and the time for such travel. This level of unprecedented demand for parks saturated many of the most popular parks, including those in the Ontario, Canada system, and in particular, Algonquin (a provincial park 2–3 hours away from Toronto). The number of cars entering Algonquin almost doubled over a 3-year period, from 28,662 in 1950 to 47,200 in 1953 (Killan, 1993).

In the USA the demand for outdoor recreation during the 1950s (Wolfe, 1964) triggered the US congressionally appointed Outdoor Recreation Resources Review Commission, which convened from 1958 to 1962, and which created 27 reports and over 4000 pages touching all aspects of outdoor recreation in the country during this time. The main objectives of the commission were to examine outdoor recreation wants and needs, determine the recreation resources needed to accommodate this demand, and develop programmes and policies (ORRRC, 1962). The final report of the Commission coordinated a vast amount of outdoor recreation literature, and developed a plan to institute concrete action. The major findings of the ORRRC are summarized as follows:

- *The simple activities are most popular.* Motoring, walking, picnicking, etc., are favourite activities regardless of age, income, education and occupation.

- *Outdoor opportunities are most urgently needed near metropolitan areas.* Such need will continue to intensify as cities continue to grow. The supply is least where the need is greatest.
- *Considerable land is now available for outdoor recreation, but it does not meet the need.* Most of the land is where people are not, and management policies are too restrictive.
- *Money is needed.* To be provided by all levels of government, federal, state and municipal.
- *Outdoor recreation is compatible with other resource uses.* Recreation must often share space with other types of human involvement (e.g. agriculture, forestry, and so on). In such cases it is critical to manage these uses for the benefit of all.
- *Water.* Is a focal point of outdoor recreation, with universal attractiveness.
- *Economic benefits.* While intangible benefits of recreation are most important, economic benefits of recreation contribute significantly at the community, regional and national levels.
- *Outdoor recreation is a major leisure time activity.* Americans will participate in one outdoor recreation activity or another on 12.4 billion separate occasions by the year 2000.
- *More research is required on the values of outdoor recreation.* Management techniques, for instance, that will help to ensure maximum public benefits.

This latter finding is indeed significant as it recognizes that land and water utilized for recreation must involve some degree of management. An important outcome of the ORRRC process involved the recommendation that outdoor recreation resources might better be managed using a system of land classification in order to accommodate a broad or narrow range of activities, much the same as articulated in the aforementioned conventions. The ORRRC developed a classification scheme (Table 3.1), which is a template that has been used and adapted time and time again in parks throughout the world.

During the 4th World Congress on Parks and Protected Areas in Caracas in 1992, researchers estimated that there were in excess of 6000 protected areas in the world. By 1997 the number had jumped to 10,401, representing 840 million ha, or 6.4% of the world's lands and waters. However, with the inclusion of parks of less than 1000 ha, the number had increased to 30,350 and 1.32 billion ha, or 8.8% of the world's land area (Green and Paine, 1997, cited in Lawton, 2001). Increasingly policy and decision makers and researchers have come to recognize the benefits that parks have to humanity but also to ecology. Despite the insistence that parks are here to support tourism, which they will continue to do, there is the realization that such activities must occur only through the protection of the resource base first. Disturbances that occur within the

Table 3.1. ORRRC Land Classification Scheme.

Class	Type of area	Description
I	High-density recreation	Areas intensively developed and managed for mass use
II	General outdoor recreation	Areas subject to a substantial development for a wide variety of specific recreation uses
III	Natural environment	Various types of areas that are suitable for recreation in a natural environment and usually in combination with other uses
IV	Unique natural	Areas of outstanding scenic splendour, natural wonder, or scientific importance
V	Primitive	Undisturbed roadless areas, characterized by natural, wild conditions including 'wilderness areas'
VI	Historic and cultural	Sites of major historic or cultural significance, either local, regional or national

Source: ORRRC (1962)

borders of parks occur for a number of reasons, including poorly conceived policies, lack of enforcement, or lack of resources. However, even the best managed parks may be ecologically 'sick' because of the substances or events that come from outside its borders. These external threats include climate change, pollution, poaching, population pressures and industry (see Chase, 1987; Lovejoy, 1992; Dearden and Rollins, 1993) and are symptomatic of problems that are of a local, regional and global scale.

Private reserves

One of the harsh realities of public park management is the significant lack of resources for effective management of visitors and the environment, especially in the less developed countries. Parks continue to operate on very tight budgets by using a number of different models to sustain themselves. Some park managers are able to capture a percentage of the revenue that is generated in their own parks, while others are required to submit their revenue into a common parks system pool or administration to be used for other general purposes. Perhaps in response to the massive underfunding of public parks in some countries, private reserves have been allowed to develop, operated in many cases by expatriates, for the purpose of habitat protection and biodiversity conservation, but also to attract tourism dollars (within the latter contributing to the former). The Monteverde Cloud Forest Reserve in Costa Rica is one such example, which has steadily increased in size, and popularity, to the point where a new entrance fee structure has been required, along with more naturalists and guides, limit on numbers of entrants and restrictions to well-marked trails (Honey, 1999).

Private reserves are those lands that are not owned by a governmental body, are larger than 20 ha in size, and are maintained in a mostly natural state (Langholz and Brandon, 2001). Alderman (1992) writes that the motivations for starting a private reserve are mixed. Some are profit-making ventures, while others are NGOs who are dedicated to habitat protection and research. Many others are hybrids, which attempt to do several things, including ecotourism, extractive industry, education and agriculture. In her research, Alderman discovered that the majority of reserves fall in the hybrid category, with 32% of reserves in her study combining tourism with extractive industry, and 25% devoted to tourism exclusively (see Table 3.2).

Fig. 3.2. Public and private reserves are both important means of protecting habitats.

Table 3.2. Types of private reserves by continent.

	Latin America		Africa		All	
Type of reserve	*n*	%	*n*	%	*n*	%
Farm and tourism	12	25.0	8	53.3	20	31.7
Tourism only	12	25.0	4	26.7	16	25.4
Research, conservation, tourism	17	35.4	2	13.3	19	30.2
Research, education, no tourism	6	12.5	1	6.7	7	11.1
Farm, conservation, education, no tourism	1	2.1	0	0.0	1	1.6
Total	48	100.0	15	100.0	63	100.0

Langholz and Brandon (2001) report that many of the newer private reserves have been developed specifically with ecotourism in mind, and cater largely to the up-scale market. These operators can afford to keep tourist numbers lower, but ensure financial viability by charging higher prices than the competition or their public counterparts. These authors suggest that there is a real conflict of interest emerging in this sector as those who are more interested in profit become tempted to cut corners, putting money first and conservation a distant second. Given the relative success of private ecotourism reserves, the sector appears to be greatly under-researched, especially when compared against the volume of literature available on public parks and protected areas. Economic, experiential and conservation-oriented data are greatly needed to better understand the overall impact that these reserves have on the ecotourism industry, especially given what appears to be an increased level of demand for the services that these reserves provide.

Resource Management

A classic definition of resources was offered by Zimmermann (1933, 1951 revised), who wrote that aspects of the environment are not resources until they are, or considered to be, capable of satisfying human needs (cited in Mitchell, 1989). Trees, fish, gold, recreational space and so on, are simply 'neutral stuff' that exist in the environment until they are perceived by humans, recognized as being able to satisfy human need, and utilized. There is a strong cultural component to resources such that some cultures value certain aspects of the environment, while others do not. Mitchell suggests that resources are thus defined by human perceptions and attitudes, wants, available technology, legal and political customs, and financial and institutional arrangements.

Resource management is a multidisciplinary field of study which focuses on the management of renewable and non-renewable resources, across sectors. At the heart of resource management is the decision-making authority: who holds the power to make decisions is quite often the entity who designs and implements the natural resource policies which are thought to be most appropriate for the setting, stakeholders and period of time. For example, the multiple-use policies designed by public landuse authorities of the 1960s were designed to produce a variety of different benefits to a large number of stakeholders from large parcels of land. Outdoor recreation was just one of many different types of land use which needed to be balanced against other forms, including hydroelectric power, fish and wildlife, timber and mining. Such a regime required the land manager to take steps and make judgements in order to balance the total production of land, according to these land uses, in an effort to continually yield optimal levels of goods and services (Knudson, 1980). Attempting to balance so many different types of consumptive and non-consumptive activities in one region became too much of daunting task. As such, the multiple-use philosophy fell out of favour because the one best solution for all groups never completely materialized. Instead compromises materialized which did little to maximize the full potential of the various land uses within the region. Land managers and stakeholder groups were never able to place their needs behind the needs of the resource base. Consequently, aspects of competition and self-interest from these groups lead to over exploitation and collapse of resources (Berkes and Folke, 1998).

In the words of O'Riordan (1971), resource management needs to be 'visualised as a conscious process of decision involving judgement, preference and commitment, whereby certain desired resource outputs are sought from certain perceived resource combinations through the choice among various managerial, technical and administrative alternatives' (p. 19). It therefore provides the means by which to analyse resource problems through social, political, legal, economic and ecological perspectives, at a number of different scales (local through to international) and time frames (past to future) (see Mitchell, 1989). Also inherent in resource management is the evaluation of various management systems, policies and practices, along with the development of alternative management approaches, which enable resource managers to better understand interrelationships between social, economic and ecological systems. This includes consideration of more contemporary management philosophies including adaptive management. This form of management allows resource managers to continually learn from feedbacks from the environment, which are thus built into subsequent management iterations. Such ongoing interplay between the environment and the management agency, through experimentation, allows for the continual modification of policy which is thus more responsive to actual social and ecological conditions (see Norgaard, 1994).

The systems approach outlined above is central to the resource management context in general and to outdoor recreation and tourism more specifically. Kreutzwiser (1989, p. 22) defines a recreational resource as an 'element of the natural or man-made environment which provides an opportunity to satisfy recreational wants'. Such resources include, for example water, mountains, theatres, sports complexes and festivals (more on this in Chapter 5). Even within recreation and tourism, however, there is a myriad of issues pertaining to different types of recreational use, and the environmental conditions for and responses from such use. These include issues related to visitor use (amount of use, inter- and intra-group differences, perceptions, expectations), effects on different resource components (e.g. soils and water), site management, and monitoring. In the past the focus of research tended to be centred on the resource itself. These days, however, there appears to be a clear consensus that the focus should centre on the resource user, along with the various environments in which they interact. The relationship between recreation/tourism user and environment is considered through the rich base of literature on impacts and carrying capacity, both of which are examined in more detail, below.

Impacts

Sterrer (1982) writes that humans, like other beings, use and modify their environment to suit their needs. Unlike other beings which change together and at the same pace, humans have introduced change at new scales of time and space, which have transformed the environment, beyond the capacity of evolution to keep pace (see also Dubos, 1976; Goudie, 1990b). The impetus for this change, writes Sterrer, is the need to seek comfort. This may be viewed as the difference between the energy acquired from food versus the energy expended in finding food. The more time saved in pursuit of the latter, the greater the opportunity to do other things. The male lion seems to have discovered the principle behind this equation, and spends the bulk of his time resting, procreating and being the first to the dinner table.

On the basis of the reasoning put forth above, tourism may be the most extreme exhibition of this comfort seeking 'instinct', predicated on tourists' travel expectations, including style of travel and level of service. The growth of the industry in the era after the Second World War lead to great prosperity and heightened levels of demand, as suggested above. For the industry to keep pace, massive hotels and associated infrastructure were developed to accommodate demand, often with a level of planning that contributed to many social and ecological problems.

Concern over the effects of tourism on the environment, however, can be traced back to the 1950s through the work of the precursor to the WTO, which held a series of discussions on the protection of natural heritage (Shackleford, 1985). This concern bridged over into the popular media in the 1960s and 1970s, highlighting destinations like Acapulco (Cerruti, 1964), Spain (Naylon, 1967) and Gozo (Jones, 1972), and the struggle between conservation and development in these regions, especially the negative aspects of large-scale hotel development (Hall, 1970). Wolfe (1966) also recognized the implications of too much development too soon, in suggesting that the time must come when an increase in aggregate travel will result in a decrease in aggregate pleasure (Wolfe, 1966).

Tourism researchers began to focus on the ecological impacts of tourism during the 1970s, spurred on by the environmental movement initiated in 1970 as well as by the realization that tourism was much more than a benign, smokeless industry. By the 1980s tourism–environment research was characterized by the identification of a wide spectrum of complex interrelationships and impacts (see Pearce, 1985). Pearce cites work by the Organisation for Economic Cooperation and Development (OECD), which documented a number of stressor activities, the primary environmental response, and the human reaction to the impact. These included: (i) permanent environmental restructuring (urban expansion, transport, facilities, and changes in land use which expand the built environment and take land out of primary production); (ii) generation of wastes (from urbanization and transportation, which include emissions, effluent discharges, solid waste and noise); (iii) effects on population dynamics, such as seasonal increases in population densities; and (iv) tourist activities (e.g. walking, skiing and trail-bike riding, which trample vegetation and soils and disturb the behavioural patterns of various species). This latter category of impact is elaborated upon in the work of Buckley and Pannell (1990), who identified the following as impacts inherent in the national parks of Australia: tracks and off-road vehicles (ORVs), trampling (human and horse), weeds and fungi, damage from boats, firewood collection, human wastes, camp sites, water pollution, changed water course, water depletion, disturbance to wildlife, damage to archaeological sites, cultural vandalism, litter, visual impacts (buildings) and noise. A more comprehensive list of impacts has been constructed by Hunter and Green (1995), who focus on the natural environment, the built environment, and the cultural environment (floral and faunal species composition impacts shown in Table 3.3 only, and the various consequences). (For a comprehensive overview of outdoor recreation impacts, see Hammitt and Cole, 1987, and Newsome *et al.*, 2002).

At its root, ecotourism is a form of travel based principally on observing wildlife in

Table 3.3. Selected impacts of tourism.

Environmental impact	Consequences
Floral/faunal species composition	Disruption of breeding habitats
	Killing of animals through hunting
	Killing of animals for souvenir trade
	Inward/outward migration of animals
	Trampling of vegetation by feet and vehicles
	Gathering of wood or plants
	Vegetation change through clearance for facilities

Source: adapted from Hunter and Green (1995).

natural habitats. Unfortunately, given the popularity of this type of activity (non-consumptive forms of outdoor recreation have been found to be on the rise in recent years (Foot and Stoffman, 1996)), and without proper management, impacts have been shown to occur, as evident, for example, in the massive scars on the African landscape from numerous vehicles which continually chase after wildlife. Although the level of impact may not be so marked as in other outdoor activities, any time people venture into the outdoors some level of impact, either direct (e.g. trampling of vegetation) or indirect (e.g. poor growth from soil compaction) is imminent. Situations where wildlife viewers have had noticeable impacts on wildlife, are as follows (Edington and Edington, 1986):

- *Habituation*. Animals such as bears, elephants and monkeys become habituated to human foodstuffs. Often the animals change their social behaviour, becoming unpredictable and aggressive – leading to the injury of tourists. This is especially true for monkeys, whose bites often send tourists to hospital because of lack of respect shown for these animals.
- *Vulnerability*. The arrival of tourists (their boats and other vehicles) often leave a species, and their eggs/young vulnerable to predation. Crocodiles, for example, are scared off by tourist boats for extended periods of time, allowing predators such as monitor lizards and baboons to raid the nest for eggs. Hyenas have been found to follow minibuses as a means of locating and robbing food from cheetah families, sometimes killing cheetah young in the process.
- *Interference with territorial behaviour*. Tourist activities often interfere with the essential processes of securing a mate and ensuring access to adequate sources of food. For example, the female Thomson's gazelle is more timid than the male and readily leaves the breeding territory when tourist vehicles approach, resulting in a tremendous impact on breeding success.
- *Parent–offspring bonds*. The movement of tourist vehicles can increase the mortality of young animals by separating them from parents at times when recognition bonds (5–6 hours depending on the species) are being established. Young which are separated before this bond is forged are often not accepted by parents, and will most certainly die or be killed by predators.

Tribe *et al.* (2000) note that although tourism research has focused primarily on the negative aspects of tourism, there are in fact a variety of positive ones. These include increased revenue, direct and indirect employment; beautification of the natural world, restoration of buildings, and increases in the diversity of leisure services, as outlined in Table 3.4. Andereck (1993) and Olokesusi (1990) underscore the positive aspects of tourism on the environment by acknowledging the many benefits of the creation of parks and protected areas. Also, Duchesne *et al.* (2000) report that with proper precautions, the caribou herd in the Charlevoix Biosphere Reserve, Canada has benefited from ecotourism. Positive impacts include heightened interest in the population, education on the role of the caribou in the preservation of biodiversity, contribution of funds towards conservation, and the prevention of poaching.

Carrying capacity

> Never has so much been said, but we sense considerable disagreement relative to the perception of how much we have learned, how useful this information has been, and how much we have yet to discover.
>
> (Burch, 1984, p. 142)

In the quote above, Burch reflects upon a real concern, and perhaps frustration, over the progression of accepted philosophies and approaches surrounding the advancement of carrying-capacity research. In one of the first papers written on carrying capacity, Lucas (1964) examined the perception of wilderness by visitors to the Boundary Waters Canoe Area in Wilderness. He concluded that all resources are defined by human perception and that different users

Table 3.4. Tourism and recreation impacts.

Facilities and activities	Impact									
	Habitat change/loss	Species change/loss	Aesthetics	Physical pollution	Soil change/damage	Noise pollution	Conflicts	Energy/ water usage	Local community	Revenue vs. costs
Car park	✖		✖			✖				✔
Cafe/shop			✖	✖				✖	✔	✔
Footpaths/roads	✖				✖				✔	✖
Information points/signage	✔			✔	✔	✔	✔			✖
Picnic areas	✖									✖
Permanent accommodation			✖	✖				✖		✔
Temporary accom./camping			✖	✖	✖			✖		✔
Toilets			✖	✖	✖			✖		✖
Visitor centre			✖	✖				✖	✔	✖
Abseiling										
Angling		✖		✖			✖			✔
Archaeology	✖		✔							
Archery										✔
Ballooning										✔
Canoeing										
Caving and potholing										
Climbing										
Children's play area	✖				✖				✔	✖
Dog walking				✖			✖		✔	
Field studies	✔									
Golf	✖		✖		✖		✖		✔	✔
Hang-gliding and paragliding										
Horse riding					✔					
Hunting	✔	✖				✖	✖		✖	✔
Husky sledding							✖			
Jet skiing				✖		✖				✔
Microlight										
Model aeroplanes						✖				

Table 3.4. *Continued.*

Facilities and activities	Impact									
	Habitat change/loss	Species change/loss	Aesthetics	Physical pollution	Soil change/damage	Noise pollution	Conflicts	Energy/ water usage	Local community	Revenue vs. costs
Motor sports				✖	✖	✖	✖		✖	✔
Mountain biking	✖				✖		✖			✖
Mushroom picking	✖	✖			✖					✔
Off-road vehicles	✖			✖	✖	✖	✖		✖	
Orienteering					✖					
Ornithology	✔	✔								
Paintball/war games				✖	✖	✖	✖			✔
Photography										
Picnicking				✖	✖					
Power boating			✖	✖		✖	✖			
Rafting										
Rowing										
Sailing and boardsailing										
Skiing	✖		✖				✖			✔
Swimming									✔	
Vehicle use				✖	✖	✖	✖	✖	✖	✖
Walking/running					✖		✖		✔	

Activities and impacts: ✔ = positive impact; ✖ = negative impact.
Source: Tribe *et al.* (2000).

such as canoeists and motor boaters will value different things in a wilderness setting. The concept of human values was also adopted by Wagar (1964), who soon recognized through his work that carrying capacity based solely on a consideration of ecological components, without human values, was doomed. These studies set the stage for a flurry of activity surrounding the carrying-capacity concept, particularly in regards to its applicability in wilderness settings.

In general, the carrying-capacity concept has a long history in the natural-resource management field which pre-dates its adoption in the field of outdoor recreation. In range management, for example, it refers to the number of animals of a species that can be maintained in a given habitat, without a degradation of the resource on which the game depend. Following this, carrying capacity for recreational purposes has been defined as the amount of use of a given kind, a particular environment can endure, year after year, without degradation of its suitability for that use (Catton, 1987). The actual damage or degradation inflicted by tourists is not straightforward, but subject to consideration along the lines of absolute numbers (the scale of tourist activity), and the relative damage per tourist, which is the efficiency with which the tourist activity is controlled (Steele, 1993). As Steele writes, 1000 tourists who each cause +1 damage, will be no worse than 200 tourists who each cause +5 damage, as depicted as follows:

Tourist damage = damage per tourist × number of tourists

In Amboseli National Park, 95,0000 vehicles per year were concentrated in only 10% of the park's area. More even distribution throughout the park created less disturbance and allowed for more vehicular traffic (an approach based on disperse-and-dilute). A concentrate-and-contain approach is said to work better in Galapagos, according to Steele, where tourists are restricted to areas (trails) of high carrying capacity only, thus eliminating them altogether in most areas of the park.

There are at least four different types of carrying capacity, including physical, ecological, facility and social, all of which are intricately intertwined. Recreational carrying capacity (social and recreational carrying capacity are often treated as synonymous) has often referred to the quality of experience that visitors will accept before dissatisfaction sets in. In more of a tourism context, it may also mean the degree of tolerance of local people to the presence of tourists, as identified by Saveriades (2000). While conceptually simple, the central theme of the concept is that recreational areas have essentially fixed abilities to accommodate use. Park 'A' is able to sustain about 500 visitors on the weekend, while Park 'B', because of its size, is able to accommodate 5000. However, such numbers have proved unreliable owing to the characteristics of the sites themselves, as well as the users. The wealth of research on the various issues confronting the carrying-capacity concept seems to fall along the following themes (Butler *et al.*, 1992):

1. There is an upper limit to the amount of use appropriate for a specific facility or resource.
2. People will readily accept some limits (e.g. seats on an aeroplane).
3. Problem concerns whether people will accept limits on vacation, or in the out-of-doors (common resources).
4. Like the Holy Grail, a magic capacity figure for an area does not exist.
5. Types of carrying capacity are not exclusive of one another (e.g. social, ecological, physical and facility).
6. The tolerance of different users groups (e.g. horseback riders, mountain bikers, and hikers using the same trail systems) to one another varies (intergroup differences).
7. At the same time, intragroup tolerances vary.
8. User encounters and levels of satisfaction vary according to a number of different variables, including space and time, user type, density, frequency and perceived crowding.
9. Sound planning and management are required.

In an extensive review of the carrying-capacity literature, Stankey and McCool (1984) wrote that (as suggested in point number 9 above) it is a management system directed toward maintenance or restoration of ecological and social conditions defined as acceptable and appropriate in area management objectives. What is perhaps most important about the statement is that it appears to be equally applicable to both socially oriented and ecologically oriented capacities (Graefe *et al.*, 1984). The diversity in types of impacts and the factors affecting impacts show similar patterns whether the focus is on contacts between users, campsite vegetation or wildlife behaviour (Kuss *et al.*, 1984). This understanding has given rise to a number of preformed planning and management frameworks, such as the recreation opportunity spectrum (ROS), limits of acceptable change (LAC), and the visitor activity management process (VAMP), used within the Canadian national parks system (see Payne and Graham, 1993, for an overview of these various strategies).

The ecological footprint

A more recent view of carrying capacity posits that the term in its conventional use is problematic in reference to humans, principally as a result of the effects of technology, preferences, and the structure of production and consumption (Arrow *et al.*, 1995). Rees (1996) writes that because we eliminate competing species, import locally scarce resources, and develop new technologies, carrying capacity is irrelevant. Globally the human population is well over 5 billion, on an earth that is decidedly disproportionate in terms of the dispensation of resources. The total net terrestrial primary production of the ecosphere currently under use by human beings is about 40% (Vitousek *et al.*, 1986). Rees writes that the carrying-capacity concept has application but only through a redefinition of the term away from a focus on maximum population (e.g. how many people can be sustained in a given area), to one which advocates maximum load (e.g. load is a function of population and the per capita consumption of population on a given area). Citing Catton (1986), Rees illustrates that in a span of 200 years, the average daily energy consumption of Americans has increased almost 20-fold. More people consuming more resources is thus the heart of the matter, with the fundamental question for economists being:

Fig. 3.3. Preformed planning and management frameworks allow recreationists to enjoy pristine environments.

Fig. 3.4. In Point Pelee National Park, Canada, built structures allow tourists to experience many of the park's environments.

> whether the physical output of remaining species populations, ecosystems, and related biophysical processes, and the waste assimilation capacity of the ecosphere, are adequate to sustain the anticipated load of the human economy into the next century while simultaneously maintaining the general life support functions of the ecosphere.
>
> (Rees, 1996, p. 3, accessed at http://www.dieoff.org/page110.htm)

Rees' solution is to examine carrying capacity in light of the maximum rates of resource harvesting, and waste generation that may be sustained, indefinitely, without an impairment of the integrity (functional and productive) of the relevant ecosystem (Rees, 1988). The corresponding inclusion of waste, suggests that for every material flow, there must be a sink to sustain these flows. For example, if a city was completely closed off to any flows the city would soon cease to function (starve and suffocate) without the input of resources and a place to put the waste. The carrying capacity of this setting would therefore be insufficient.

The concept of an ecological footprint, the calculation of the amount of land needed to produce the resources which we consume, and absorb the wastes we produce (Rees and Wackernagel, 1996), provides a rough indication of the natural capital needs of any population. For example, this may include amount of crop and grazing land, wood, built environments, fossil energy, amount of seascape, and so on. Rees (1996) calculated that the people of the Netherlands, because of their dependence on external ecoproductivity, require a land area more than 15 times larger than their current country to support their population's consumption of energy. The Netherlands, along with Japan, is recognized as generating huge unaccounted ecological deficits.

The implications of the ecological footprint to ecotourism are far-reaching, in allowing researchers to focus not only on the impacts of a specific region alone (e.g. Toronto), but rather to consider the movement of tourists outside their home ranges. It would then be necessary to measure the per capita ecological footprint of a traveller to Costa Rica, and compare that to his or hers at home, including whether the footprint increases or decreases in size when travelling. Such information would allow researchers to compare ecotourists with local people (e.g. what is the ecological footprint of a Costa Rican versus a Canadian visiting Costa Rica), in quantitatively measuring the effects of these different groups. Although quite provocative, at least from a tourism standpoint, economists have criticized the ecological footprint in recent years on the basis of its inability to have practical and policy applications. Consequently it may best be conceived as a heuristic device which provides the grounds for further investigation of impacts and capacities.

The precautionary principle

The precautionary principle has generated a great deal of interest among academics and government officials as one of the more contemporary views of human–environment interactions. The concept appears to have emerged in Germany during the 1970s and 1980s at a time of industrial restructuring and modernization (Boehmer-Christiansen, 1994). In general, the concept underlies an emphasis on preventative action by governmental officials, such that any damages to the natural world would be avoided in advance. O'Riordan and Cameron (1994) write that precaution adopts the perspective that people can be earth protectors and stewards, and that this may only happen if people anticipate possible problems and take care through everyday practices. These authors define the precautionary principle as 'a culturally framed concept that takes its cue from changing social conceptions about the appropriate roles of science, economics, ethics, politics and the law in pro-active environmental protection and management' (p. 12). In this regard, precaution has been extended to include six basic concepts, including: (i) preventative anticipation; (ii) safeguarding ecological space; (iii) restraint adopted is not unduly costly; (iv) duty of care, or onus of proof on those who propose change; (v) promotion of the cause of intrinsic natural rights; and (vi) paying for past ecological debt.

Hunt illustrates that one of the fundamental issues facing the precautionary principle is the fact that uncertainty is prevalent in the decisions that are made about the environment. This uncertainty may be viewed from two very different perspectives. The first and lesser version accepts that environmental discharges cause harm, which science cannot necessarily prove over time (with links to the burden and proof of harm). The second, and more radical version posits that scientific knowledge is indeterminate as regards the effects of humans on the environment. The implications of this perspective lead us to query the level of harm that a discharge might cause as well as whether we need the process which causes the harm. In addressing these concerns, Hunt suggests that the precautionary principle should prompt decision makers and various other groups to be open to the following three priorities. First, to include a wider participation in groups producing scientific advice; second to allow these groups to have equal decision-making power; and third to encourage the debates about scientific research and various procedures to be open and publicly accessible.

Discourse analysis

In talking face-to-face or through various media, people are involved in discourse. Discourse is thus about communication and meaning. These discourses may be very specific, as in language, or very general as in the social process of communication (Lemke, 1995). More specifically, discourse analysis deals with interpretations regarding the speaker, his or her spatio-temporal location, as well as to other variables that identify the context of the communication (Benveniste, 1971). As such, it is a shared method of understanding the world.

Once the domain of logico-semantic analysis, more contemporary approaches to the concept have evolved in many different disciplines allowing for widely varying conceptual schemes. In tourism, discourse might help to clarify meaning as regards the traveller vs. tourist, tour brochures and carrying capacity. This latter concept, as we have seen, can be interpreted in different ways, in different circumstances, by different people and at different times (Mowforth and Munt, 1998). In general, the theory appears to be moving in two directions: one that discusses discourses as sources of frameworks for power which direct the powerless, the other which looks at how discourses are communicative devices that define reality and new futures. This latter perspective has been adopted by researchers keenly interested in environmental politics, especially in regards to policy issues. For example, Bunce (1998) writes that the movement supporting the preservation of farmland has been controlled by the intersection of popular and professional discourses. He writes that the mainstream farm voices (many of whom resist restrictions on their development rights) are barely detectable in the farmland preservation movement. Instead, it is those from the urban fringe who view farmland as a physical symbol of a mix of ideologies and values. The division between these two main players has severely limited the ability of decision makers to implement policy within a broader rural land-use perspective.

Discourse, according to Dryzek (1997), conditions the way people define, interpret and address environmental issues. This is important because, as we have seen in the previous example, it demonstrates that people attach different meanings to the same issue, largely depending on what they have experienced, and to which stimuli or influences they have been subjected. From the environmental standpoint, there are at least two perspectives with regard to the information that the public is exposed to and uses to make informed decisions. The first is that the media continue to shield governments and corporations from the responsibility of being accountable for their actions that have proven harmful to the natural world (Lee and Solomon, 1991). The second is that the principal watchdog responsible for reporting the wrongdoings of government and industry (environmental non-governmental organizations; ENGOs) has not yet learned how to play the game well enough. In addressing the shortcomings of environmental advocacy, Cantrill (1996: 76) notes that:

> Few of the environmental vanguard recognize that the 'facts' about ozone depletion, toxic waste, and the overconsumption of limited resources do not stand alone in the minds of most people; rather, 'facts' are interpreted in light of preexisting notions about the world and are modified by all manner of information-processing biases. Thus, the shortcomings of human cognition often preclude the worth of many logical arguments to protect our planet.

In addition, discourse analysis, from the environmental standpoint, provides a meaningful way with which to view the Dominant Social Paradigm. However, as Dryzek (1997) suggests, this perspective must be quite diversified in its approach. For example, it can be reformist or radical; or it can be prosaic or imaginative. In the case of the former, the political economy of society, as established by industrial interests, is seen as a fairly structured template. Actions that are required to resolve issues do not lead to a new kind of society, but rather are the type which have been defined by industrial interests. In the latter case (imaginative), departures seek to redefine society through an approach that considers the environment as being at the heart of the problem, rather than lying outside the realm of influence (see Table 3.5). The four-cell matrix of Table 3.5 illustrates a number of different ways with which to classify environmental discourses. All four reject industrialism, but they do so at very different levels, with very different points of departure. These are as follows (Dryzek, 1997, pp. 13–15):

- *Environmental problem solving*. Defined by taking the political-economic *status quo* as given but in need of adjustment to cope with environmental problems, especially via public policy. Markets may put price tags on environmental harms and benefits; while the administrative state may institutionalize environmental concern and expertise in its operating procedures.
- *Survivalism*. Continued economic and population growth will eventually hit limits set by the Earth's stock of natural resources and the capacity of its ecosystems to

Table 3.5. Classifying environmental discourses.

	Reformist	Radical
Prosaic	Problem solving	Survivalism
Imaginative	Sustainability	Green radicalism

Source: Dryzek (1997).

support human agricultural and industrial activity. The limits discourse is radical because it seeks a wholesale redistribution of power. It is prosaic because it can see solutions only in terms of the options set by industrialism.

- *Sustainability*. Defined by imaginative attempts to dissolve the conflicts between economic and environmental values. The concepts of growth and development are redefined in ways which render obsolete the simple projections of the limits discourse. However, without the imagery of apocalypse that defines the limits discourse, there is no inbuilt radicalism to the discourse.
- *Green radicalism*. It is both radical and imaginative. Its adherents reject the basic structure of industrial society and the way the environment is conceptualized therein in favour of a variety of quite different alternative interpretations of humans, their society, and their place in the world. Given its radicalism and imagination, it is not surprising that green radicalism features deep intramural divisions.

Of the four outlined above, the sustainability discourse appears to be the one most firmly embraced by tourism researchers, which makes discourse analysis of particular use for ecotourism. If ecotourism is to continue to espouse an ecological approach to the tourism industry, it must do so from a clearly defined perspective. Discourse, as a method of analysis (although not fully elaborated upon here), may provide a means by which to better organize the meaning of ecotourism, in reference to the involvement of multinationals, community development, policy development, aboriginal interests, as well as better defining differences that exist between hard- and soft-path ecotourists and the service providers who support these different tourist types. As such, ecotourism is probably perceived very differently among the various stakeholders involved in the ecotourism industry, a fact which might be quite effectively demonstrated through discourse analysis. Dryzek reminds us that environmentalism is composed of a variety of discourses, in complementary and competing frameworks. The disentanglement of these may occur in light of: (i) basic entities whose existence is recognized or constructed (the ontology of a discourse); (ii) assumptions about natural relations (cooperation, competition, hierarchies); (iii) agents and their motives (actors who are individuals or collectivities); and (iv) key metaphors and other rhetorical devices ('spaceship', 'machines', 'organisms', and so on).

Norms

The concept of norms has been defined by McDonald (1996) as 'evaluative rules that involve some levels of shared group agreement or consensus, and focus on what is appropriate behaviour, social, and environmental conditions for a given situation' (p. 1). A 1996 edition of the journal *Leisure Sciences* focused on the concept of norms as a more contemporary manner with which to examine the various issues discussed in the section above on carrying capacity. The concept, therefore, appears to have an especially strong link to outdoor recreation, where there is conflict between different users for the same resources. The classic outdoor recreation example is the canoeist who has different expectations, sensitivities to density, and feelings of crowding from other users of a resource such as anglers or river rafters. Management must be aware of these various group-oriented perspectives and act accordingly. Fennell (1997) found that canoeists and white-water rafters on the remote Clearwater River, Saskatchewan, each had different inter-group and intragroup use encounter expectations (as well as different expectations in regards to encounters with float planes). Table 3.6 shows the average upper limit encounters of these groups per day. Canoeists said that their upper limit for encounters with other canoeists was, on average, 4.3 as compared to 6.0 for rafters. Canoeists were also willing to put up with 1.4 float plane encounters per day, as compared to 1.8 for white-water rafters. Finally, the table shows that canoeists identified a normative limit of 3.4 rafters per day, while rafters said the upper limit of other rafters per day was 8.2.

Table 3.6 Encounter expectations for canoeists and rafters

Encounter variable	Mean scores	
	Canoeists (n = 14)	Rafters (n = 26)
At what number would you say there were too many of the following *for me* on the river?		
Canoeists ...		
upper limit per day	4.3	6.0
Float planes ...		
upper limit per day	1.4	1.8
Rafters ...		
upper limit per day	3.4	8.2

Source: Fennell (1997).

One of the more comprehensive papers in the *Leisure Sciences* edition, mentioned above, was one by Shelby *et al.* (1996), who examined the norms concept from perspectives of history and application. These authors suggested that the norms concept is based on the work of Jackson (1965), who used an impact accessibility curve (IAC) to describe different norms for different groups. The various curves which may be placed on the IAC describe social norms in terms of average individual evaluations. Shelby and his colleagues used the following example (Fig. 3.5) to examine three different types of activities in terms of the level of acceptability of different numbers of encounters.

The horizontal axis on the figure represents impacts increasing from right to left, while the vertical axis represents the evaluations, with those which are positive on top of the curve, and negative ones below (apart from neutral ones which may also occur). More specifically, the authors identify a number of conditions which help to examine differences between groups. These include: (i) the optimum condition, characterized by the most positive evaluation; (ii) the range of tolerable conditions, which include evaluations above the neutral line; (iii) intensity of the norm, which is the relative distance of the curve above or below the neutral line; and (iv) agreement or crystallization, which

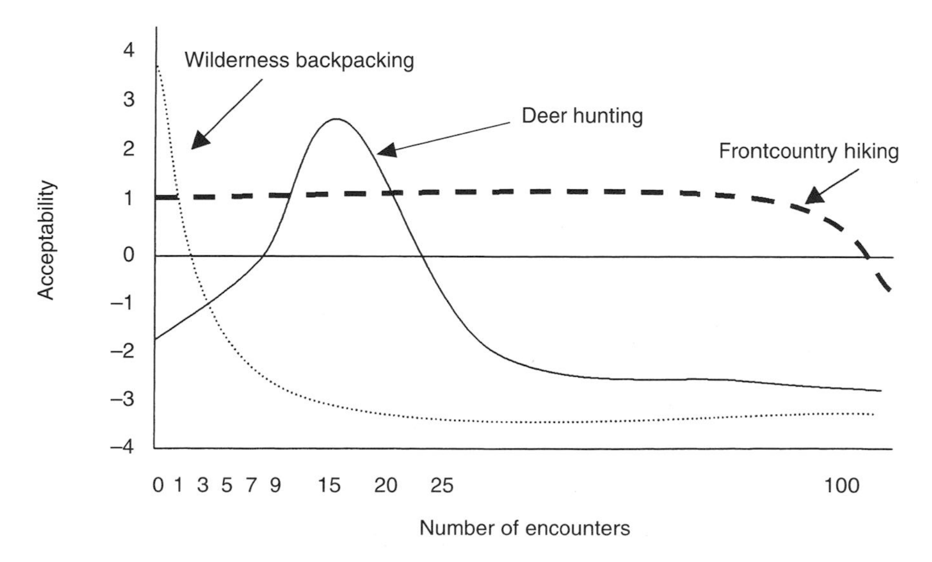

Fig. 3.5. Hypothetical norm curves. (Source: Shelby *et al.*, 1996.)

is the variation among evaluations at each impact level. In the example above, the wilderness backpacker has an optimum condition of zero, the range of tolerable conditions is low, the intensity is high and the crystallization is high. Conversely, for the frontcountry hiker the optimum condition shows many encounter levels, the range of tolerable conditions is higher, the intensity is lower, and crystallization is also lower. The authors illustrate a number of other types of norms including those for different activities, different settings, the same setting, ecological impacts, aesthetic impacts and stream flows, for different recreational users.

The utility of the norms concept for ecotourism lies in its ability to examine encounter levels, aesthetics, and so on, in regards to different types of ecotourists, such as hard path and soft path, as well as ecotourists compared with other types of tourists. In addition, other variables that are more specific to ecotourism may be compared across groups, including conservation group affiliation, impacts, consumptiveness, education, ethics and responsibility, and so on.

Community Development

In a recent article on the greening of the Amazon economy, Romero (2000) examined a number of different economic enterprises being considered for this region. He concluded that one of the most pressing issues surrounded the conflict between rainforest preservation and the need for people to feed and school their children. The latter becomes much more important than the former, it was suggested, because it is an issue of survival, rather than one of preservation in an environment that is seemingly endless and bountiful. The principal case study used in the article involved the marketing of guarana, which is a bright red, highly caffeinated berry with the potential to be used as an aphrodisiac in soft drinks. One of the world's largest soft drink manufacturers has been closely involved and it was considering the use of this ingredient in one of their products. Although the level of involvement of this company was not disclosed, a familiar scenario in international development is unfolding: the overwhelming influence of multinational corporations who dominate the manner in which resources are harvested, marketed and distributed. In such cases, control is often removed out of the hands of local decision makers and entrepreneurs and placed into the hands of the dominant entity, with devastating impacts on the social, economic and ecological aspects of the community. The literature that has emerged on ecotourism, sustainable tourism, alternative tourism, green tourism and responsible tourism has attempted to address this imminent concern by advocating the need to give control back to those who often stand to lose the most: local people in peripheral settings.

Tourism–community issues

Tourism as a presence in a less developed country may be viewed differently by local people than other export industries, such as bananas, coffee, or the one mentioned above. Although these have a distinct presence within the community the presence may not be overwhelming as that of tourism which attracts lots of people who may be distributed widely throughout the region. What may transpire is a relationship which progresses from one that is very positive to one which is antagonistic (see Doxey, 1975). Local inhabitants are often forced into a power struggle over resources with a transient population of tourists who share space, who come for a short period of time, who are often treated preferentially, and who later leave. The benefits of the industry, as outlined by Butler (1991), are rarely uniform, with advantages going to those involved in the industry, and costs borne by those who have no stake in the industry at all.

Unfortunately, as suggested by Wearing and Neil (1999), the power struggle that is inherent in many tourism case studies occurs within a number of different political levels, and it is one of tourism's ironies that those who are most removed from the tourism setting are those who ultimately determine where to develop, how to develop, and who benefits. Indeed, the element of control inherent in the aforementioned discussion strikes to the very heart of issues on tourism development. This is supported by Clark and Banford (cited in Wearing and Neil, 1999), who write that there is no rational reason why the community should not decide on what level of tourism development is appropriate for their region, and the level of changes they are prepared to absorb. Scheyvens (1999) writes that instead of focusing almost completely on the economic dimensions of this power struggle, we should also turn our attention to a consideration of social matters. She has developed the ecotourism empowerment framework (Table 3.7) as a means by which to identify the various economic, psychological, social and political empowerment and disempowerment signs which lead to various impacts on a community.

The United Nations has defined community development as 'a process designed to create conditions of economic and social progress for the whole community with its active participation and the fullest possible reliance on the community's initiative' (United Nations, 1959, p. 6). A more contemporary view of the concept identifies environment as

Table 3.7. Framework for determining the impacts of ecotourism initiatives on local communities.

Economic empowerment	Ecotourism brings lasting economic gains to a local community. Cash earned is shared between many households in the community. There are visible signs of improvements from the cash that is earned (e.g. improved water systems, houses made of more permanent materials).
Economic disempowerment	Ecotourism results in small, spasmodic cash gains for community. Most profits go to local elites, outside operators, government agencies. Few individuals/families gain direct financial benefits from ecotourism, while others cannot find a way to capitalize because they lack capital and/or appropriate skills.
Psychological empowerment	Self-esteem of many community members is enhanced because of outside recognition of the uniqueness of their culture, their natural resources and their traditional knowledge. Increasing confidence of community members leads them to seek out further education and training opportunities. Access to employment and cash leads to an increase in status for traditionally low-status sectors of society (e.g. women, youth).
Psychological disempowerment	Many have not shared in the benefits of ecotourism, yet they may face hardships because of reduced access to resources. They are thus confused, frustrated, disinterested or disillusioned with the initiative.
Social empowerment	Ecotourism maintains or enhances the local community's equilibrium. Community cohesion is improved as individuals and families work together to build a successful ecotourism venture. Some funds raised are used for community development purposes (e.g. to build schools or improve roads).
Social disempowerment	Disharmony and social decay. Many in the community take on outside values and lose respect for traditional culture and for elders. Disadvantaged groups (e.g. women) bear the brunt of problems associated with the ecotourism initiative and fail to share equitably in its benefits. Rather than cooperating, individuals, families, ethnic or socio-economic groups compete with each other for the perceived benefits of ecotourism. Resentment and jealousy are commonplace.
Political empowerment	The community's political structure, which fairly represents the needs and interests of all community groups, provides a forum through which people can raise questions relating to the ecotourism venture. Agencies initiating or implementing the ecotourism venture seek out the opinions of community groups and provide opportunities for them to be represented on decision-making bodies.
Political disempowerment	The community has an autocratic and/or self-interested leadership. Agencies initiating or implementing the ecotourism venture treat communities as passive beneficiaries, failing to involve them in decision making. Thus the majority of community members feel they have little or no say over *whether* the ecotourism initiative operates or *the way* in which it operates.

Source: Scheyvens (1999).

a key element within community development. Christenson *et al.* (1989), for example, define it as: 'a group of people in a locality initiating a social action process (i.e. planned intervention) to change their economic, social, cultural, and/or environmental situation' (p. 14). The concept of planned intervention in this latter definition is important, as it suggests that social change is a result of the application of human knowledge, capital and technology to help in solving the problems of a particular individual or group. As such, this intervention is purposefully directed change for the betterment of society which is a fundamental and driving force of the community development process (Christenson *et al.*, 1989; see also Drake, 1991).

The concept of community development has been explored at least as far back as the 1940s, through projects that were initiated by the Ford Foundation in India (Voth and Brewster, 1989). While community development may be viewed as a programme (focus on the programme rather than the people involved) or a movement (cause to which people become involved), in tourism it has most often viewed as a *process* or *method* (see the work of Sanders, 1958). In the former case, change occurs through a series of stages: (i) from a situation where a few elite make decisions for the community, to one where the people themselves make the decision; (ii) from minimum to maximum interaction and cooperation; and (iii) from the use of external resources/specialists to one where local resources are used. In the latter case, *method*, community development is a means to an end. It uses process in order that the will of those using the method may be carried out.

Many definitions of ecotourism focus specifically on the notion that ecotourism must be a vehicle for positive community development (see Fennell, 1999). These definitions state that control must rest in the hands of local people and that the benefits of ecotourism must be realized by local people over the long term. The ecotourism literature supports a number of different recipes for involving the community and planning for community development. A number of these are included in Table 3.8.

Social capital

Although not talked about extensively in the literature on ecotourism, social capital has generated a good deal of interest among sociologists and theorists in other fields (including recreation). Social capital has been described as the glue that holds society together (Svendon and Svendson, 2000), the ingredients of which consist of social organization (social networks, contacts and trust) and norms of reciprocity (based on shared values) which help to facilitate the coordination and cooperation for mutual benefit (Warner, 1999). Coleman (1988) refers to social capital as the ability of people to work together for common purposes within groups and organizations. More elaborately, Bordieu (1986) defines social capital as:

> the aggregate of the actual and potential resources which are linked to possession of a durable network of more or less institutionalized relationships of mutual acquaintance and recognition – or in other words, to membership in a group – which provides each of its members with the backing of the collectively-owned capital.
>
> (pp. 248–249)

Two forms of social capital include: (i) integration, including intra-community ties; and (ii) linkage, which refers to extra-community networks. Both are seen as being of importance for economic development initiatives, such as tourism. At the heart of the issue of social capital and community development is the element of benefit. On one hand, community members are expected to contribute to the community, in order to also receive certain benefits (an embedded perspective where the sources of norms and group reciprocity are central to the perspective). As regards ecotourism, there are a number of stakeholder groups which can, or rather should, be called upon to make comprehensive ethical decisions for the good of the community. These include the tourism industry, government, NGOs, universities, the community, and other national and international agencies with an interest in tourism development. While at times the forces which are charged with the task of planning and developing ecotourism are often at odds

Table 3.8. Sample of recipes for achieving community development through ecotourism.

Brandon (1993)	Horwich *et al.* (1993)	Williams (1992)	Drake (1991)	ARCDC (1997)
Local participation's role	Plan at village level	Grassroots planning, driven by local people	Determine role of local participation in project	Organize local citizens and government and educate them on ecotourism planning and benefits
Empowerment	Local integration	Understanding of ecotourism markets	Choose research team	Use methods that involve the public in planning
Participation in project cycle	Broad, legal, local empowerment	Inventory of region's resources to determine suitable areas for ecotourism	Conduct preliminary studies on conditions in community	Coordinate efforts with existing planning initiatives/groups
Creating stakeholders	Use existing resources	Goals and objectives sensitive to culture and ecology	Determine the level of local involvement	Evaluate methods and opportunities for training local people for ecotourism services
Link benefits to conservation	Appropriate scale	Creation of vision statements	Set up appropriate participation scheme (sharing of information)	Create support for the ecotourism plan by promoting ecotourism's benefits
Distributing benefits	Sustainability	Establish a formal tourism board to work with operators and public	Initiate dialogue and educational efforts (building consensus)	Create a vested interest for local people to protect natural resources
Involve community leaders	Local needs/conservation are primary		Collective decision making	
Using change agents	Professional must contribute		Develop action plan and implementation scheme	
Understand site-specific conditions	Conservation is a viable development strategy		Monitor and evaluate	
Monitor and evaluate	Conscientious tour operators and investors			

because of the emphasis on their own needs, deficiencies and problems (such as those identified above), those community initiatives which are strongest are those that take, value and mobilize the skills, capacities and talents of these various constituents (see Kretzmann and McKnight, 1993).

Perhaps as a result of the realization of the foregoing discussion, a fundamental pillar of social capital is the focus on trust and values that promote social cooperation and social stability. For example, Newton (1997, p. 576) suggests that 'social capital focuses on those cultural values and attitudes that predispose citizens to cooperate, trust, understand and empathize with each other – to treat each other as fellow citizens, rather than as strangers, competitors, or potential enemies.' However, the principles of social capital may not be articulated or implemented evenly across cultures. For those societies defined as high trust (e.g. Japan, Germany), norms and values are structured such that they can subordinate individual interests to those of larger groups. As Fukuyama (cited in Arai, 1999) writes, out of shared values comes trust, and trust may be transferable into measurable economic activity. However, in lower trust societies such as China, Italy and France, the family constitutes the basic unit of economic organization. The construction of larger economic entities is constrained because of the focus on the family (Arai, 1999). As such, the state must intervene as a catalyst for small- and large-scale economic activity.

One perspective or method which has been proposed, as an off-shoot of the social capital theory, is the entrepreneurial social infrastructure (ESI) model developed by Flora and Flora (1993), and further refined by Flora (1998). Like social capital, ESI is built on a foundation of trust, social networks and norms, and designed to encourage collective action to achieve tangible goals, including sustainable development and well-being within the community (see Fig. 3.6). The ESI is based on three main components: (i) legitimacy of alterations; (ii) mobilization of resources; and (iii) network qualities.

- *Legitimacy of alternatives*. Flora (1998) writes that respect must be shown for multiple points of view in the community. Strength may be drawn from people with different backgrounds and experiences, especially those which require limited commitment to the group. There is the acknowledgement that communities will differ in their acceptance of diverse ideas and people. Community politics must be depersonalized so that individuals are not at risk in their reputations and livelihoods. Finally, the manner in which issues are discussed and decided upon are just as important as the outcomes.
- *Mobilization of diverse resources*. ESI is enhanced when resources from inside or outside the community can be readily accessed. Mobilization includes those in positions to invest in the community, the commitment of taxes to community betterment, and the innovations which permit the channelling of resources with the community. An example of this type of resource mobilization occurred in Shetland where significant amounts of money generated through the offshore oil industry have been put in trusts to stimulate entrepreneurialism and build facilities and infrastrucure to better the lives of Shetlanders.
- *Network qualities*. Networks can serve to include and exclude, and to consolidate and share power. Community networks are most successful if they are diverse, inclusive, flexible, horizontal (linking those of similar status) and vertical (linking those of different status, particularly local organizations/individuals with agencies who have access to resources).

This tool may be a fruitful means by which to draw on the alternatives, resources and networks available within a community, and to effectively ensure that ecotourism lives up to its billing as a meaningful form of community development. Like so many other phenomena, it is not necessarily the concept itself which guarantees accessibility, equity, balance and prosperity, but rather the processes that are operationalized to ensure that the philosophical intent of the vehicle is truly represented in a community situation. In ecotourism we have perhaps fallen victim to this mentality, without attempts to secure the appropriate mechanisms which would help to increase our chances of success.

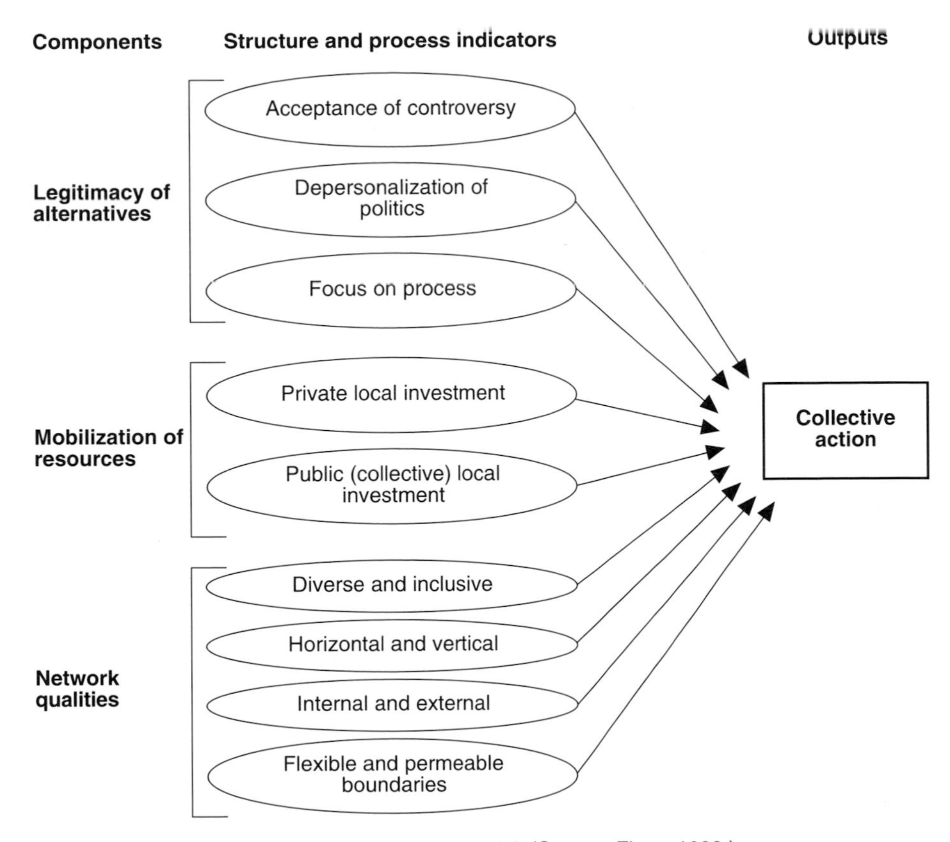

Fig. 3.6. The entrepreneurial social infrastructure model. (Source: Flora, 1998.)

Inducing participation by members of the community is an important step in ensuring that ecotourism is accepted and integrated with current and future development initiatives. The sooner the connection is made between development plans and citizens, the more likely it will be to succeed. In fact, in order to be consistent with the goals of ecotourism as a form of sustainable development, it is critical to gain the local support of citizens. Cases do exist whereby ecotourism development plans were imposed on local communities with disastrous consequences. In Tres Guarantias, Mexico, for example, the top-down development approach was implemented without proper social integration and few benefits realized by local people (MacKinnon, 1995). In Panama, the configuration of an imposed 'ecotourism' industry placed Kuna Indians in low-pay jobs, and with little control. As a result of this the Kuna forced investors out of the region and established an industry based on local control, small size, and equity and dialogue (Chapin, 1990). Maintaining open lines of communication between stakeholders is critical in developing successful local ecotourism industries. Paul (1987) suggests that citizen involvement can take many forms, including:

- *Information sharing*. Investors and designers share information with local citizens in order to facilitate collective (e.g. conservation groups) and individual action.
- *Consultation*. The public is consulted on many issues at many times during the project cycle.
- *Decision making*. The public becomes involved in making decisions about project design and implementation.
- *Initiating action*. Local citizens actively take the initiative regarding decisions about ecotourism development.

While some citizens will want to become involved at one level or another, others may not want to be involved at all. Furthermore, the invitation to have local participation in planning decisions may be difficult to organize, it may slow the process of planning, add to the cost of the project, and also involve more people than are required. However, the benefits may be numerous, including the development of leadership in the community, increases in community commitment, an opportunity to tap into a pool of latent human resources, innovation, and empowerment. Some techniques for inducing public participation are as follows (adapted from Ontario Ministry of Tourism and Recreation, 1985b), with the associated most effective uses of these techniques on the basis of a select few variables (see Table 3.9):

- *Public hearing.* A formal, controlled approach that is often used by government; it usually deals with narrow issues, often with a legal twist; verbal and written submissions are accepted.
- *Community forum.* Not as formal as a hearing; it allows for an exchange of information but usually in a controlled way.
- *Vote for referendum.* A political method for resolving an issue by a majority decision; to be successful the issue must be clearly understood; action may require more than a simple majority.
- *Large public meeting.* An informal gathering, open to everyone; it usually proves to be a good exchange of information.
- *Small group meeting.* A structured meeting usually with members of one organization or interest group to explore in depth their concerns.
- *Community conference.* A flexible, informal gathering that can be used for different purposes in a wide variety of situations; one staff person can work with volunteers over a period of time to plan, organize and run the event; outside resource people can be used to stimulate discussion; it can be broken up into small discussion groups for immediate feedback.
- *Mass media.* Can be used to create awareness, present information, define choices or gather information; the quality of the response obtained may be suspect.
- *Expert panel.* Can be used with a large public meeting to provide accurate information; it can also be used for gathering, testing and clarifying information.
- *Survey and questionnaire.* A commonly used technique for gathering information; the data collected depend on the questions asked; the respondent remains passive unless the question is left open-ended; accuracy depends largely on the sample selected to respond.
- *Personal interview.* A slow, time-consuming technique; the response depends on the sample chosen and the questions asked.
- *Participant observation.* Method can be used at almost every meeting; those present are invited to express their opinion or reaction to whatever is being presented or discussed.
- *Workshops.* Somewhat similar to small group meeting but participants will become more actively involved during the session; it is an excellent method for developing a site plan.
- *Simulation and role playing.* Two sophisticated techniques that have limited use without high levels of skill and practice; role playing is excellent for working on attitudinal problems.
- *Advisory group.* The technique may allow too little active participation and, probably, no decision-making authority; the members can be given much greater responsibility provided that it is mutually agreed upon.

Citizens will have to decide which level of participation is best for them, the project, and for the community. From the perspective of the ecotourism service provider, it is important that as many linkages are drawn to the community as possible, not only for the support of the ecotourism operation itself, but also from the programme side. A tremendous pool of community human resources may be tapped into, which may enable the operator to provide enhanced services, and win support from sides of the community. This, according to Falk and

Table 3.9. Public participation techniques.

Technique	Scope		Method provides			Staff time	Approx. cost	Most effective use
	Broad	Specific	Public awareness	2-way communication	Level participation			
Public hearing	L	M	L	L	L	M	M	Make policy
Community forum	L	M	M	M	L	M	M	Resolve conflict, exchange information
Vote or referendum	M	H	H	L	H	L	H	Test consensus
Public meeting	M	M	H	L	L	M	L	Test reaction
Small group meeting	L	H	L	H	M	M	L	Give information, gather ideas, build trust
Community conference	H	H	H	H	H	L	L	Many and varied
Mass media	H	H	H	L	L	L	H	Generate interest, test reaction
Expert panel	L	H	L	M	L	L	L	Gather information, evaluate
Survey or questionnaire	M	H	M	L	M	H	H	Test attitudes, test opinions
Personal interview	M	H	L	H	L	H	H	Gather information, test assumptions
Participant observation	L	M	L	L	L	M	L	Gather information
Workshop	H	H	M	H	H	H	M	Educate, evaluate, test alternatives
Simulation or role playing	L	H	L	M	L	H	L	Provide information, resolve conflict
Advisory group	M	H	L	H	L	L	L	Sounding board

H, high; M, medium; L, low.
Source: Ontario Ministry of Tourism and Recreation (1985b).

Kilpatrick (2000), translates into *trust* within the community, *interconnectedness* between various members and levels within the communities' hierarchy, and the provision of *opportunities*, which all contribute to social capital. It is worth noting two points which have been raised by Wyatt (1997) as regards community: (i) community is an achievement, not a given; and (ii) a community is essentially what its members intend it to be. Ecotourism has the potential to aid the process of community building, especially through efforts to maintain control and to be as inclusive as possible.

Although generally accepted as a worthwhile tool, the social capital concept has been criticized regarding its usefulness to communities that are marginalized or peripheral. Some youth, for example, who are invested in and who become the brightest tend to be those who end up leaving their communities behind. In their wake, they leave resentment, feelings of failure by other youth, and disjointedness, leading to a situation of social deficit rather than social capital. Furthermore, many traditional communities are hundreds of years old and the Western ideals put forth by social capital are at odds with the traditional mechanisms that may be employed for social organization.

Tourism Ethics

Ethics has been defined as 'inquiry into the nature and grounds of morality where the term morality is taken to mean moral judgements, standards, and rules of conduct' (Taylor, 1975, p. 1). Our attitudes and behaviours are founded on belief systems which are in turn shaped by the environment in which we live, including the influences from family, friends and media. As such, there is a good deal of support to suggest that ethics, which includes values and conduct, are a result of the learning process (see Rest, 1975). Collectively, attitudes and beliefs of individuals determine the culture of organizations, and more broadly of communities and societies. In corporations, for example, business practices (which may or may not be ethically based) define the standards which guide the internal and external behaviour of the entity (Hunt *et al.*, 1989).

Ethics has emerged as a central area of research in many disciplines over the course of the past decade, including medicine, sport, law, business and environment. In these disciplines theoretical and applied ethics have been debated strongly, with the consensus that both are instrumental in providing solutions to problems which are very complex. The same cannot be said for tourism, however, despite the overall magnitude of the industry as well as the sheer number of interactions which occur from stakeholders (i.e. tourism brokers, including hotel managers, operators, and so on; many different tourist types; and local residents) who represent a variety of social, economic and ecological interests. It is the interactions between these groups that is so much the subject of debate in the tourism literature, which have contributed to a vast array of very different impacts. As a case in point, Arlen (1995) found that of 34 ecotourism operators contacted, 27 stated that they do not give environmental concerns a high priority. Other examples cited include operators suggesting that comfort was the critical concern for the company, and its clients, marketing of ecotourism sites unseen, and the capture of wild animals for the photographic convenience of ecotourists.

Unfortunately, as discussed by Wheeler (1994), the determination between right and wrong behaviour in tourism is made on the basis of intuition rather than formal reasoning. He suggests that the principal means driving ethical behaviour – codes of ethics – have had little overall effect on the industry, with more confusion than appropriate direction. This sentiment, however, is still very much open to debate. And while there are few articles on tourism codes of ethics, especially ones driven by appropriate ethical theory, it may suffice to take stock of the literature in business ethics and other disciplines, until such time as tourism researchers have explored the area adequately.

Codes of ethics have been developed by a number of different stakeholder groups, including governments, NGOs and tourism operators. Often these are local (i.e. site specific), but many have also been written as cosmopolitan statements (i.e. for tourism in general) which direct tourists, operators, or local people to either act or not act in accordance within the normative views related to the site. (See the work of Mason and Mowforth, 1995; and Malloy and Fennell, 1998, for extended lists of codes of ethics.) Table 3.10 is an example of a code of ethics that has been developed for Costa Rica specifically and Latin America in general (Blake and Becher, 2001).

Theoretical ethics

Although quite different, normative approaches to ethics, such as deontology and teleology, have long been considered as meaningful ways in which to guide decision making. Briefly, deontology is a theoretical perspective which posits that acts must or must not be done regardless of the consequences of their performance or non-performance (referred to as 'right' behaviour). 'Right' behaviour then is a function of one's motivation and ability to follow pre-established rules or doctrines. Using the example of the code of ethics in Fig. 3.10, each and every one of the statements is deontological in nature (i.e. they instruct the actor to behave in one way or another). Using the fifth guideline as an example: 'Wildlife and natural habitats must not be needlessly disturbed', the agent is directed to not needlessly disturb wildlife and natural habitats. However, what might be more helpful, is more information which describes what the consequences would be of the actor's performance or non-performance. This consequential element is the cornerstone of the teleological school of normative ethics, which focuses on the ends of behaviour. From this perspective, the rightness or wrongness of any act (which may be viewed as 'good' behaviour) depends entirely upon its consequences. The teleological extension to the example used above might read as follows:

> Wildlife and natural habitats must not be needlessly disturbed. Wild animals should not be fed as this alters their diet and behaviour. Furthermore, some ecosystems such as coral reefs are particularly sensitive, and excessive human intrusion (stepping on coral heads, turbidity, and so on) may cause irreparable damage.

In this case, 'alters their diet and behaviour' and 'excessive human intrusion ... may cause irreparable damage' is information which provides the agent with much more information in deciding whether to engage in certain actions. Malloy and Fennell (1998) write that it is unfortunate that most codes of ethics are deontological in nature, thus not providing the tourist with some critical information about the implications of his or her actions.

This is the premise under which Fennell and Przeclawski (2002) worked in attempting to establish a model of ethics for the variety of tourism interactions which may occur within a community. This research followed from the work of Fennell (2000a), who suggested that ethical dilemmas were derived from interactions, largely based on the situation, but also on the assumptions and beliefs which these various groups maintain (about

Table 3.10. Code of ethics for sustainable tourism In Latin America.

1. Tourism should be culturally sensitive
2. Tourism should be a positive influence on local communities
3. Tourism should be managed and sustainable
4. Waste should be disposed of properly
5. Wildlife and natural habitats must not be needlessly disturbed
6. There must be no commerce in wildlife, wildlife products, native plants, or archaeological artefacts
7. Tourists should leave with a greater understanding and appreciation of nature, conservation, and the environment
8. Ecotourism should strengthen conservation efforts and enhance the natural integrity of places visited

each other, for example). Inappropriate behaviour by one or more groups may lead to social and/or ecological impacts, and these negative experiences are then fed back into the basic assumptions that these individuals hold towards others. Figure 3.7 expands on this model, by focusing on the types of evaluation which agents may make in the determination of sound ethical judgements. These may be normatively based judgements (i.e. deontological or teleological), or existential in nature. (Briefly, existentialism is a more subjective and introspective theoretical perspective positing that acts are right, based on the free will and responsibility of the individual. The integrity of the individual is secured by being true to oneself through self-determined, authentic action.) Certain situations may compel the ecotourist to override certain universal laws (deontology) in being true to oneself, but also to other people and the environment. The interaction between these theoretical perspectives and the various actors within a tourism destination (including the environment) are illustrated in the figure, with a graphic

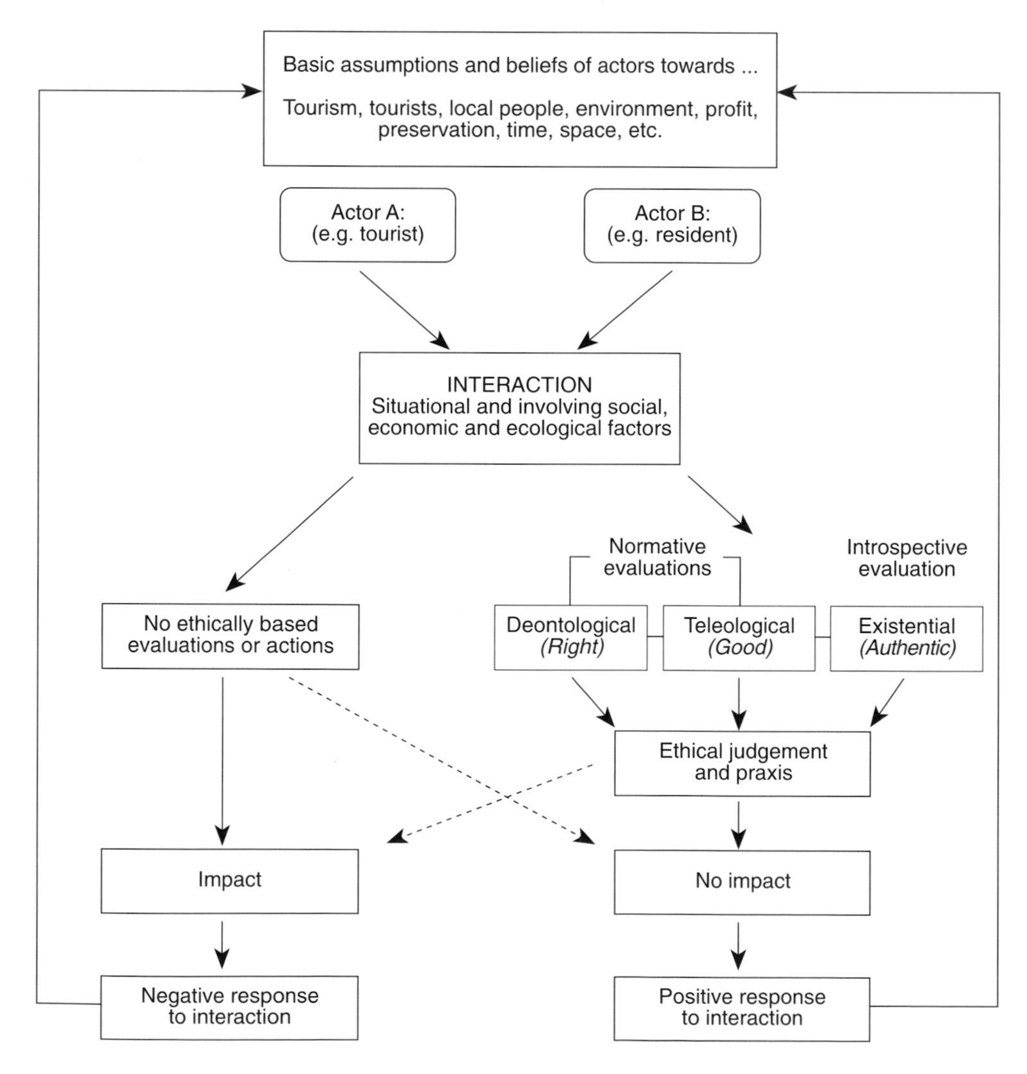

Fig. 3.7. Tourism ethics – interaction framework. (Adapted from Fennell, 2000a.)

description of how such theories may aid in avoiding impacts, as suggested by Fennell and Przeclawski:

> for those actors who avoid making ethically based evaluations or actions, the chances of creating impacts are higher (indicated by the solid black line) than those who do (indicated by the broken line). Similarly, the act of making ethical judgements does not eliminate the possibility of creating an impact, but one's chances of avoiding impacts are thought to be much stronger if they employ normative and introspective evaluations in situations which demand them.

In a relatively short period of time, tourism and ethics have grown from a focus which has been decidedly hospitality driven, to one which has explored very specific aspects of tourism and ethics (see Ahmed *et al.*, 1994; Hultsman, 1995; Walle, 1995; Malloy and Fennell, 1998; Fennell and Malloy, 1999). It is an area that will command a great deal more effort and respect from decision makers and researchers if we are to fully embrace the reported need to curtail the many negative issues that currently confront the ecotourism industry.

Conclusion

This chapter has touched upon a number of elements that are central to the setting in which ecotourism takes place. Although at times broad in scope, the discussion has attempted to be inclusive of a number of themes and considerations which are key aspects of the industry. These include brief analyses of the environment as a system, ecotourism supply (including biogeography, nature reserve design, and public and private reserves), the positive and negative impacts of tourism, carrying capacity, norms, community development and ethics. The message brought forward suggests that in order for ecotourism service providers to be successful, they must first be sensitive to the social and ecological conditions of the local and regional environments in which they interact.

4

Programming

The people who bind themselves to systems are those who are unable to encompass the whole truth and try to catch it by the tail; a system is like the tail of truth, but truth is like a lizard; it leaves its tail in your fingers and runs away knowing full well that it will grow a new one in a twinkling

(Ivan Turgenev)

Introduction

This chapter introduces the concepts of planning and programming, which together provide the necessary ingredients for the development of well conceived recreation and tourism experiences. Different types of planning are briefly examined as they apply to tourism and recreation, particularly with reference to some of the most frequently used planning strategies. The chapter follows with a discussion of the concept of recreation programming and its importance to the field of recreation. The basic philosophical structure of programming is explored from the context of mission and vision statements, and goals and objectives. It should be noted that goals and objectives are examined in this chapter to emphasize the importance of the mission and philosophy in programme development. Otherwise, goals and objectives are often considered in later planning stages to enable the programmer to gather key information on needs and assets in assisting in the design of the programme. The chapter follows with an examination of different programming theories and strategies that are commonly used in the field of recreation programming, as well as a discussion of the programming model developed for this book. This model involves a systematic approach to programme planning, based on needs and assets, followed by programme design, trip planning, implementation and evaluation. This process is built from an understanding of the philosophical influences which exist in programme planning, which in turn are affected by a number of influences of the broad environment. The chapter concludes with an examination of the programmer, a discussion of tour operators and the nature of a tour.

Planning

Planning involves arranging, charting a course, designing, preparing and plotting, all with an eye to the future. All people plan to varying degrees, especially with regard to leisure pursuits (e.g. social functions and interactions), but also with respect to our long-term future: jobs, children, houses and retirement. In these we look to the short- and long-term future, establishing goals and objectives, selecting a course of action, and judging results. Just as individuals plan so too should organizations for the purpose of anticipating external perturbations, streamlining their internal organizational structure, being competitive, and keeping employees happy. Defined, planning is a 'dynamic cyclical process that allows interaction between the people who are involved and their resources. Those participating are able to determine, implement and evaluate a course of action that is designed to help them achieve desired goals and objectives' (Ontario Ministry of Tourism and Recreation, 1985a, p. 9). The definition underscores the notion that planning is dynamic, cyclical and interactive, meaning that the efforts of those involved in the planning process feed back into the system creating the opportunity to realize objectives, and thus the ability to create positive changes through successive planning iterations.

One of the positive benefits of the economic down-turn of the 1980s was that it forced public and private organizations (especially governments) to look more seriously at planning. Indeed they were forced to as a result of huge debts which were accumulating throughout the 1970s, and a public that demanded a great deal more fiscal accountability – a legacy that continues to the present. Before the 1980s, planning in recreation and tourism was very much a reflection of the times: the feeling that resources and funds were unlimited in supporting a growing number of essential and non-essential programmes. As the bottom fell out of this model in the 1980s, resources became fewer and governments, and consequently many recreation providers, were asked to become more accountable for their programme expenditures. Ultimately many programmes (social, recreational and so on) were simply eliminated because they were either too expensive to run or simply not profitable.

In tourism, planning appears to have undergone a significant transformation from a perspective which was decidedly economic in focus, to one which at present appears to be more holistic. These different and distinct stages which have emerged over the course of the last few decades include: (i) the Boosterism approach, characterized by an emphasis on the maximization of benefits wherever possible; (ii) the economic approach, which positioned tourism as a means by which to generate economic development (the huge tourism developments in the Mediterranean and Caribbean of the 1960s and 1970s were representative of this stage which emerged to be out of step with the social and ecological conditions of these regions); (iii) the physical/spatial approach, which attempts to minimize tourism's negative environmental impacts; and (iv) the community approach, which sees tourism as political and social and best developed through local control (Getz, 1987, cited in Simpson, 2001). To this list we can add a fifth, the sustainable tourism approach which is based on the amalgamation of the latter three approaches (see Inskeep, 1991).

According to Inskeep, planning approaches that are comprehensive, systems-based, integrated, or strategic, enable planners to understand better how tourism fits within the context of the community or region, thus allowing the planning scheme to be both implementable and relevant. As one of the most broadly based forms of tourism planning, comprehensive planning focuses on all aspects of the tourism region, including institutions, the social realm, and the ecological realm, in a manner which is more holistic in scope. It is a cooperative venture, usually led by the municipality, to: (i) help the municipality make rational and informed decisions about the development of future tourism and recreation services; and (ii) help avoid gaps in services and the duplication of facilities and programmes (Ontario Ministry of Tourism and Recreation, 1985b). Systems planning allows decision makers to examine tourism as a set of interrelated parts, and planning emphasizes these parts through various systems-analysis techniques. Its strength lies in its ability to identify a number of different aspects or phenomena of a region – social, economic, and ecological – allowing for a more complete understanding of how these systems work independently and as part of a larger whole. The integrated approach to planning, which is related to the preceding two strategies, is based on the notion that tourism needs to be planned and developed as part of the overall pattern of development within the region. In his discussion of integrated planning, Hall (2000) writes that tourism planning by nature is very complex. He suggests that the problems within society have a habit of becoming interconnected between different spheres of social life. The problem for tourism is that although there is the acknowledgement that planning is essential, especially in our efforts to develop sustainable tourism, it is difficult because tourism is socially diffuse, and suffers from an absence of a clear control agency (Hall, 2000). Finally, strategic planning is a process undertaken by specific agencies to develop a broad plan of action in dealing with major issues, opportunities, needs and performance. It includes the setting of a path, guiding the allocation of resources, to achieve long-term objectives (see Schermerhorn, 1996). This planning approach is designed to help the agency to re-assess the fundamental reasons for all of their programmes and policies, as well as gauging their effectiveness in meeting participant needs (see Hall and McArthur (1996, 1998) for applications of strategic planning to heritage tourism). More specifically, strategic planning is based on the following principles derived from the management literature (Simpson, 2001, p. 13):

1. Develop an agreed portfolio of critical stakeholder values.
2. Use critical values to articulate a broad vision for the future.
3. Establish generic tools which will contribute realization of vision.
4. Adopt an exhaustive evaluation of current status as a basis for future action.
5. Establish specific objectives to bridge the gap between current status and generic goals.
6. Assign priorities, responsibility and control systems to monitor implementation effectiveness.

In addition to the various approaches outlined above, tourism planning may occur at a number of different scales, as illustrated below (Inskeep, 1991). Included are some of the key characteristics and functions of these levels:

- *International planning*. Concerned mostly with international transportation, the flow of tourists among different countries, complementary development of major attractions and multi-country marketing strategies.
- *National planning*. Tourism policy, the physical structure of attractions, infrastructure, accommodation, tour routes, marketing strategies, education and training, and impact analyses.
- *Regional planning*. This relates to planning for a province/state and includes many components mentioned in national planning, but more specific.
- *Sub-regional planning*. Again, more specific than regional planning, but not as detailed as development area planning.
- *Development area land-use planning*. Includes resorts, resort towns, urban tourism and tourist attractions. Indicates the specific areas for accommodation, shops, facilities, parks, roads, sewerage and so on.
- *Facility site planning*. Very specific for individual buildings or complexes, such as hotels or facilities.
- *Facility design*. Architectural, landscaping, and infrastructural designs and engineering specifications for resorts, restaurants, attractions and facilities.

Many authors write that although valued by industry stakeholders (principally government), the reality is that strategic planning does not actually occur at local, regional or national scales (see Middelton and Hawkins, 1998). Added to this is the notion that even though planning has taken place, there is no guarantee that it will actually lead to successful outcomes. This is perhaps due to the multifaceted nature of tourism, which provides a very challenging environment in which to isolate the huge number of social, economic and ecological variables that dictate the ebb and flow of the industry.

Recreation Programming

> The program is what recreation services are all about. All else – personnel, supplies, areas and facilities, budgets, public relations – exists primarily to see that the program occurs and that people enjoy participating. Planning is the tool that makes programs happen.
>
> (Russell, 1982, p. ix)

From this quote, we begin to see the importance of employing a programme-planning approach to the delivery of recreation services. While people will always engage in leisure pursuits that are spontaneous and relatively unstructured, they may also engage in any number of activities that require a great deal of structure and planning (e.g. organized travel, a swim class, a scuba-diving course). In the latter case, the onus for generating recreational satisfaction switches from the individual him or herself to the agency responsible for presenting the programme. Accepting this responsibility, service providers must endeavour to organize and plan their activities and actions for the purpose of maximizing participants' experiences. This may be accomplished by adopting a recreational programming approach. Programming is a process that entails following many simple but well planned principles, which are not only geared towards the satisfaction of participants, but also to better understanding how the agency may be accountable to itself in addressing issues related to capital, equipment, effective use of time, behaviour of employees or clients. Simply, but effectively stated, programming is the process of organizing resources and opportunities for other people for the purpose of meeting their leisure needs (Searle and Brayley, 1993, p. 214). These needs may be educational, cognitive, affective, social or psychomotor, and the satisfaction of such enables these individuals to live healthier, happier lives.

Although a systematic programming approach has been used in recreation for many years, it has altered significantly over time. Simplistic methods have given way to an approach that is quite specialized in terms of philosophy, content and the realization that different populations have different programme requirements. This perspective is summarized by Kraus (1997, p. 15):

> All this means is that many recreation programmers have moved sharply away from their earlier, traditional functions in carrying out simple after-school or summertime activities within a narrow range of age-related programs in day camping, sports and games, arts and crafts, swimming, and nature studies. Instead, programmers are responsible for serving a much broader range of participants in activities involving more innovative programs relating to special interests, personality development and living skills, health and fitness, family and gender-related concerns, travel and tourism, adventure and 'risk' activities, performing arts, and special services for those with physical, mental and social disabilities.

In general the inherent strength of the systematic approach to programme planning is that it is one integrated process, broken down into a few broad units, which typically include philosophy, planning, implementation and evaluation. Depending on the model chosen, these units will specify certain aspects off the programming process. However, because there are literally dozens of tasks which need to be accomplished along the way, certain tasks will need to be done in a logical order, concurrent to others, or even in reverse order (owing to the nature of the programme). This means that although there is a fairly common process to follow, certain tasks are not necessarily exclusive to one broad unit over another. For example, in some models risk management is placed in the programme implementation unit, whereas in others it may be part of the planning process of the programme. Furthermore, it should be noted that recreational programming occurs in just about every organization and sector, including the public sector (municipal, e.g. parks and recreation departments; provincial, e.g. tourism and parks departments; and federal, e.g. national parks service), the private/commercial sector (ski resorts, residential camps, health spas, zoos), and the not for profit sector (e.g. YM-YWCA). More specifically, the who, why, when, and hows of programming are nicely summarized in Table 4.1, and adapted to the particular focus of this book.

Unfortunately, the programme planning model is often absent in the development of tourism services. While a good deal of effort is often placed into business planning (i.e. getting the finances of the operation in order), it is the systematic planning for recreational activities themselves that is often lacking. However, it is not only the applied situations which demand a greater sensitivity to programming planning, but also tourism research. Few if any tourism models, such as the ones discussed in Chapter 1, discuss pro-

Table 4.1. The who, why, when, and how of programming

Who does it?	Recreation and tourism agencies and departments Those in organizations who provide programme services
Why do it?	To improve the quality, quantity and effectiveness of services based on defined needs To ensure the effective use of all available resources To help to avoid unnecessary duplication or gaps in services
When should it be done?	Usually done on an ongoing, cyclical basis It is probably being done, on an informal basis at least, in most organizations. As needs change, or as the level of service drops or participation drops, chances are that it is time to formalize the process
How is it done?	Needs assessment and problem identification Goal and objective setting Implementation strategy Delivery of programme services Evaluation

Source: Adapted from Ontario Ministry of Tourism and Recreation (1985b, p. 5).

gramme planning as an integral aspect of the tourism experience. This is indeed unfortunate, as one might suggest that the programme, and how it is offered, is the one key element that weighs heavily on whether the service will be successful or not (success defined here as meeting the needs of participants). The integration of programme planning in tourism research may be demonstrated in reference to the five major phases of an outdoor recreation experience, as developed by Clawson and Knetsch (1978) (see Chapter 1). Although spatial and temporal by nature, this model is very much an experiential one in recognizing that tourists have different perceptions during the anticipation, travel to site, on-site, travel home from site, and reflection stages of the trip, which ultimately contribute to the satisfaction of the overall travel event. Programming, as discussed earlier, is a service which, at its heart, is concerned with generating positive experiences for participants through an array of services which cater directly to the tourist's needs. Applied to ecotourism, this means that to be an effective service provider, the operator will increase his or her chances of success by: (i) tapping into the anticipation stage through effective advertising; (ii) ensuring that the travel to the site is comfortable and enjoyable; (iii) exceeding the ecotourist's expectations on site through a variety of educationally based experiences; (iv) that the trip home is safe and enjoyable, with perhaps some follow-up literature on the attractions of the trip; and (v) that the reflections are so meaningful that the tourist will either want to return to the site, continue travelling as an ecotourist to other sites, or highly recommend the site to his/her friends and associates.

Apart from what appears to be very few sources which link programmes and products (some of which will be discussed later), the programming term appears infrequently in the tourism literature. In one case, Tribe *et al.* (2000) use the word 'program' to refer to the approach adopted in the third stage of their Environmental Management System. Programme, in the context of how they use this term, refers to the setting of a formal agenda for action. This includes: (i) assessing current environmental management; (ii) establishing and implementing a programme for an environmental management system; (iii) prioritizing significant impacts; (iv) setting environmental targets; and (v) establishing a programme to achieve environmental targets. Although environmentally based, this is a case which emphasizes the word programme in a systems context, which is consistent with the approach used in delivering tourism 'recreational' services, as the focus of this book.

Programming Ideology

Philosophy

The development of an ecotourism philosophy (for the industry, for a service provider or for programming) demands the examination of some basic assumptions regarding the place of ecotourism in the context of the environment, the nature of human activity, the nature of human nature and the nature of human relationships (Malloy and Fennell, 1998). These may further evolve into a series of other questions related to the meaning of leisure for individuals and for society as a whole. Kraus and Curtis (1990) write that the philosophy which leisure service organizations develop is a direct reflection of the acceptance of leisure as an important part of life and the extent to which it meets human needs. These needs may relate to accessibility, happiness from involvement in the pursuit, the consideration of a means–end relationship, rights of the individual, freedom, as well as equity (see Edginton *et al.*, 1998). An agency's philosophy, then, is a direct reflection of what the entity values.

Hunt *et al.* (1989) suggest that values help define the core of people, including what they dislike, love, and the sacrifices which they make to attain their goals. Within the organization, values both convey a sense of identity to its members, and guide decision making. Underlying these values are corporate ethical values, according to Hunt *et al.*, which help to establish and maintain standards and the right course of action. Plummer (1989) writes that we are in the midst of a shift in the values which we hold to be true. He says that more traditional values are being replaced by newer values which underscore the importance of technology, pluralism, and so on, in a vastly changing, postmodern world (see Table 4.2). While it is not necessarily fair to suggest that these values point to a hedonistic outlook on life for the individual, they do suggest that leisure, in whatever form, is at least part of the movement towards self-actualization of the individual.

Researchers have identified a series of core values which appear to underlie many of the basic feelings and behaviours that are common within and between different populations. Kahle (1983), for example, has determined that fun and enjoyment in life, and excitement, self-respect, warm relationships with others, sense of accomplishment, self-fulfilment, security, sense of belonging and being well respected, are fundamental values which are cosmopolitan in nature. The first six of these are internally motivated values (with the top two being hedonistic), while the last three are externally motivated ones. While it would be interesting to see how ecotourists compared and contrasted with other populations on these items, there is perhaps a series of other core environmental values which would be articulated by ecotourists over other types of tourists, as being fundamental to the travel

Table 4.2. Changing values.

Traditional values	New values
Self-denial ethic	Self-fulfilment ethic
Higher standard of living	Better quality of life
Traditional sex roles	Blurring of sex roles
Accepted definitions of success	Individualized definitions of success
Traditional family life	Alternative families
Faith in industry, institutions	Self-reliance
Live to work	Work to live
Hero worship	Love of ideas
Expansionism	Pluralism
Patriotism	Less nationalistic
Unparalleled growth	Growing sense of limits
Industrial growth	Information/service growth
Receptivity to technology	Technological orientation

Source: Plummer (1989).

experience. These may include reducing/reusing/recycling, harmony, exploration, multiculturalism, preservation of landscapes, biodiversity conservation, integrity, learning, service and knowledge. In regards to this last item, knowledge is an important value because information changes at an incredible rate where the uncertainties of today become the frivolities of tomorrow. What this means for the manager of a tourism business is that in today's highly competitive marketplace, one must continually attain and operationalize knowledge in order to be competitive but also to be wary of the social and ecological implications of one's actions in corresponding to appropriate local and cosmopolitan rules of corporate conduct.

Not surprisingly, it would seem logical to assume that ecotourism service providers and ecotourists share many of the same values – an essential factor for successful ecotrips. From the organization's standpoint, such values must be freely communicated through all levels of the firm, but also to ecotourists as well as members of the community, who might also benefit from the positive environmental values of the service provider. This means that the service provider needs to take on a very active leadership role in the communication of appropriate social and ecological values. It is worth mentioning that profit may also be a value of the service provider, along with many of the aforementioned ones. While this does not have to be outwardly advertised, it does mean that perhaps more value should be placed on the things that will perhaps guarantee profit, including good service, equity, preservationist attitudes and so on.

Mission and vision statements

> Eco-X will create and promote the highest quality possible ecotourism opportunities and preserve and enhance parks and natural areas to enrich the lives of tourists and local people (Mission).
>
> Eco-X will be a leader in the development of the region's best ecotourism opportunities where tourists may enrich their lives through first-rate programmes. Eco-X will always treasure our heritage and natural resources and will help preserve natural areas for the children of the future (Vision).

Both mission and vision statements are extremely important as vehicles by which to articulate the philosophical direction of an organization. While the former focuses on the current state, the latter makes a projection into the future based on some forecasted reality. Both should be bold and inspiring, and, to be effective, must communicate direction to customers and workers within the organization. As regards the mission, Campbell *et al.* (1990) suggest that each statement should have the following four elements:

- *Purpose*. Defines why the organization exists.
- *Values*. Declares what the organization believes in and its moral principles.
- *Strategy*. Explains the organization's position *vis-à-vis* the competition and its competitive advantages.
- *Standards and behaviour*. Sets out how the organization's staff operates, their competencies and the way they act.

In essence, all of the various things that the organization will plan on doing will emanate from the mission statement. Just as people's philosophical positions are subject to change, an agency's mission statement may too. Kraus (1997) writes that the mission should be reviewed periodically in light of changing conditions, needs and agency capabilities.

On the other hand, vision statements are forecasted realities which have yet to come about (Snyder *et al.*, 1994). To these authors, the vision acts as an inner force compelling a leader to action. This long-term goal is something which defines leaders and which catalyses others to follow. For the organization, the vision is:

> a target at which the leader directs the organization's energy and resources. With an eye on that vision, the leader keeps the organization moving forward in spite of personal, organizational, and environmental challenges and resistance, including a fear of failure; suspicion, derision, or sabotage from superiors, peers, or

subordinates; competitor attacks; government obstructions; economic disruptions; and any and all external events

(p. 74)

Both mission and vision statements are thus essential in communicating the organization's intent. By advertising the mission and vision statements, as many operators are wont to do these days, in brochures, on walls or over the web, the service provider is creating a conduit into the inner workings of the agency, their philosophical orientation, and their intentions. These statements of purpose are operationalized through the development of a set of goals and objectives, which establish direction for the service provider in a slightly different, but related way.

Goals and objectives

Kraus (1997) writes that goals and objectives, although often used interchangeably, are different in what they aim to accomplish. Goals are designed to reflect the mission of an organization, and may be viewed as purpose(s) which are often communicated as statements which emanate from the mission statement. On the other hand, objectives are much more specific, measurable, and attainable, depending on the nature of the service provider. They are the steps required to reach specific goals, and are written in a straightforward way, according to the actions to be achieved as well as the outcomes that are expected over a given period of time. Objectives will quite naturally flow from the various goal or purpose statements which are a direct reflection of the agency's overall mission. Table 4.3 is a graphic representation of the relationship that exists between mission, goals and objectives, established here on the basis of three distinct tiers.

Table 4.3. Mission, goals and objectives.

Mission statement		
Goal 1	Goal 2	Goal 3
Objective 1	Objective 1	Objective 1
Objective 2	Objective 2	Objective 2
Objective 3		Objective 3

Objectives have been described as being of two types. The first type are programme (also called operating, production, input or implementation) objectives. These relate to the means (materials, leaders, facilities, etc.) used to operate the programme, the amount of energy expended in the programme delivery, or matters related to client participation. As such, there may be categories of programme objectives in the same way there are different goals that stem from the mission statement. The second type of objective is the performance (otherwise known as outcome) objectives. These are the measurable end products of the programme, the return expected from the effort expended, or the skills, attitudes and knowledge gained from participation, and must also directly reflect the initiatives set out in the mission and goal statements. The following are examples of programme and performance objectives:

Programme objectives:

1. To increase awareness of the endangered tree frog within the community by the end of the current year.

2. There should be a 10% increase in the number of participants in the programme within 2 years.

Performance objectives:

1. Participants will be have identified 75% of the animals listed on the identification list.

2. All of the participants will enjoy the outdoor component of the field day.

Service providers must be aware of the fact that certain types of performance objectives – on the basis of how they are written – are more easily measured than others. This is the case in the two examples written above. It is a relatively straightforward task to examine one's animal list at the termination of the programme to see the percentage of animals identified. Not so with performance objective two, because human behaviour, especially the affective domain, is a much more difficult construct to evaluate because of its intangibility. In attempting to resolve this potential problem, Edginton *et al.* (1998) suggest that emphasis should be placed on the quantification of intangibles such as enjoyment, excitement and so on. This may be accomplished through the use of evaluation instruments that attempt to gauge satisfaction on the basis of numerical scales, which will be discussed in Chapters 5 and 10.

Programming Theories

The programming processes that service providers develop are hinged not only on what individuals and organizations hold to be true (which help to develop the philosophical orientation of the operation), but also on some of the narrowly or widely accepted ideas about human behaviour. These theories help to explain the underlying reasons why people engage in leisure pursuits, and perhaps why they do not, in addition to the reasons why service providers endeavour to provide opportunities for these individuals at all. An understanding of why people participate in leisure pursuits such as tourism is important as a means by which to formulate different approaches to programme delivery. Some of these theoretical perspectives are outlined below.

Symbolic interactionism

This theory, which is grounded in the work of C.H. Cooley, Margaret Mead and E. Goffman, attempts to explain the behaviour of individuals in certain situations and under certain conditions. It relates to the shared meanings that individuals, such as tourists (in a group), communicate and interpret during the process of interaction. This may involve relationships that exist between tourists within the group, tourists to locals, tourists to operators and tourists to the environment in which they are a part. For example, tourists and local people may interpret meanings much differently and thus react differently to the same situation or phenomenon. As outlined by Searle and Brayley (1993), programmers must allow for the participant to help shape the overall experience and allow an element of freedom in regards to choice. This clearly links to the concept of positive affect, a leisure-defining concept which posits that people are most happy when they feel that they have had some control over their recreational experiences. Ecotourism operators would do well to include elements of positive affect in their programmes. This many include allowing some freedom to make choices about activities, time spent involved in these activities, and perhaps some control over the amount of free time, instead of being continuously directed from one facility or attraction to another.

Social exchange theory

Searle and Brayley (1993) write that people enter into relationships and situations (such as recreation and tourism programmes) seeking rewards. These relationships are sustained if rewards are valued, if costs do not exceed benefits and if they evolve over time. This theory is said to substantiate the need for employing good assessments of the needs of participants. If participants are not receiving good value for their money, then they will discontinue their relationship with the provider. The implications for the ecotourism industry are varied, and touch upon aspects of quality service provision, and accreditation and certification, especially in relation to guide training, environmental education and facilitation.

Optimal arousal and flow

Optimal arousal has been defined as a psychological construct referring to the level of mental stimulation at which physical performance, learning or temporary feelings of well-being are maximized (Smith, 1990a). Physical performance is linked with levels of arousal, such that as arousal increases, performance increases. However, too much arousal (stress) leads to decreasing performance, just as too little arousal stifles performance (boredom). People seek an optimal state of arousal in their participation in order to maintain a state of sensoristasis or homeostasis. Csikszentmihalyi's (1975) concept of flow, although related to optimal arousal, refers more to the transcendental state that recreationists, surgeons, writers, for example, reach in their participation. Flow describes a state of experience that is engrossing, intrinsically rewarding, and outside the parameters of worry and boredom (Csikszentmihalyi and Csikszentmihalyi, 1990). For an activity, if skills are greater than challenges, boredom emerges; while if challenges are greater than skills, anxiety occurs. However, if these two (skills and challenges) are in balance the participant is able to: (i) concentrate on the performance; (ii) centre their attention; (iii) experience a feeling of transcendentalism; (iv) exercise control over their actions; and (v) fulfil their goal; with the goal of maintaining the flow state (autotelism).

Intrinsic/extrinsic motivation

As outlined in Chapter 1, there are two important ways in which to define a leisure experience. One is by way of perceived freedom, which relates to the lack of constraint in participating freely in the activity. The other is intrinsic motivation, which refers to the internally generated intention to act without any other apparent motive (Smith, 1990). Intrinsic motivation is thus hinged upon self-expression, self-development, competence, self-realization (Mannell and Kleiber, 1997), as well as the ability or need to participate in an activity for its own sake. As a consequence of the fundamental differences that exist between extrinsic and intrinsic motivation, researchers have been able to differentiate between types of outdoor recreation enthusiasts based on this motivational premise. For example, Ewert (1985) discovered that experienced mountain climbers were motivated by intrinsic factors such as exhilaration, challenge, personal testing, decision making and locus of control; whereas, inexperienced climbers did so for extrinsic reasons, including recognition, escape and social motives.

Personality

'Personality is the sum of the unique psychological qualities of an individual that influence a variety of characteristic behavior patterns ... in relatively consistent ways across different situations and over time' (Zimbardo, 1979, p. 467). We come to understand the personality of people by their actions. From the definition above, behaviour is a result of the psychological qualities of a person which remain fairly consistent over time. These fairly stable tendencies allow researchers to compare people with others, such as in determining social competence or why some have the necessary skills it takes to interact with people while others do not. Mannell and Kleiber (1997) suggest that there appears to be some agreement in the academic community that there are five main personality factors. These include: extroversion, agreeableness, conscientiousness, neuroticism and openness to experience. The first and fifth factors seem to have the most applicability to ecotourism, and tourism in general. Plog's (1972) allocentric tourist (those travellers who seek novel, adventurous and unstructured experiences) has been linked to extroversion, as has Zuckerman's (1979) sensation-seeking scale, which identifies a personality trait which strives for novel, varied, complex and risky experiences.

Life stage

Kelly and Godbey (1992) write that the life course consists of a variety of different transitions. Members of a cohort go through life experiencing many of the same events at the same ages. At the same time, the individuals of these cohorts appear to have the same needs, at greater or lesser intensities, than others who are at different periods of their life (Patterson, 1991). Development theories such as Erikson's (1963) psychosocial crisis, or Kohlberg's (1984) moral stage, help us to understand their needs better (Table 4.4).

Patterson (1991) notes that in each of the stages in Table 4.4, individuals develop behavioural patterns which are the foundations for all cognitive behaviour. For the recreation and tourist service provider, it is important that he or she has a general understanding of the behavioural patterns of these

Table 4.4. Stages according to Erikson and Kohlberg.

Stage	Erikson	Kohlberg
Infancy	Trust vs. mistrust	Premoral
Early childhood	Autonomy vs. doubt; initiative vs. guilt	Obedience and punishment (stage 1); reciprocity (stage 2)
Late childhood	Industry vs. inferiority	Good child
Adolescence	Identity vs. role diffusion	Law and order
Young adulthood	Intimacy vs. isolation	Social contract
Middle age	Generativity vs. stagnation	Principled
Old age	Integrity vs. despair	–

Source: Erikson (1963) and Kohlberg (1984).

different population groups. This is especially important for programme development, which must accurately reflect the needs of these groups, but also to match available resources with these needs. Consequently, Patterson (1991, p. 24) has developed a breakdown of characteristics for the following major life stages, as they relate to recreation and programme planning (Table 4.5). In general, studies have found that leisure activity levels decline with age, especially those which demand high levels of intensity. Furthermore, the elderly are much less likely to experiment with new activities, and the tendency to seek novelty, especially through the adoption of new leisure activities (see Mannell and Kleiber, 1997).

Constraints

Constraints to participation in leisure activities include 'any factor which precludes or limits an individual's frequency, intensity, duration or quality of participation in recreation activities' (Ellis and Rademacher, 1986, p. 33). Constraints have generally been discussed along three dimensions: (i) structural, such as a disability, lack of money, transportation or facilities, which would enable participation; (ii) interpersonal, such as being unable to find others to participate with, including friends or family members; and (iii) intrapersonal or psychologically based constraints, which may include a lack of interest (attitude or mood), or having a low level of energy. Owing to the cost of many ecotours, there will be those who will be unable to participate because of a lack of money (structural constraint). An interpersonal constraint may be a family member trying to prevent someone from travelling to a particular destination because of political or environmental concerns. Finally, an example of an intrapersonal constraint might include an individual, possessing a lower level of education, who refuses to travel as an ecotourist because of the perception that ecotourism is too much the realm of the educated, which would thus prevent him from enjoying such an experience.

Programming Strategies

Due to the multi-sector (public, private and not-for-profit sectors) extent of the recreation services delivery system, the number and type of programmes that are available to the public are quite diffuse. Not-for-profit agencies, for example, who may have different values than a for-profit agency may develop programme strategies that are geared more towards accessibility, equity and involvement. In so doing, the not-for-profit may be less concerned about developing services that are going to be participated in by large numbers of people or that make substantial sums of money. Comparatively, the public sector may also choose to emphasize other values and goals in the development of their programmes which may reflect the profit motive. The same holds true for ecotourism, where all three sectors are involved in the development of ecotourism programmes,

Table 4.5. Groups, ages and recreation characteristics.

Group	Age	Characteristic
Preschool	Under 4	Dependent on others, short attention span, self-centred, major motor development, and desire for immediate reward.
Primary	5–7	Developing social relations and additional motor skills, imagination and exuberance, develops self-control, becomes industrious.
Intermediate	8–12	Perseverance, diligence and competency develop, play is serious business.
Early adolescence	13–15	Rapid physical growth, onset of puberty, sometimes awkward and self-conscious.
Youth	16–19	Develops individual identity, discovers talents. Strong likes and dislikes, group memberships, sets goals and strives for independence.
Young adult	20–29	Seeks meaningful relationships, makes commitments, may pursue outdoor and risk activities to test competency.
Adult years	30–49	Major productive years, child-rearing and other social responsibilities. Looks to recreation for activity, diversion, status and autonomy.
Pre-retirement	50–64	Reduction of intensity of some needs, secure, enjoys social outings, leadership oriented.
Early retirement	65–79	Recreation replaces work as reason for being. Can be active or passive, depending on health. General physical decline and more difficult to stimulate.
Later maturity	80 and over	Contingent upon health, can be a rewarding but semi-passive way of life. Those who prepare for retirement and later maturity can prevent despair associated with loss of independence and isolation.

Source: Erikson (1963) and Kohlberg (1984).

despite the fact that the bulk of these operations are for-profit entrepreneurs. Unfortunately there is a dearth of research on the programme offerings of these three sectors from a comparative sense, which restricts our understanding of these different perspectives. Government agencies (public), museums (not-for-profit) and for-profit operators are all active players in the ecotourism industry, and all have developed programmes that are profit-oriented.

In recreation, these different sectors have tended to adopt one of three different conceptual perspectives from which to guide the decision making in terms of values, profit and education, as well as a number of other variables. The perspective adopted is critical because it forces the service provider into making programme decisions that are based on the selected service delivery paradigm. These three are outlined below, with further elaboration in Table 4.6.

1. *The direct service delivery or social planning approach.* The service provider uses his or her own professional training and expertise to diagnose, intervene and provide services. The expert is thus responsible for developing the programme based on what he or she feels is best for the participants.

2. *The marketing approach.* The oft quoted adage: 'to sell Jack Jones what Jack Jones buys, you have to see Jack Jones through Jack Jones' eyes', is a reflection of this approach. Here, the delivery of a service occurs through an assessment of what participants want (Crompton, 1985). Programmers attempt to match the wants of the client with available programme resources, such as money, equipment, people and time. An interesting spin on the marketing concept is one advocated by Edginton *et al.* (1998) who suggest that a better term is *social* marketing. This latter terminology is selected because: (i) social marketing connotes an emphasis on the delivery of experiences rather than tangible products; and (ii) the main thrust of a leisure services organization should promote the wise use of leisure, environmental ethics and human dignity, over efficiency, profit and other goals.

3. *The indirect or community development model of planning.* This model assists individuals in initiating programmes to meet their needs. From this end, the programmer acts as a facilitator, enabler or collaborator, who strives to help individuals identify their problems, needs, and the resources which may be required to address these problems and needs. This is perhaps a model that might be more strongly embraced by the public sector, who may value the role as facilitator, teacher or partner.

The social planning approach, illustrated above, places the ecotourism programme planner in the role of expert. This approach is based on the feeling that the needs of the ecotourist are best met on the basis of the service providers professional opinion. This model is less market driven and more the hallmark of a traditional community or municipal parks and recreation perspective, which is basically: 'this is what we feel the community needs and this is what we will provide'. The social marketing approach is driven more by the dynamic that exists between demand and supply. Such an approach is more profit driven, and therefore reliant on the customer, his or her needs, and the financial outcomes of the service. This is the basic model for the provision of ecotourism services as regards the tourism 'industry'. On the other hand, the community development approach puts the service provider in the role of facilitator or enabler, where the client is seen as a partner rather than customer or consumer. It should be noted that public, for-profit and not-for-profit organizations may adopt any one of the aforementioned models of service provision. Some not-for-profits, such as the YM-YWCAs, which historically adopted a model based on accessibility and equity, have more recently adopted a marketing-based approach to remain viable and competitive in the marketplace. This fundamental shift in organizational strategy has made such organizations more competitive in the marketplace, but at the expense of accessibility.

Programming Approaches

The social planning, social marketing, and community development strategies outlined above have stimulated the evolution of a tremendous array of different programming approaches. Programme approaches are

Table 4.6. Strategies for delivering leisure services.

	Strategy		
Function	Social planning	Social marketing	Community development
Goals	Problem-solving orientation to community needs, service	Satisfying customer needs, profit	Self-help, teaching process skills, promotion of democratic values, service
Basic strategy	Fact gathering followed by rational decision making regarding the distribution of resources	An analysis of needs, integration and initiation of marketing mix to meet these needs	Assisting individuals to determine and solve their own problems
Professional roles	Programme planner, fact gatherer, analyst, programme implementer	Analyst, planner, implementer, promoter	Enabler, catalyst, coordinator, teacher, value clarifier
Sector	Public, quasi-public or private	Discrete target markets	Public, quasi-public
Conception of population served	Consumers (we are the experts; people consume what we diagnose is good for them)	Customers, consumers, guests	Partners

essentially ways in which programmes are conceived and delivered. They are numerous because there are a number of ways in which to present activities, facilities, and other programme aspects to participants. Some of these approaches fit completely within the context of a particular strategy (e.g. the approach is directly representative of the social marketing strategy), while others are more diverse in terms of their programming intent and application. They are important to mention here because they allow students and practitioners to appreciate the fact that there is no one best way to develop a programme, but rather many – depending on the strategy and philosophy of the organization. With this in mind, the following 28 approaches are briefly illustrated, along with the author of the approach, as well as the date it appeared (adapted from Carpenter and Howe, 1985). The task for the ecotourism service provider is to assess which of the following 28 are most applicable to his or her particular operation, and the various programmes that are offered (which may in fact differ from one another). This task will be made easier in Fig. 4.1, which links these approaches with the three programme strategies discussed above.

Danford and Shirley (1970):
1. *Traditional.* Relies on what has worked in the past. 'We've always done it this way, it's tried and true'.
2. *Current practice.* Programming based on current trends. 'If it's new and everyone is doing it, it must be good … lets do it'.
3. *Expressed desires.* Basis is the expressed needs of the client. 'We'll do what ever the people want us to do'.
4. *Authoritarian.* Programmer uses only personal knowledge in determining programme outlook. 'I'm a trained professional, I know what the people want'.

Tillman (1973):
5. *Reaction.* Responds to individuals and groups influential to the decision-making process of the organization. 'The loudest voice gets the action.'
6. *Investigation.* Fact-finding methods are used to determine the behaviour and needs of participants. 'A needs inventory is required to determine client concerns.'
7. *Creative plan.* An interactive relationship between participants and professionals. 'We use people's demands and interest inventories in deciding upon programmes.'

Murphy (1975):
8. *Cafeteria.* Clients select programmes from a list of offerings prepared by the agency. 'Our programme consists of a variety of opportunities for all ages, interests and abilities.'
9. *Prescriptive.* The professional diagnoses the needs of the participant and then provides services accordingly.

Edginton and Hanson (1976):
10. *Trickle-down.* Programmes are initiated at higher levels, and trickle-down to the consumer. 'If this is what the administration wants us to offer, then this is it.'
11. *Educated guess.* Planning is based on someone's hunch they will meet community needs. 'I've got a hunch that this is what the clients want.'
12. *Community leadership input.* Advisory boards represent and input the concerns of the public. 'Since the advisory board represents the needs of our participants, we'll do what they want.'
13. *Identification of need.* Data is collected on participants to determine programmes (e.g. demographics, behavioural).
14. *Offer what people want.* The voiced opinion of participants. 'After talking to our clientele, we offer them whatever they want.'
15. *Indigenous development.* Moving straight to the people in their own neighbourhoods, encouraging them to plan their own programmes.
16. *Interactive discovery.* No subordinate relationship exists between agency and participants. 'We meet face-to-face with our clients and decide together what we will offer.'

Kraus (1977):
17. *Socio-political.* Response to pressure groups. 'We have to watch-out for these groups, they are in-touch with city hall.'

Kraus and Curtis (1990):
18. *Quality of life/amenity.* Recreation as an

important community service that is pleasurable and constructive and which contributes to the physical/mental health of participants.

19. *Marketing*. Recreation is viewed as a service or product that has economic value and that should be designed and aggressively promoted to yield the maximum financial return.

20. *Human services*. Based on a holistic and humanistic view of human life. Strives to make all recreation programmes highly purposeful and to link them to other community-based programmes of a health-related, vocational, counselling or other social service nature.

21. *Prescriptive*. Recreation as designed to accomplish specific goals. Prescriptive planners urge that recreation be applied as a distinct form of therapy, carefully formulated to achieve treatment outcomes for patients and clients.

22. *Environmental/aesthetic/preservationist*. This approach links several groups in society who are concerned with preserving the natural environment along with ensuring that government takes responsibility of the environment. It is found chiefly among recreation and park managers who serve in resource-based agencies.

23. *Hedonist/individualistic*. Stemming from the dramatic shift in social values and the 'me generation', this approach stresses the pursuit of pleasure as a primary focus of leisure programmes, along with the promotion of individual forms of creative expression, rather than mass programming in familiar, stereotyped activities.

Farrell and Lundegren (1993):

24. *Programming by objectives*. The success or failure of the programme is a function of a series of stated objectives which guide implementation and evaluation.

25. *Programming by desires of participants*. An approach that links the actual expressed desires of participants, and the methods of attaining this information, to actual services.

26. *Programming by perceived needs of participants*. An approach that attempts to anticipate the needs of participants, without the input of participants.

27. *Programming by cafeteria style*. The provision of several different activities/services, from which customers can choose.

28. *Programming by external requirements*. Programmes are developed using normative standards developed by professional organizations, which the organization can use to evaluate their programmes.

Perhaps the easiest way to digest all of the various programme approaches listed above is to link them up with the aforementioned Strategies for Delivering Leisure Services (Table 4.6) along a continuum (Table 4.7). It should be noted that many of these approaches are still open to some interpretation, especially from the perspective of private, public or not-for-profit service providers, and their place on the continuum in the following figure should be viewed as being somewhat flexible. Furthermore, some approaches are very similar in intent, e.g. 'Expressed desires' and 'Investigation', simply because they have been developed by different individuals. For the ecotourism service provider who is working within a social marketing structure, 'Investigation', 'Identification of need', 'Marketing', 'Environmental/aesthetic/preservationist', 'Programming by objectives' and 'Programming by external requirements', are perhaps some of the approaches that might appeal to the organization. Furthermore, some of the programming approaches outlined above interface with the tourism planning approaches identified by Inskeep (1991). Inskeep writes that tourism planning should be integrated and based on principles of sustainable development. Such an approach suggests that tourism must be planned and developed as an integrated system with all other types of development occurring within the region. The sustainable-development approach identified by Inskeep values natural and cultural resources of the destination, and maintains them on a permanent basis for use by future generations; while the community-based approach to tourism planning maximizes the involvement of local community in the planning and decision-making processes of tourism.

Table 4.7. Continuum of programme approaches.

Authors	Programming approach													
	Social planning						Social marketing							Community development
	Programmer as Expert ←												→	Programmer as Enabler
Danford and Shirley (1970)	4	1			2					3				
Tillman (1973)									5		6		7	
Murphy (1975)		9				8								
Edginton and Hanson (1976)			10		11				12	13	14	15	16	
Kraus (1977)								17						
Kraus and Curtis (1990)		21		20	18		19		22		23			
Farrell and Lundegren (1993)					26	27	28	24			25			

Adapted from Edginton *et al.* (1998) and DeGraaf *et al.* (1999).

The Programming Model

Many of the books on recreational programming are organized according to a general model which starts with needs assessment, then moves to planning, implementation, and ends with evaluation. Evaluation feeds back into the needs assessment stage, and the process continues over again. The systematic nature of this procedure provides the needed clarity in addressing many critical steps which must be satisfied to develop successful programmes, as stated earlier in this book. The process adopted for this book is quite similar to these other works, in moving from needs assessments to evaluation. However, because ecotourism involves the facilitation of outdoor experiences in remote settings and over extensive periods of time, three programme design steps have been developed, including structure (Chapter 6), preparations for the field (Chapter 7), and leadership and risk (Chapter 8) for the purpose of emphasizing the importance of these elements in the programme design phase (see Fig. 4.1). Much of the discussion in these chapters is therefore of an applied or practical nature consistent with the programming focus at this stage of the cycle.

This general model is heavily influenced by a level (tier 2) of what has been termed 'philosophical influences'. The inclusion of philosophies, values, theories and strategies in the general programming model, appears to be consistent with the general literature on recreation programme planning. The place of 'philosophical influences' in Fig. 4.1, however, occurs at a different level than in other texts. This has been done to emphasize the importance of these influences to the overall programme planning process. It has also been set apart from the general programming cycle to suggest that ecotourism must be guided by a philosophy or ethic which in turn should influence how ecotourism programmes are developed across the board. Ecotourism service providers do not approach the industry from the same philosophical perspective, which is why there appears to be such a difference in the offerings of programmes from one provider to the next.

At the very root of programming and service provision, providers will be influenced

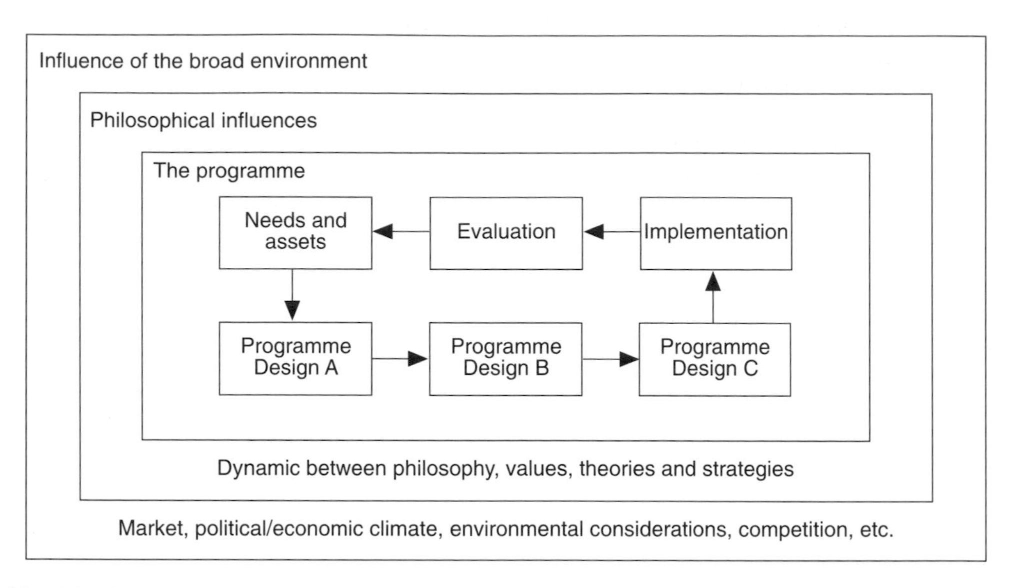

Fig. 4.1. Ecotourism programming framework.

by a number of market, political, economic, environmental and industry-related issues. Externalities such as politics and global weather patterns may periodically dictate the number and type of individuals who will visit a region. Similarly, the service will have to compete in an industry that continues to experience significant levels of growth from one year to another. These influences of the broad environment will not necessarily affect the philosophical influences of the ecotourism operator – they may or may not – but rather these are important operational considerations which will in every case have some type of influence on who, what, where, when, why and how things will be accomplished, if at all. As such, while each of the three levels influence each other, the process is not necessarily an iterative one in the sense that one must positively influence another for programming to occur. These are simply levels of influence which exist, and which in turn influence the programming process spatially and temporally.

The Programmer

In their research on ecotourism operators in western North America, Fennell and Malloy (1999) found that the average size of ecotourism companies was 5.9 individuals, with as few as two people working in the company. What this means is that employees of the operation will need to 'wear many hats'. There are numerous tasks that must be completed in getting the programme(s) up and running – and maintained over time – which include administration, public relations, community liaison, marketing, promotion, evaluation, programme design, establishment of goals and objectives, needs assessment, leadership, budgeting, facilitation, teaching and environmental education, scheduling and so on. All of these tasks may be shared in smaller companies. However, if there is someone charged with the responsibility of programme development, many of these will fall on his or her shoulders. Consequently, such middle management positions are important as they act as a conduit between front-line tasks and upper management.

Understandably, such responsibility can be overwhelming, but it is critical to the success of the operation. The recognition of the level of involvement that goes into such a job has prompted researchers and practitioners in the field of recreation to develop a series of guidelines that provide professional guidance to the programmer or service provider (at a broader level). These guidelines have been developed almost exclusively to ensure that programming philosophies and practices are put in place in order to guarantee that the best interests of participants are foremost in the minds of employees (adapted from Edginton *et al.*, 1980):

- *Placing the needs of the participant first.* Ensuring that the whole organization is devoted to meeting this goal.
- *Commitment to the ideals of ecotourism.* The entrepreneur should be consistent with the philosophical tenets and most up-to-date approaches of the industry.
- *Protection of the participant's rights.* The relationship between participant and provider should be based on trust, respect, and confidence.
- *Acquisition of adequate and appropriate knowledge before engaging in professional activities.* Should have sufficient education or work experience to enable the programmer to carry out the responsibilities of the job.
- *Practice of the highest standards of professional service.* The provider should strive to be consistently excellent day in and day out.
- *Continuous upgrading of professional knowledge, skill and ability.* The programmer must be motivated to pursue learning and advancement, and to keep abreast of current trends and concerns.
- *Operating ethically and equitably.* Responsibility to be honest, forthright and direct in his or her relationship with the client, and avoid being coerced, bribed or have one's integrity compromised.
- *Maintaining a collaborative relationship with the participant.* Maintain a two-way dialogue between the provider and participant. Establish a complementary rather than subordinate relationship with the participant.
- *Self-regulation.* It is the responsibility of the profession to establish performance guidelines/standards. Operators must regulate their behaviour accordingly.
- *Contributing to the development of the profession and other professionals.* The professional ecotourism operator must endeavour, unselfishly, to share knowledge, skills, abilities for the betterment of the field as a whole.

These guidelines underscore the importance of a more ethical and humanistic approach to the delivery of services. The customer is valued as more than simply a means to a financial end. Valuing and practising equity, high standards, collaboration with others, commitment, knowledge, and protection of rights is critical for success, but also in terms of the communication of such values within the operation. This overall intrinsic commitment to the ecotourism industry will quite naturally get absorbed by tourists and perhaps ultimately lead to an enhanced profile in the industry.

Programming Tip 4.1.

The use of flashcards.

Ontario-based Niagara Nature Tours makes use of laminated flashcards to pass around to tour participants on-site. These flashcards have pictures and descriptions of flora and fauna that are known to be found along the trail. The cards dissuade leaders and participants from picking up and disturbing plants and animals for the sake of interpretation, and are both visually appealing and scientifically correct, and thus almost always sufficient to address the curiosity of the most inquisitive ecotourists.

Tour Operators

According to the Saskatchewan Tourism Education Council (1997), a tourism small business operator is an individual who is: (i) accountable for planning, decision making and management of day-to-day and long-term operations; and (ii) operates a business in one or more of eight identified sectors of the tourism industry. These sectors include accommodation, adventure tourism and recreation, attractions, events and conferences, food and beverage, tourism services, transportation, and travel trade. One of the key aspects of the definition above, is that operators (e.g. ecotourism service providers) may operate a business in one or more of the sectors, depending on the focus of the business enterprise.

Weaver and Oppermann (1999) write that the primary role of a tour operator is to provide a package of services for the consumer, including accommodation, transportation, restaurants, attractions, and so on. They do this through the purchase of transport, accommodation and other services which they combine and sell, either directly or indirectly, to a particular type of consumer. This list of responsibilities may further include producing and paying for brochures, making confirmations, writing up tour vouchers, paying commissions, handling financial transactions, coordinating all passenger manifests, as well as advertising, sales, promotions and other operations (Mitchell, 1992). For the consumer, the benefits of a package tour are many (see McKercher, 1998), including: (i) the elimination of a great deal of involvement (e.g. select route, find partners, purchase food, and so on); (ii) lower costs, in time and finances; and (iii) lower risk by picking an operator who has been successful at conducting the tour over a period of time. For the operator, the benefits include an opportunity to expand the market base, the ability to maintain a consistent quality on each trip, and the reduction of risk through the development of stable systems, which will also promote growth (McKercher, 1998).

In such a specialized market, ecotourism operators target a niche which may be characterized according to their interests in certain types of activities (e.g. wildlife viewing), the setting(s) (e.g. New Zealand's South Island), and perhaps on other aspects of the experience, such as the type of transportation included in the ecotour (e.g. bicycle touring). Specialized operations are said to have many more advantages over mass-market companies. These operators are able to access smaller hotel units which may be more abundant, especially those which may be adjacent to protected areas (and which are restricted to a certain size). Furthermore, the specialized tour operator may be better able to alter arrangements around transportation, accommodation, and the food and beverage industry, in situations which warrant such changes. Table 4.8 outlines a number of perceived differences between large-scale mass-market operations and smaller scale ones.

Table 4.8. The perceived differences between large mass-market tour operators and small specialist operators.

Variable	Large-scale	Small scale
Environmental impact	High	Low
Economic impact	Low per head expenditure	High per head expenditure
	Benefits few	Benefits spread widely
	High leakage	Low leakage from community
Socio-cultural impact	High	Low
Host–community relations	Formal	Informal
	Institutionalized	Personal
Commitment to destination	Low	High
Types of tourist	Generally insensitive	Sensitive to/interested in destination, specifically
	Disinterested in destination	.

Source: Swarbrooke (1999).

The reader may note why an ecotourist may be more attracted to small-scale operations which place a great deal more emphasis on, for example, lower environmental and sociocultural impacts.

While the advantages are clear, small specialized tourism operators are confronted with a number of obstacles not experienced by the larger, mass-tourism operator. These include fewer vacation options in fewer destinations, fewer distribution channels, as well as reduced power to negotiate contracts with local suppliers. In this latter case, the mass-tourism operator is able to secure lower prices for their customers, with the spin-off of leaving less money in the local economy. A further danger relates to the problems of restricting oneself to one particular travel region. This 'all your eggs in one basket' approach increases the chances that certain major external events, such as a natural disaster or political instability, will severely cripple the operation. Given the problems stated above, Swarbrooke (1999, p. 279) writes that the smaller tour operators are able to survive by employing the following measures:

- Selling holidays at prices that generate a higher profit margin per customer than mass-market vacations.
- Relying heavily on repeat purchases by loyal customers.
- Offering high levels of personal service.
- Providing niche market, specialized products that are not offered by the mass-market operator.

Other Issues

A trend which appears to be on increase is the involvement of outfitters in ecotourism, beyond their conventional job of providing the equipment necessary for an activity or experience. Tims (1996) writes that outfitting traces its roots back to the early explorers where people provided a service (tents, horses and so on) to others in their struggles against nature. More recently, outfitting has progressed to the point where it has taken on professional status through America Outdoors, and the Professional Guide Institute (PGI). The PGI provides training for people in the areas of wildlands heritage, backcountry leadership, interpretation, and outfitter operation, with the mission to: 'identify, enhance and disseminate the natural interpretive and educational resource of the outfitting industry so that outfitters and guides can offer the highest quality of experience to the public' (Tims, 1996, p. 177). Clearly, the PGI operates with a mandate that involves guides as a natural ingredient to providing effective backcountry experiences. An example is Algonquin Outfitters in Ontario, Canada, which provides the gear for people to participate in outdoor recreation (the conventional role of the outfitter), but has also become involved in the actual trip planning and guiding for participants who want their experience as a part of an overall package (the contemporary role of the outfitter). (See section on Professional Development in Chapter 8.)

Programming Tip 4.2.

Use of operators and guides.

As an international service provider, Quest Nature Tours will employ local operators to handle many of the on-site arrangements at the destination (e.g. hotels and transportation). At the time of the trip, Quest will send down its own leader to work with a local guide (usually two for every 16 participants), both of whom are experts on the natural history of the region. The use of reliable local guides is key – they help out with language as well as the many conditions and situations that occur on a trip. Quest will also send along extra guides to places like Antarctica and Galapagos for experience. This type of training enables Quest to rely on a bank of experienced guides for future trips.

As suggested earlier in the book, the not-for-profit sector has become an active player in the ecotourism industry. In a study on ecotourism operators, Higgins (1996) found that 17% of those sampled were not-for-profits, while similar results were generated by Weiler (1993). There appears to be a natural cross-over between ecotourism and not-for-profits, as many of these groups have a vested interest in parks or communities, and are often in search for other funding mechanisms. Ozark Ecotours is an example of a not-for-profit community development corporation run by the Newton County Resource Council, in the USA. The mission of this organization is to 'provide ecological and cultural interpretation to visitors which promotes an appreciation and protection of the natural resources and cultural heritage of Newton County' (www.ozarkecotours.com/ecotours.html). The commitment shown to ecology and community is demonstrated through training workshops on impact monitoring and responsible travel for guides, as well as for members of the community. This organization actively seeks methods for community economic development which limit the harmful side effects of ill-planned tourism. Much of the work of the Newton County Resource Council involves education. For example, their guide handbook explains, in detail, a number of aspects associated with the running of ecotours, as identified below (see Northwest Arkansas Resource Conservation and Development Council, Inc., 1997):

1. *Staging period*
   - Greet people as they arrive for the ecotour.
   - Make best first impression. Smile, be friendly.
   - Try to be the first to introduce yourself; greet everyone; spread your attention around.
   - Size up the group and make sure everyone has what they need.
   - Make sure all forms have been filled out and payments made.
2. *Introduction*
   - Introduce the assistant guide, where appropriate.
   - Introduce them to your concept of ecotourism and specific guidelines.
   - Set the tone by generating interest and enthusiasm.
   - Announce the itinerary and when you expect to return/end for the day.
   - Repeat important information about the trip, e.g. degree of difficulty, equipment, clothes.
   - Make time for a last minute restroom break.
   - On the bus: incorporate an ice-breaking activity or other activities/discussion.
   - Orient audience to theme by distributing information packets or describing routes.
   - Get to know your group, especially their names.
3. *Body*
   - During narrative stops, ensure that everyone has caught up to the group before starting.
   - Give people something to look for or think about between stops.
   - Look for or capitalize on teachable moments at every opportunity.
4. *Conclusion*
   - On return, reinforce trip theme; summarize; talk about highlights; laugh.
   - Relate lessons to their own experiences or lifestyles.
   - Ask for feedback; ask them to fill out an evaluation form.
   - Suggest how they might get involved in local efforts; recommend other ecotours they might consider taking, whether they be yours or another company.

These elements are very specific to the tour experience itself. As such, they provide a great deal of practical information about what must be done during the face-to-face phase of the ecotourism experience. At a more conceptual level, however, Holloway (1994) writes that there are a number of other elements to consider in developing tours but also understanding the nature of different markets. These include: (i) the motivation for touring, such as holidays, business, study, or religion, and is dependent upon the needs of tourists (education about nature and natural resources may form the basis for travel, in the case of ecotourists); (ii) the characteristics of tourists, including an analysis of tourist nationality, gender, age, lifestyle, life cycle stage, social class, occupation, personality, interests, and so on; (iii) mode of tour organization, which may be independent, or inclusive, which involves having the operator organize most of the major components of the tour including transportation, hotel, food, activities and other attractions and services; (iv) the composition of the tour, which consists of the makeup of the trip itself, including the form of travel (air, sea, land), the style of accommodation (guest house, B&B, ecoresort), and any links between the hotel and lodge (taxi, bus, and so on) or other components of the tour; and (v) the characteristics of the tour, including domestic and international travel, the kind of destination selected and the length of time spent on the trip (1 full day or 21 days).

This more broadly based perspective forces the ecotour operator to extend his or her thinking to include many of the most essential aspects of the tour. Consequently, the operator must ask many questions during the early stages of planning in attempting to understand how his or her operation will take shape and satisfy the many expectations that ecotourists will have. Some of these issues will be dealt with in greater detail in the chapters that follow, starting with the motivations and characteristics of tourists, which is dealt with in the next chapter.

Conclusion

The focus of this chapter was on introducing many essential principles that have been instrumental in recreation programme planning. Logically, programme planning stems from the consideration of agency philosophies, as well as mission and vision statements, and goals and objectives which reflect the overall ideology of the organization. These, particularly mission and vision statements, should be made available both within the organization and to ecotourists in an open and accessible manner. The chapter also discussed a number of theories that have been used in leisure studies, as a means by which to understand better the motivations of ecotourists: why they participate and what benefits they expect to accrue from their participation. Furthermore, the chapter referenced a number of programming approaches, which programmers use, and which fall within three general programming strategies, including the social planning approach, the community development model and the marketing approach. These were used to provide a framework from which to examine a series of programming approaches that have long been used in recreation programming, and which may be used for the purpose of developing ecotourism programmes. Along with a discussion on the tour operator, many of the essential roles of the programmer were discussed, especially as regards the importance of maintaining an ethical, and values-based approach to interactions with clients.

5

Needs and Assets

> Human behaviour can be genuinely purposive because only human beings guide their behaviour by a knowledge of what happened before they were born and a preconception of what may happen after they are dead; thus only human beings find their way by a light that illumines more than the patch of ground they stand on.
>
> (Medawar and Medawar, 1977)

Introduction

An understanding of: (i) the needs of participants, and (ii) resource assets, is critical both for the decision to start a new programme, and for the choice to continue with a programme after it has been offered. Needs and assets are not static, but rather change over time. Consequently, a periodic assessment of these helps to keep the programme timely and efficient. The ecotourism service provider will be confronted with a pool of needs (see Knowles, 1980), which derive from participants, the organization itself and the community. All of these will need to be filtered according to the overall requirements of the ecotourism operation, as well as the various sources of need. Based on the philosophical orientation of the service provider, such needs will be operationalized through a series of programme priorities, which will provide the basis of the ecotourism programme. In this chapter the needs assessment is discussed as the key vehicle by which one can collect important information on the specific needs of ecotourists. This is addressed through a brief overview of the steps involved in conducting needs assessment studies. In addition, attractions are examined as fundamental motivators for travel. An inventory of attractions is provided, which acts as an example of how service providers might go about the task of identifying critical features to include in their programmes. The chapter concludes with a discussion of the business plan as a key aspect of the programme, with particular attention paid to PESTE and SWOT analyses.

Motives (Needs)

One of the fundamental ways in which to differentiate between tourist types has been through an understanding of needs or motives. This is particularly true in the case of ecotourism, where a number of authors have sought to understand how the ecotourist differs from other traveller types (Eagles, 1992). Motivation is derived from the Latin word *movēre*, meaning 'to move', and have been described as internal factors which arouse, direct, and integrate a person's behaviour (Murray, 1964). Because of their 'internal' nature, motives cannot be observed, but are rather inferred from one's behaviour. That is, we must look at one's actions in order to understand what it is they are after.

Mannell and Kleiber (1997) developed a model of motivation based on the following four aspects: (i) motives, (ii) behaviour, (iii) satisfaction, and (iv) feedback (as illustrated in Fig. 5.1). These authors write that needs are synonymous with motives in that they are physiological and socially learned. Need emergence (motive) is said to create disequilibrium in a person. This disequilibrium is caused by either a lack of something or a desire for something. The model further denotes that a certain action or set of actions (behaviour) will restore equilibrium. If the actions are not satisfying, the negative feedback results in a modification of the activity or stopping of the activity altogether (perhaps being substituted by another). The relationship between these various elements of the model is shown in the following hypothetical situation:

1. As a result of being overwhelmed at work, the individual is motivated to relax (need).
2. Travel to a destination for 1 week (behaviour).
3. Still tied to work (phoning and e-mailing) leads to dissatisfaction.
4. Feedback.
5. Still the need for more relaxation (need).
6. Stay away for an extra week with no tie to work (behaviour).
7. Satisfaction.

Lundberg (1976) suggests that what tourists identify as motivations maybe reflections of deeper seated needs which the traveller does not understand, may not be aware of, or may not wish to articulate. Given the sheer number of needs expressed by an individual (perhaps 30 or 40 different needs), Iso Ahola (1980) wondered whether people constantly assessed these at different times and places for each activity participated in. He, like Lundberg, felt that researchers must look deeper as to why people behave the way they do, as illustrated in Fig. 5.2 on the levels of causality of leisure behaviour. Iso Ahola noted that the list of motivations which recreationists and tourists report, such as freedom and independence, are simply superficial explanations that are shaped by deeper needs. These include perceived freedom (feeling free to engage in an activity) and competence (feeling competent at the activity), which are very strong indicators of feelings of leisure. In turn, motives are shaped by the need for optimal arousal and incongruity, meaning that too much stimulation, or too little, is damaging to an individual. Too little stimulation leads to boredom, while too much often leads to stress (see discussion on optimal arousal and flow in Chapter 4). Thus, individuals search for the right level of arousal under the situational influences and social environment in which they are interacting. Finally, the need to participate in a recreational activity, like travel, is said to be a function of one's personality which is in turn a function of inherited characteristics (biological dispositions) and how the person has been socialized. Consequently, decisions on travel such as when to go, for how long, and what to see are driven by many very basic internal drives which may be quite difficult to articulate.

In a subsequent study, Iso Ahola (1982) wrote that travel needs (motivations) were purely psychological in nature (as opposed to being sociological, which has been

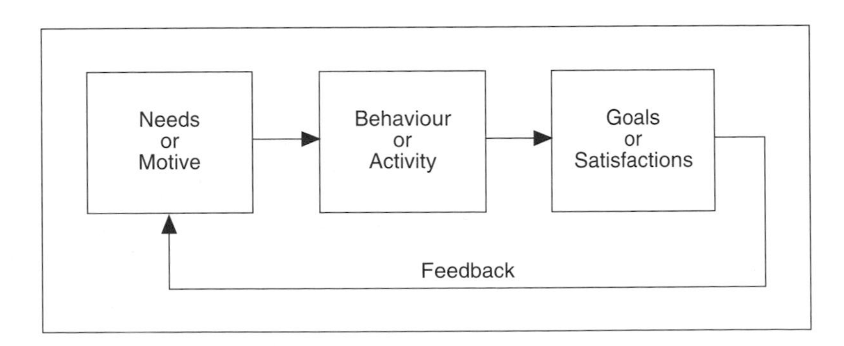

Fig. 5.1. Model of motivation. (Source: Mannell and Kleiber, 1997.)

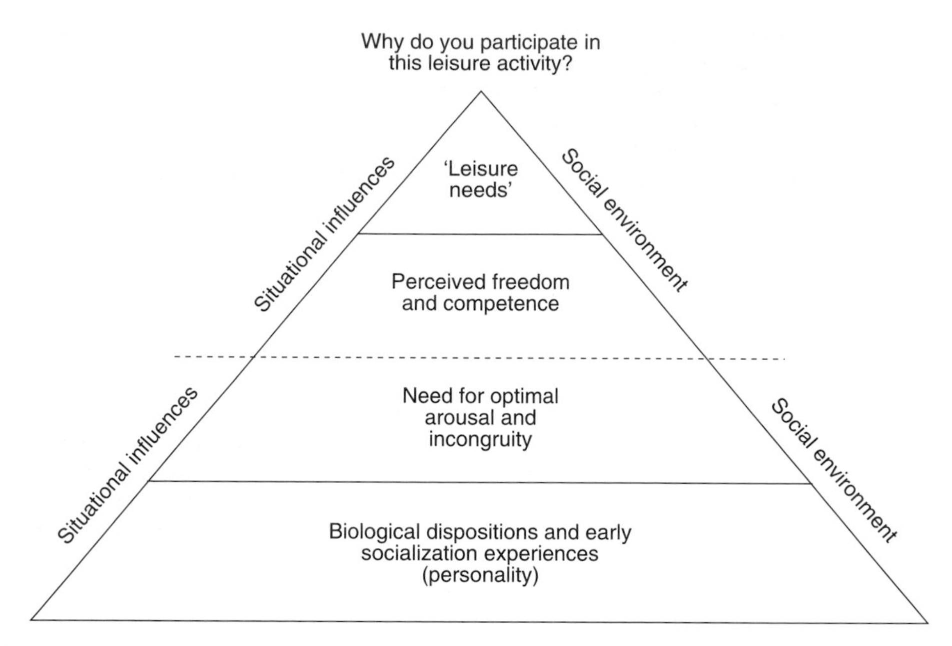

Fig. 5.2. Levels of causality of leisure behaviour. (Source: Iso Ahola, 1980.)

reported in other research). He argued that people travel for two main reasons: (i) to seek intrinsic rewards (novelty), and (ii) to escape the everyday environments (avoidance). These two motivational forces simultaneously influence people's travel choices. (Tourism, he argues, may be more escape-oriented.) Using the example of escaping, a person may leave behind personal troubles or failures, or interpersonal relationships between co-workers, family members or neighbours. The amalgam of these elements is a four-cell matrix, where a single tourist could theoretically go through one or all aspects of the model during the course of one trip.

Needs Assessment

The first step, and one of the most important in the process of developing recreational programmes, is the identification of participant needs which the service provider will seek to isolate and later satisfy. Indeed, it should be the prime directive of service providers to strive to meet or exceed the expectations of participants through the programmes that they develop (McCarville, 1993). In consideration of the foregoing discussion on superficiality of needs identification, it is still a responsibility of service providers to attempt to tap into the needs of participants for the purpose of developing effective programmes and generating meaningful experiences. The vehicle that has been used to accomplish this end in the recreation field is the needs assessment, which incorporates any number of different survey or measurement tools to explore needs. (Tourism planners have often employed tourism strategies or tourism marketing plans which focus on aspects of demand from generating regions, as well as on the state of affairs within destination regions, including capacity, attractions, and impacts (Veal, 1992). The extent to which these effectively address recreational or tourism need, in the manner addressed in this book, is open to debate.)

Quite simply, the needs assessment is 'the process by which the program planner identifies and measures gaps between what is and what ought to be' (Windsor *et al.*, 1994, p. 63; see also Gilmore and Campbell, 1996). Such needs may be identified for an individual, but more typically for groups of people (grouped by gender, age, geographic location, and so on), which is an approach much more easily accomplished (see the discussion on encounter norms in Chapter 3). Table 5.1 addresses some of the reasons why practitioners conduct needs assessments (adapted from DeGraaf *et al.*, 1999), and these are categorized under service orientation, the desire for quality, and programme management. For example, under service orientation, the needs assessment may allow tourists to provide valuable input into the programme; it may allow for new ideas in creating innovative programming; it may allow the service provider to serve as many people as possible; and it helps the programmer stay in tune with psychological and physiological needs of participants.

Depending on the time, situation, resources, and experience of the programmer, a number of steps will be followed in conducting the needs assessment, including the following general catregories: (i) defining need; (ii) identifying study subjects; (iii) asking the right questions; (iv) collecting the data; (v) analysing and interpreting data; and (vi) reporting and using the data for planning. Each of these will be covered in more detail below.

Defining need

Despite the implicit assumption that ecotourists represent one homogeneous sub-set of tourists, which may be true but only to a degree, one of the certainties that ecotourism service providers must contend with is that participants will have a number of different needs, both within and between groups (i.e. within a tour and between different tours). Such needs may include requests for more educational material in the programme, less physical involvement in the

Table 5.1. Why conduct needs assessments?

Reason	Description
A. Service orientation	
Tourist input	Programming decisions may be influenced positively by allowing tourists the opportunity to share their views
Solicit new ideas	Allows for creativity and innovation in programme planning
Inclusion	Enables the provider to serve as many people as possible, regardless of ages, sex, ability, race, religion, education
Meet real needs of people	Helps the programmer stay in tune with the physiological and psychological needs of participants
B. Desire for quality	
Professional commitment	Operators must strive to be ethically based and to exceed standards which may be set for the industry/region
Resource allocation	Helps to inform the operator on where s/he needs to focus (e.g. more staff, better transportation, and so on)
Accountability	Being responsible for our actions and who and what they impact (tourists, parks, local people, other land users)
Increase profits	Important element to ensure that the company makes profits while ensuring tourist satisfaction
C. Programme management	
Manage duplication	Duplicating services within or between operations is a waste of resources and leads to unhealthy competition
Address safety	How to better use facilities, transportation, and better understand hazards in very dynamic environments
Plan a variety of programmes	Help us design and implement the best possible programmes for the specific targets that we cater to
Develop/meet individual goals	To help providers and participants to meet their goals, such as skill development and environmental education

day-to-day initiatives of the group, a higher level of service quality, a lower guide-to-participant ratio, less time transporting between natural areas, more time spent in natural areas, and so on. Need, however, as suggested earlier, is not an easy construct to understand. Following the work of Bradshaw (1972), Edginton *et al.* (1998) suggested that people with different sets of values, have different reference points for determining needs. Needs can thus be classified in a number of different ways, including normative, felt, expressed, comparative and created. These are discussed in greater detail below (and adapted for the purpose of this book):

1. *Normative need.* These needs refer to objective standards that have been established by experts in the field. For example, these standards may include the number of campsites, car parks, parks, or open spaces that are required for an area. The province of Ontario, for example, implemented a set of standards for the development of their recreation class parks on the basis of population. They used a 3-hour time limit to define camping supply for Metropolitan Toronto by using averages for camping use from past studies. On the basis of 1.18 camper visits/person/year they were able to determine the number of recreation class parks which would meet Toronto's demand (within this 3-hour radius).
2. *Felt need.* These needs are defined as the desires that an individual has but has not yet expressed actively. These learned patterns of behaviour are based on what people think they want to do, rather than what they are actually doing or have done in the past. As Edginton *et al.* (1998) suggest, this type of need is developed further or intensified by mass communications which expand the person's knowledge of the world beyond his or her root of existence.
3. *Expressed need.* These are simply felt needs that have been put into action – the fulfilment of one's needs. While this type of need is important in development of programmes (e.g. preferences, tastes and desires of participants), a strategy based solely on this type would preclude the development of other programmes which would not meet all of the various needs of participants.
4. *Comparative need.* In this approach, ecotourists would compare the types of opportunities available from one ecotour operator with those available from another, in the selection of an appropriate service. However, this also means that one operator could compare his/her offerings for a particular population, and the behavioural outcomes of these programmes, with other operators. Edginton and his colleagues stress that caution must be practised when using this approach in assuming that what works well in one situation will also be effective in another.
5. *Created need.* Clever marketing and promotion create or induce a need where none existed before. Consequently, the inducement is supported and identified as a need by clients, who in turn are awakened to a need heretofore unrealized.

Programming Tip 5.1.

General versus specific trip content.

Quest Nature Tours takes pride in developing general nature tours within the destinations of their choice. These trips focus on all aspects of natural history instead of one or two main attractions, such as birds or rafting. In this way the participant is open to experience as much of the flora and fauna of the destination as possible, with a heavy emphasis on learning. The aspect of learning is also valued by Niagara Nature Tours, who suggest that information should not be 'lectured', but rather 'offered'. A leader may be one of the most knowledgeable in a field, but if they cannot pass information along in an interesting way the information may be lost on tour participants.

Identifying study subjects

The identification of study subjects might typically range from the identification of a few ecotourists to many ecotourists, selected randomly, or on a convenience basis, and who are thought to be representative of the type of information sought. Such information might include an assessment of differences related to particular geographical regions (Europeans, Australians and so on), gender, income, age and education. It may also be that service providers make a habit of allowing all participants to express their needs in one way or another. In all likelihood this will be a function of the number of ecotourists taking the programme(s) as well as the approach to the needs assessment. Short one-page questionnaires are easy to administer, while lengthy surveys may involve a broader approach centred around mailing a large number of surveys to past participants. In the latter case, this may include drawing an appropriate sample of subjects if numbers are too large (sampling is discussed further in the section on 'Collecting the data').

Asking the right questions

In conducting needs assessments, Henderson and Bialeschki (1995) suggest that organizations should address some or all of the following five types of question, in their efforts to ask the right questions (adapted for the purpose of our discussion on ecotourism). These questions may be targeted to the past, present, or future, and include (see Appendix 1 for an example of a needs assessment survey):

1. *Experience/behaviour* (e.g. How many times did you travel as an ecotourist last year?)
2. *Opinion/value/attitude* (e.g. What did you like most and least about the last trip you took to Costa Rica?)
3. *Emotional* (e.g. Did your interactions with the tour guide give you a positive feeling about how important ethics are to the company?)
4. *Knowledge* (e.g. How informed is the group on the various ecological processes taking place in a park?)
5. *Background/demographics.* Questions on age, gender, household income, level of education, occupation, marital status, number of children, race or ethnicity.

Hudson (1988) writes that leisure needs are often difficult to uncover because needs assessment surveys generally rely on interest questions instead of 'need' questions. Needs assessment instruments often contain a long list of recreation activities asking tourists to indicate which of these activities would best suit them. Hudson says that a problem with this approach is that it does not enable the provider to understand the underlying motives of the respondent (refer back to the discussion on needs earlier in the chapter), particularly in regards to interest in one activity over another. A way to avoid falling into such a trap, is to ask people their recreation or tourism environment preferences. Questions are thus geared to group dynamics (big or small groups), structured or unstructured activities, or relaxing or stimulating activities. Figure 5.3 contains an example of such a question (Hudson, 1988, p. 19).

Collecting the data

Service providers will want to pay close attention to the type of data they are wanting or willing to handle. An interview format demands a much different type of data collection technique than a self-administered questionnaire. The types of questions asked, therefore, and how they are asked, have obvious implications on the results, but also on the techniques which need to be employed to understand these results. Briefly, there are two main approaches to data collection (or research, in general), including qualitative research methods and quantitative approaches. In the case of the former, analysis involves:

> breaking up, separating, or disassembling of research materials into pieces, parts, elements, or units. With facts broken down into manageable pieces, the researcher sorts and shifts them, searching for types, classes, sequences, processes, patterns, or wholes. The aim of this process is to assemble or reconstruct the data in a meaningful or comprehensible fashion.
>
> (Jorgenson, 1989, p. 107)

Please respond to each set of words based on your immediate reaction to the various styles of participation in leisure activities. A LEISURE ACTIVITY is anything you do for fun and pleasure (e.g. jogging, reading, swimming).

GENERALLY, I LIKE TO PARTICIPATE IN LEISURE ACTIVITIES WHICH ARE:

	Strong	Moderate	Neutral	Moderate	Strong	
Active						Passive
Safe						Risky
Novel						Repetitive
Simple						Complex
Hard						Easy
Creative						Routine
Spontaneous						Planned

Fig. 5.3. Recreation environment preferences. Source: Hudson (1988).

This approach, through focus groups or interviews, allows the researcher to describe experiences in words (Henderson and Bialeschki, 1995). This method is particularly useful in allowing for an in-depth understanding of a particular problem or issue. Focus groups, for example, enable the practitioner to examine a group of individuals who possess certain characteristics and who provide qualitative information through focused discussions (Krueger, 1988). Focus groups are especially useful to companies who are conducting market research, but the technique may also be used before a programme begins, in the middle, or at the conclusion. Such groups may consist of, for example, just males or females, 18- to 35-year-olds, seniors, the economically disadvantaged, and so on. They are typically small, because some individuals may choose not to disclose information in larger groups, and also because larger groups become more unwieldy to facilitate. The facilitator must attempt to pose questions in such a way as to maximize the amount of interaction from participants. The open-ended aspect of this approach lends itself well to initiating further probes into particular issues or details. Information, or data, is recorded through tape recorder, note pad, or both, and transcription of the data provides the researcher with the opportunity to examine trends or specific statements that are representative of key themes.

The Delphi technique is a method that allows professionals to gather information from a number of experts on the perceptions of policy or future courses of action. While the research may examine an issue that is fixed in time, the technique is also well suited to longitudinal analyses (Kraus and Allen, 1997). Experts in the field, who remain anonymous, are asked to complete a survey indicating their views or perceptions on a particular issue. The first step in the process is to develop a survey that highlights a key issue. This is followed by the selection of experts who comment on the issue, followed by a tallying of their results for the purpose of examining agreement and disagreement. The experts then receive a second survey underscoring the items of disagreement, and the reasons for initial judgements. These steps are repeated until consensus is reached (Siegel *et al.*, 1987). As suggested earlier, this technique is excellent for soliciting responses on key policy issues. Subjects for scrutiny using Delphi may include the development of indicators for sustainable tourism, the development of codes of ethics for an attraction, or the ways in which to induce community–industry cooperation.

In using quantitative approaches, methods are employed which generate data in the form of numbers. These numbers may be measured, assessed and interpreted with the use of mathematical or statistical manipulation (Alreck and Settle, 1985). The self-administered questionnaire is an example of a survey tool that yields quantitative data (more on this to follow).

Although, as suggested above, a number of methods or tools may be used to collect data for needs assessments, three types have been frequently used. These include: (i) mailed or self-administered questionnaires; (ii) telephone interviews; or (iii) face-to-face interviews. A number of questions related to each of these methods are addressed in Table 5.2, including the type of information sought, sample, and the time frame, as well as the various conditions in which each should be employed. Although not mentioned in Table 5.2, the service provider may elect to conduct a self-administered survey on-site. This is an effective way to collect information on the pros and cons of the programme, while allowing the operator to save substantial sums of money that might otherwise be spent on stamps and envelopes.

Table 5.2 suggests that not all methods are appropriate depending on the conditions under which the needs assessment is taking place. And, as suggested in the last variable under 'Time', needs assessments and other evaluations should never be undertaken if the service operator does not have the time to think about the research that might need to be done, which includes asking the right questions. The exercise will be a massive waste of resources if time and effort have not been put into asking the most effective questions. Furthermore, all information is important, but the most meaningful will emerge from the application of solid research techniques using a few well established methods. The following, although detailed, is an example of the Total Design Method (TDM) developed by Dillman (1978). It is a widely accepted method which is quite effective in generating meaningful results in *mail surveys*. The construction of the TDM is below (Table 5.3), and is followed by the implementation strategies for the TDM in Table 5.4. (NB:

Table 5.2. If a survey is to be done, what type is best?

Type of information	Face-to-face interview	Telephone survey	Mailed survey
Type of information			
If you have ...			
A complicated series of questions to ask	Yes	No	Sometimes
Many open-ended questions	Yes	Possibly	Rarely
Many alternatives to be considered or rated	Possibly	Rarely	Yes
Do you need to know ...			
Responses to specific questions	Yes	Yes	Yes
Deeper attitudes and feelings	Yes	Rarely	No
Sample			
Do you need ...			
Answers from people who might not respond themselves	Yes	Yes	No
Answers from people who are likely to respond	Yes	Yes	Yes
Answers from a lot of people	No	Possibly	Yes
Do you have ...			
Those who won't grasp print	Yes	Possibly	No
People not wanting anyone to know what they think	Rarely	No	Yes
People who speak other languages	Possibly	Possibly	Possibly
Time			
If you lack ...			
Trained interviewers or the time to train people	Never	No	Yes
Time to code responses or people to do it for you	Never	Rarely	Rarely
Time to think about the research you are doing	Never	Never	Never

Adapted from Abbey-Livingston and Abbey (1982).

Table 5.3. Construction of the TDM Mail survey.

1. The survey is designed as a booklet, approximately 16.5 × 21 cm (6.5 × 8.25 inches).
2. The survey is typed on normal sized paper and photo-reduced to fit into the booklet, thus providing a less imposing image.
3. The booklet is printed on white paper and is slightly lighter than normal paper for lower mailing costs.
4. No questions are printed on the first page, which is used for an interest-getting title, a neutral but eye-catching illustration, and any necessary instructions.
5. Similarly, no questions are allowed on the last page (back cover). It is used to invite comments and express appreciation to the respondents.
6. The most interesting topic-related questions comes first; potentially objectionable questions next; followed by demographic questions.
7. Special attention is given to the first question, which should apply to everyone and be easy to answer.
8. Each page is formulated with great care in accordance with these principles:
   - Lowercase letters are used for questions and uppercase for answers.
   - To prevent skipping items each page is designed so that whenever possible respondents can answer in a straight vertical line instead of moving back and forth across the page.
   - Overlap of individual questions from one page to the next is avoided, especially on back to back pages.
   - Transitions are used to guide the respondent much as a face-to-face interviewer would warn of changes in a topic to prevent disconcerting surprises.
   - Only one question is asked at a time.
   - Visual cues (arrows, indentations, spacing) are used to provide directions.

Source: Dillman (1978) *Mail and Telephone Surveys: the Total Design Method.* John Wiley & Sons, New York.

Table 5.4. Implementation of the TDM survey.

1. A 1-page covering letter is prepared, and explains:
   - that a socially useful study is being conducted;
   - why each respondent is important;
   - who should complete the survey;
   - a promise of confidentiality.
2. The exact mailing date is added on to the letter, which is then printed on the sponsoring agency's letterhead stationery.
3. Individual names and addresses are typed on to printed letters in matching type and the agency's investigator name is individually signed with a blue ballpoint pen using sufficient pressure to produce slight indentations.
4. Surveys are stamped with an identification number, the presence of which is explained in the covering letter.
5. The mail-out package, consisting of a cover letter, survey, and business reply envelope is placed into a monarch-size envelope on which the recipient's name and address have been individually printed (address labels are never used) and first-class postage is affixed.
6. Exactly 1 week after the first mail-out, a postcard follow-up reminder is sent to all recipients of the survey.
7. Three weeks after the first mail-out, a second covering letter and survey is sent to everyone who has not responded.
8. Seven weeks after the first mail-out, a second letter complete with another covering letter and replacement survey is sent by certified mail.

Source: Dillman (1978) *Mail and Telephone Surveys: the Total Design Method.* John Wiley & Sons, New York.

some of the methods employed will now have changed due to developments in technology, but the basic mechanics of survey design remain unchanged.)

Depending on the size, scope and methods used, a sampling procedure may need to be employed for the needs assessment. Sampling is the process of examining part of one population (e.g. all the ecotourists visiting an island) in order to represent the whole population. It is understandably a very ambitious and costly task to survey each and every individual of a population, which may number in the thousands. The sample, however drawn, is thus a more economical means of gathering information, while at the same time ensuring that the sample is representative of the overall population. For example, on the basis of a population of 10,000 individuals, a representative sample would be 370; a population of 1000 would require a sample of 278; a population of 500 ecotourists would require a sample of 217; and a population of 200 would require a sample size of 132. The relationship between population and sample is more effectively shown in Table 5.5.

Scaling down to a manageable size is the big advantage of sampling. However, in scaling down, the researcher may set the stage for bias and error. In eliminating such problems, the researcher may wish to use a number of techniques to guarantee reliability (freedom from random error), and validity (measuring what is supposed to be measured), which have been summarized in Table 5.6 (from Neuman, 1997, p. 205). The reader will note two distinct types of sampling in the table. Non-probability sampling, as described by Alreck and Settle (1985), as that which deviates away from the random selection of respondents or some other design where the probability of inclusion of individual sampling units (e.g. tourists) is equal and known. Alternatively, probability sampling is where every element in the population has either an equal probability of selection, as with random sampling, or has a given probability of being

Table 5.5. Table for determining sample size for an evaluation.

Population	Sample	Population	Sample	Population	Sample
10	10	220	140	1,200	291
15	14	230	144	1,300	297
20	19	240	148	1,400	302
25	24	250	152	1,500	306
30	28	260	155	1,600	310
35	32	270	159	1,700	313
40	36	280	162	1,800	317
45	40	290	165	1,900	320
50	44	300	169	2,000	322
55	48	320	175	2,200	327
60	52	340	181	2,400	331
65	56	360	186	2,600	335
70	59	380	191	2,800	338
75	63	400	196	3,000	341
80	66	420	201	3,500	346
85	70	440	205	4,000	351
90	73	460	210	4,500	354
95	76	480	214	5,000	357
100	80	500	217	6,000	361
110	86	550	226	7,000	364
120	92	600	234	8,000	367
130	97	650	242	9,000	368
140	103	700	248	10,000	370
150	108	750	254	15,000	375
160	113	800	260	20,000	377
170	118	850	265	30,000	379
180	123	900	269	40,000	380
190	127	950	274	50,000	381
200	132	1,000	278	75,000	382
210	136	1,100	285	100,000	384

Source: Krejcie and Morgan (1970) as cited in Henderson and Bialeschki (1995).

Table 5.6. Types of sampling.

Non-probability	Probability
Haphazard. Select anyone who is convenient	*Simple.* Select people based on a true random procedure
Quota. Select anyone in predetermined groups	*Systematic.* Select every *k*th person (quasirandom)
Purposive. Select anyone in a hard-to-find population	*Stratified.* Randomly select people in predetermined groups
Snowball. Select people connected to one another	*Cluster.* Take multistage random samples in each of several levels

Source: Neuman, 1997.

selected that is known in advance and used in analysis to assess significance.

Data analysis

The data analysis stage involves organizing data, coding data and data analysis. Once surveys are received they must be treated so that when they are coded, through whatever means, the researcher is able to match a series of numbers on a spreadsheet with an actual survey. While this sounds overly simplistic, the proper treatment of surveys when they arrive will ensure that numbers may be cross-checked or further analyses may be completed. Data analysis may occur on many different levels, from univariate statistics, to sophisticated multivariate computations. It may be that the service provider only requires some basic information about the population being studied. In such a case, frequency counts and percentages, or measures of central tendency such as mean, mode and median, may suffice. Furthermore, the analysis of this data may be done without the aid of computer, or it may be done using advanced statistical packages, such as SPSS. Larger organizations, with many more clients, may wish to invest in software packages to generate different statistics, but also for the purpose of handling data over long periods of time. Small organizations may not wish to invest significant amounts of money on statistical software packages that yield sophisticated results.

It is beyond the scope of this book to elaborate on the tremendous array of parametric and non-parametric statistics that may be used to analyse data. In general, the selection of techniques will be based on the types of questions asked (e.g. nominal, ordinal or interval level data) as well as the type of information sought. Students and practitioners would be wise to consult any number of good books which have been designed to cater directly to research and evaluation in the tourism and recreation and leisure studies fields, some of which have been mentioned above. A book that is particularly user-friendly is one written by Gray and Guppy (1999). This resource is succinct, humorous and informative.

Reporting and using the data for planning

The data, which may have taken a great deal of time and money to generate, must be shown to have a meaningful application to the organization. As suggested above, the importance of placing sufficient time into the construction of good surveys, which ask the right questions, to the right population, and use the right measurement tools, cannot be emphasized enough. Having accomplished this, it is also worthwhile to place a great deal of time and effort into the production of a professional report, which is not only informative in content, but also aesthetically pleasing. This means taking the time to report the data through the use of appropriate written and graphic presentation techniques. The following is an overview of some of the key elements that should be included in a needs assessment or evaluation report.

- *Cover.* Includes the title of the report, the author, funding agency (if applicable), the date and the name of the group for whom it is prepared.

- *Covering letter*. This is optional, but may identify important information related to the research that may not necessarily be included in an Executive Summary.
- *Table of contents*. This page(s) will list all of the main sections and sub-sections of the report, including the appendices and the page numbers where these can be found. A list of tables and figures will also be helpful.
- *Executive summary*. This section highlights the purpose of the study, key aspects of the population or event, and some of the key findings of the research.
- *Acknowledgements*. An important, but often overlooked aspect of the report. It should recognize the important help of respondents, colleagues, sponsors and other individuals who aided, in any respect, the researcher in his or her efforts to complete the work.
- *Introduction*. May include a discussion of the 'setting' and context in which the research has taken place. After a brief discussion of the background to the study, the researcher should include a statement of purpose as well as objectives, terms of reference, and/or list of problems. This may also include a discussion of the various theories or philosophies incorporated, as well as a list of various issues that need to be examined.
- *Methodology*. This section should include a discussion of subjects as well as information on the setting, data-collection techniques and response rates.
- *Results*. This section includes the data in written and graphic form. Tables and figures are a welcome addition to this section in order to provide information in a synoptic form. Tables must be crafted to 'stand alone', meaning that all pertinent information about the table's contents should be included. The availability of graphics packages makes the development and inclusion of pie graphs, line charts, 3-D area charts, scatter charts and so on, very straightforward.
- *Conclusions*. This section may simply discus the findings of the previous section, or elaborate on these findings through a discussion of the results and their implications.
- *Recommendations*. Based on the results of the study, the programmer is given the opportunity to identify a number of recommendations to ameliorate the problems that have been identified through the research.
- *Appendices*. This section will contain information that does not easily fit into the main body of the report, but nevertheless is required for clarification or for further information.

The needs assessment is thus a key aspect of the programme planning cycle, which allows the service provider to address some fundamental questions related to the organization and participants. These include: (i) the types of programme that ecotourists want; (ii) the attitudes and opinions of ecotourists regarding the programmes that are currently being offered, and the level of service; (iii) the programmes, of a number of possible alternatives, that are the most popular and financially successful; and (iv) the changes needed to make the current programmes better. It is a mechanism that allows the operator to examine current and ideal conditions, imbalances in the provision of services, agency values, effective use of resources, and long- and short-range planning, in an attempt to be accountable to financiers, as well as to the participants. In regards to this last point, operators cannot assume that a population of 55- to 65-year-olds is going to want the same type of experience as a group of 25- to 35-year-olds. Consequently, the operator must shoulder the responsibility of knowing as much as possible about the population to be served, including how much time to spend in the field, interest in recreational activities during 'free time', benefits from participation in certain type of activities, and differences on sociodemographics (e.g. gender, education, income).

Attractions

Tourism attractions have long been a topic of discussion among tourism scholars, especially in regards to how they should be defined (see Lew, 1987; MacCannell, 1989). In general, attractions have been examined on the basis of a number of characteristics related to space, management, experience and perception. Spatially, they include areas, objects, or people, that are either sedentary (fixed, like a building) or transitory (e.g. birds). This space is identified by a marker(s), such as signs, and is a component of some overall management system within a holiday / geographical region. Furthermore, there appears to be some type of formal or informal relationship between the tourist (who gains positive or negative experiences from viewing the attraction) and the attraction, which must absorb different types and levels of impact. The perceptual element relates to the fact that the attraction must be perceived as such by both management (e.g. national parks agency or museum administration) and tourists alike.

There is little doubt as to the importance of attractions as tourism pull factors. Attractions such as beaches, casinos, rainforests, people and so on, are the focus of numerous different promotional tools, such as brochures, which induce people to travel to a destination. For ecotourists, there is the combination of factors that draw individuals to a destination such as the Galapagos Islands, including the historical significance of the islands, the flora and fauna, and perhaps the geographical distinctiveness of the region. As alluded to above, attractions differ across a number of dimensions, allowing them to be classified according to a number of different characteristics. Figure 5.4 illustrates how attractions may be categorized on the basis of culture, nature, events, recreation and entertainment. The first two attraction clusters, cultural and natural, form the basis of ecotourism experiences, although there is some hesitancy in including cultural attractions as primary ecotourism attractions, simply because they do not focus on the natural history aspects of ecotourism (see Fennell, 1999). Nevertheless, many operators include cultural themes in their ecotours, including sites like Macchu Picchu in Peru, Mayan pyramids in Mexico, Guatemala and Honduras, Viking settlements in Newfoundland, Canada, and the monuments of explorers (e.g. the monument to Speke in Uganda in recognition of his finding the source of the Nile).

Natural attractions, as outlined in Fig. 5.4, are not so different from some of the undeveloped resources that are identified below, such as weather, surface materials, and flora and fauna. The difference perhaps lies in how the resource or attraction is perceived and used. Resources may be critical in enabling participation (the characteristics of space, good weather, topography, water); whereas attractions are the focus of participation (e.g. a ski hill) and will have been presented according to the definition of attraction outlined above (e.g. it is marked, managed, and so on). Nevertheless, developed resources such as of highways, facilities, sewerage, buildings, and so on, and undeveloped resources (below), provide the necessary conditions that allow for the tourism industry to exist. Chubb and Chubb (1981) outline seven different types of undeveloped resources, which are identified as being of importance to outdoor recreation and ecotourism planning, as well as to participant satisfaction. Although independent, these may be envisioned as integrated, as it is the combination of environmental elements that provides the basic foundation of our experiences. These undeveloped resources are as follows:

1. *Geographic location*. The characteristics of space that determine the conditions, in association with other variables, for participation (e.g. skiing).
2. *Climate and weather.* Determined by latitude and elevation relative to large landforms, mountains, ocean currents and high-altitude air currents. Along with geology, climate is the prime controller of the physical environment, affecting soils, vegetation, animals and the operation of geomorphological processes such as ice and wind.
3. *Topography and landforms*. The general shape of the surface of the earth (topography) and the surface structures that make some geographical areas unique (landforms). A landform region is a section of the earth's surface characterized by a great deal of homogeneity among types of landforms.

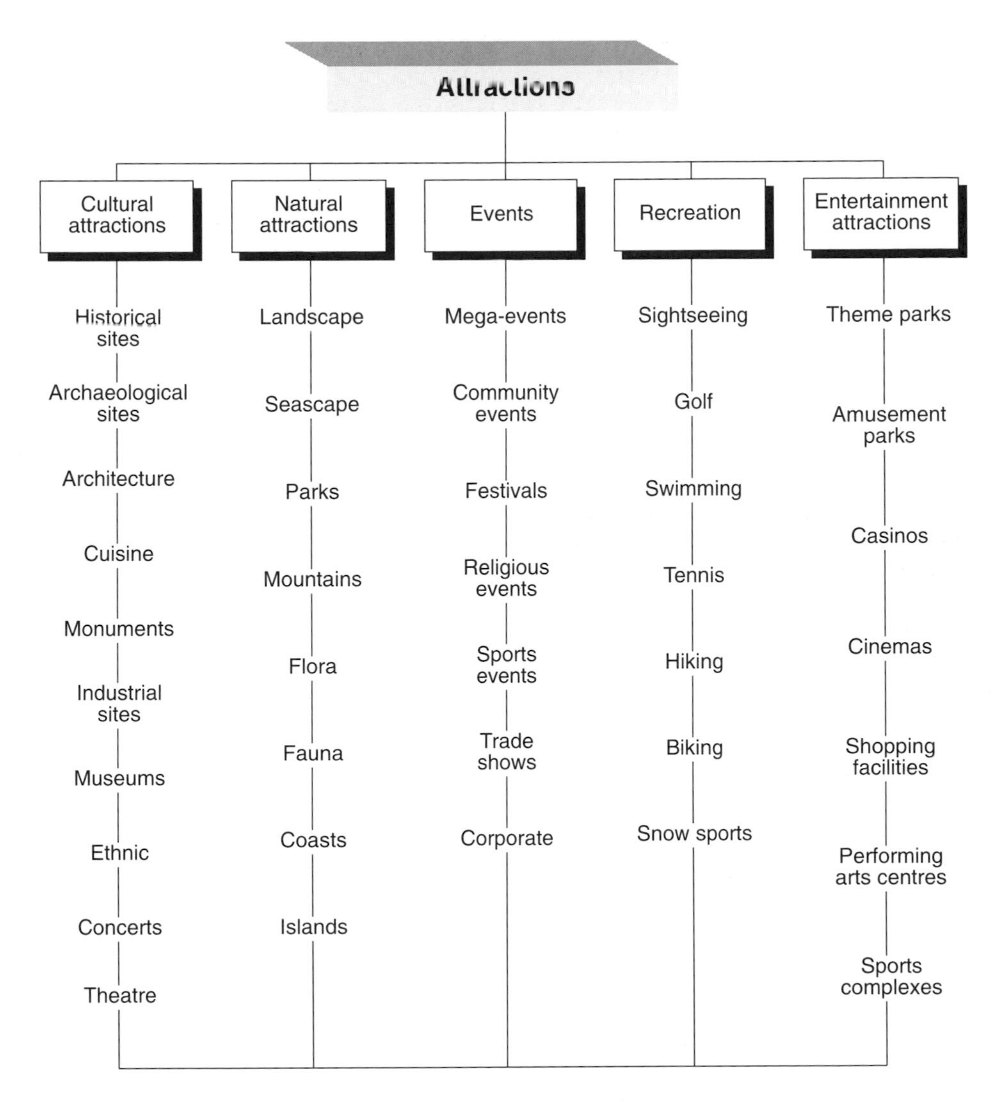

Fig. 5.4. Overview of attractions. (Source: Goeldner *et al.*, 2000.)

4. *Surface materials.* The nature of the materials making up the earth's surface, including rocks, sand, fossils, minerals, soil, sand, and so on.

5. *Water.* This substance plays a critical role in determining the type and level of outdoor recreational participation in ocean and freshwater settings.

6. *Vegetation.* Vegetation refers to the total plant life of an area. Ecotourism is dependent on plant life directly (e.g. tourists taking pictures of unique plant species), or indirectly (e.g. trees acting as a wind barrier for skiers).

7. *Fauna.* Animals play a significant role in tourism activities that are both consumptive and non-consumptive. Consumptive activities are those which take more of a utilitarian or controlling approach to wildlife (e.g. fishing, hunting, etc.), while non-consumptive activities are those which have a softer impact on the resource base (e.g. birdwatching).

Ecotours take place in many of the most beautiful areas of the planet. The attractiveness of these regions, however, should in no way be confused with the concept of attraction. Husbands (1983) drew a relationship between tourist space and tourist attraction in attempting to address the problem of the measurement of natural attractions. He viewed the concept of attraction, as many researchers do, with the actual levels and patterns of tourist visitation. Attractiveness, on the other hand, has characteristics akin to a sense of place. The concept of place carries with it a subjective value, and is therefore difficult to measure in a qualitative sense. This feeling is captured in the following quote by Dearden and Sadler (1989, p. 3):

> The visual and aesthetic qualities of the landscape in which we live, work and play make up a fundamental dimension of everyday experience. Our response to them is layered and complex, often difficult to personally capture and articulate, let alone to measure or translate into planning or educational terms.

The implications of Husbands' work suggests that the marketing of a destination cannot be based on attractiveness alone, but must also consider its relative attraction among a set of competing destinations. Marketing research relies on quantifiable data on the attractions of a destination to aid in understanding which factors are important to tourists when they evaluate attraction choices (Fodness, 1990).

Fig. 5.5. Natural attractions, such as seabirds in Shetland, are particularly attractive to tourists.

Inventories of attractions and resources

Just as the needs of participants change from year to year and from programme to programme, so too do the base of attractions and physical, natural, recreational and sociocultural resources, which form the basis of ecotours. In view of this, the service provider may wish to undertake an inventory of attractions and resources on a regular basis in order to document these changes, but also to ensure that he or she is taking advantage of the best of what a region has to offer. In general, such an inventory is a systematic compilation of attractions and resources that occur within a tourism destination region, and which will help to demarcate the spatial and temporal boundaries of the ecotour. It is the task of the operator not only to list these potential attractions and resources, but also to identify any special significant factors related to these attractions, which would allow the operator to generate greater value in programme offerings.

A comprehensive list of general natural, cultural, human and tourism industry attractions and resources is included in Table 5.7 (general categories adapted from Patterson, 1997), and a more specific list is found in Table 5.8. Table 5.7 includes the 'Presence' or 'Availability' of a resource, the 'Quantity' of the resource (e.g. waterfalls), as well as the 'Quality' of the resource for the purposes of the ecotour. It also includes a means by which to subjectively assess the ability of the resource to absorb the demands of a tour, which could include selected composite

Fig. 5.6. Cultural attractions may be a central or peripheral aspect of the ecotour, depending on the needs of the population served.

measures, such as temporal and spatial aspects of the tour, levels of impact, levels of use, transportation, and so on. The operator may also record a subjective assessment of the future potential of the resource. The latter table is a more refined version of the general inventories, allowing the programmer to record more pertinent information on specific attractions. The example provided in Table 5.8 is birds, where the programmer would simply list the bird species, or those of special interest, which are endemic to the region, and other bits of information related to seasonality, sitings, the ability of the bird to withstand numbers of ecotourists (e.g. accessible, shy, aggressive, etc.), as well as any other notes that which would help the programmer to plan. Recreational activities have not been included owing to limitations of space, but might include white-water rafting and canoe rental shops, theatres, bicycle rentals, and so on. The inclusion of recreational activities should be strongly considered to reflect the interest and abilities of participants (more on this later).

In short, the operator must actively search for any and all resources that will aid in the delivery of the programme. Once completed, the information yielded from this process will be an essential component of the programming planning stage. From the natural resources standpoint, the various habitat types documented may yield a range of different opportunities to view wildlife. Bogs and swamps will be habitat for a number of bird species which will not be encountered in other areas (sand dune areas). Also, the operator should search for individuals or groups – human resources – in the community who may be able to help in the interpretation of the environment. Every community has people who are knowledgeable about the natural and human history of the region, and who may be willing to share this information with ecotourists. For this reason, the identification of a number of specialists (through universities or local conservation groups) will help put variety into programmes, with the added benefit of building strong presence and support in the community.

The Business Plan

The decision to enter into business as an ecotourism operator may be based on many

Table 5.7. General resource inventory.

Resource	Presence/ availability	Quantity (No.)	Present quality (Rank 1–10)	Capacity to absorb tour (low, med, high)	Future potential (low, med, high) and notes
Natural attractions					
Beaches	No	–	–	–	Nothing available, not appropriate
Birding	Yes	Many types	8	(medium)	Migratory and residents species (high)
Karst landscapes	etc.				
Temperature					
Rainfall					
Average sunshine					
Desert environment					
Forests					
Fungus					
Geological formations					
Hiking trails					
Islands					
Lakes					
Lichen					
Mammals					
Mountains					
Nature trails					
Parks/protected areas					
Plants					
Rivers					
Waterfalls					
Wilderness areas					
Cultural/historical attractions					
Archaeological sites					
Ethnic culture					
Indigenous sites					
Interpretive centres					
Museums					
Festivals					
Special events					
Traditional lifestyles					

Table 5.7. *Continued.*

Resource	Presence/ availability	Quantity (No.)	Present quality (Rank 1–10)	Capacity to absorb tour (low, med, high)	Future potential (low, med, high) and notes
Accommodation					
B&B					
Campgrounds					
Hostels					
Hotels/motels					
Ecolodges					
Restaurants					
Ethnic					
Fine Dining					
Speciality (health, theme)					
Interpretive services					
Bus tours					
Dramatic presentations					
Guided walks					
Talks or video					
Transportation					
Airline					
Stock					
Boat/ferry					
Bus					
Helicopter					
Rail					
Infrastructure					
Communication services					
Medical					
Police					
Sewerage					
Human resources					
Bilingual guides					
Trained guides (first-aid)					
Interpreters					
Community leaders					
Like-minded comm. groups					

Table 5.8. Specific resource inventory.

Bird species	Seasonality	Sitings	Capacity to absorb tour (low, med, high)	Notes
List bird species that might be encountered on tour				

factors, including love of the outdoors, a wish to interact with people, the need to help conserve the environment, or perhaps to make money while doing something they truly love. All of these are excellent motives. However, moving from passion into a strong and viable business venture takes time, energy and a great deal of planning. This sense of passion has been well articulated by Timmons (1994) who writes that the 'typical' entrepreneur is characterized by commitment and determination, leadership, opportunity obsession, tolerance of risk, creativity and motivation to excel. However, going into business also takes money, which may be attained more easily on the basis of a solid business plan.

Financiers, whether they be banks, venture capitalists or other groups, will need to see the steps that the service provider is prepared to take in making the business a success. Dolli and Pinfold (1997) suggest that the majority of capital in new businesses usually comes from the entrepreneurs themselves (through the assistance of banks), friends and relatives, or what are referred to as 'angels'. Angels, according to the authors, are those wealthy individuals and companies who have experience in the tourism industry and who have the means by which to invest in new ventures. These individuals usually find the operator, rather than vice versa, through their connections with friends, business associates, or personal searches for investment opportunities. Taylor *et al.* (1998) write that the tourism sector typically relies on the owners' own funds, as an internal source of funding, and debt financing from banks, venture capitalists, and government grants, as an external source of funding.

Patterson (1997) has suggested that a business plan outlines, step-by-step, one's business activities, marketing plans and financial forecasts, usually for a period of up to 5 years (see Table 5.9). In addition, it will identify any problems or constraints and the solutions required to remedy these problems. These will be organized through the development of a series of objectives, and the time frame required to achieve them. Accordingly, it will be reflective of the mission and vision statements of the entrepreneur and act as a chart from which the operator guides his business through the implementation of appropriate management

Programming Tip 5.2.

The link between funding and business plans.

Without a credible business plan it is usually impossible to persuade any of the external sources to commit funding to a new venture, even in situations where considerable financial security is available. In any event, the production of a business plan is considered to be essential even if no external funding is required.

Source: Morrison *et al.* (1999, p. 105)

strategies. In short, it is a statement in words and numbers of what you want to do and what you need to get there (Industry Canada, n.d.). McKercher (1998, pp. 31–34) explains that business plans are developed to serve a number of functions, including:

- *A tool for integrating ideas.* Takes the information gathered during the research phase and refines it into a discrete, coherent document.
- *A decision-making tool.* Allows the entrepreneur to test new or existing business in a low-cost, low-risk environment. Decisions may be made in regard to price, capacity, trip size, service level and so on.
- *An analytical tool.* Allows the existing business to be analysed in an unbiased way, by stepping back and fully analysing risks that are often ignored.
- *A 'flight plan'.* The business plan is a dynamic document that maps out the future and identifies a possible direction to follow.
- *A tool to measure performance.* It provides a mechanism for measuring the growth of a business and for keeping the business on track to reach its targets.
- *A tool for raising finance.* It demonstrates to various individuals or groups that a possible business opportunity exists and that lenders have a strong chance of getting their money back.
- *A tool for assessing ideas.* Finally, and perhaps most importantly, strategic analysis through the business planning process will allow operators to identify potentially fatal flaws before they destroy the business.

The business plan is an essential planning tool which involves research, an analysis of alternatives, and the identification of problems and opportunities which will aid in the development of the business (see Fig. 5.7). The plan itself need not be overly elaborate, but rather should be short and crisp, well organized, identify a target market, demonstrate benefits of the service, be forward looking, not overly optimistic in regards to sales, identify potential problems, be management oriented and demonstrate how investors will make money (Schilit, 1987). (See Margerison, 1998, for an elaborate overview of business planning in the tourism sector, with special emphasis on financial considerations, including cash flow forecasts, profit and loss accounts, and balance sheet projections.) Furthermore, business planning may continue well on into the life of the operation through the implementation of feasibility studies. These studies may be used when new ideas for programmes develop (e.g. a new canopy tour), to assess their financial viability, but also to allow financiers the opportunity to assess the value of the new programme. The feasibility study involves doing research to make an assessment of the market, financial and operating logistics of the programme. (See Saskatchewan Tourism Education Council, 1997, for an overview of tourism feasibility worksheets.) Despite its strengths, the business plan is *not* a process that will guarantee the development of meaningful programmes and satisfy the various needs of a variety of different participants.

PESTE and SWOT analysis

The outline of a business plan, above, illustrates the tremendous number of variables which must be considered in initiating a business venture. Some of these are quite observable, and easy to control, while others are more difficult to anticipate. Two planning tools that will help in understanding the various external and internal catalysts and constraints to the business include PESTE analysis and SWOT analysis.

PESTE is an acronym for Political, Economic, Social, Technological, and Environmental analysis, and involves the many broad considerations that affect the tourism industry environment (e.g. the political environment), and which are outside the control of the service provider. For example, the terrorist attacks of 11 September 2001, in the USA have had catastrophic effects on the tourism industry. The extreme decline in travel, especially by air where there have been massive lay-offs, has had a number of ripple effects on all sectors of the industry. Other indirect events which have had various

Table 5.9. Business plan outline.

- Title Page
- Table of Contents
- Executive Summary (a short summary of the highlights of the plan)
- Background
  - Statement of history of the project (if applicable)
  - The product or service (what are you planning to do)
  - General statement regarding the intended mission of the venture
  - Discuss the industry (economic considerations, competition, environment)
- Management
  - Describe the management team – highlight education, experience, skills, functions
  - Résumé(s) highlighting these characteristics (in the appendix)
- Business goals and objectives
  - General goals (at least two) and objectives (two for each goal) and required initial investment (e.g. business size, production levels, performance levels)
- PESTE and SWOT analyses
- Marketing strategy
  - Research (primary and secondary)
  - Target market(s) (common characteristics and benefits sought)
  - Marketing mix (product or service, price, place, promotion, for each market)
- Operations
  - Site/location/layout, facilities/equipment, engineering/design, labour, process/production planning, suppliers/purchasing/material control/inventory, quality control, handling/delivery, customer service policies, resource management /environmental standards)
- Financing
  - Establish capital requirements and make feasibility projections (includes sales forecasts here or in a separate section)
  - Determine sources of funding (for capital and operating costs)
  - Plan the accounting and bookkeeping systems
  - Determine financial evaluation measures
  - Also include: initial salary/salaries and salary/salaries within 5 years
  - Include a balance sheet, income statement, break-even analysis, and cash-flow projection for 2 years
- Human resources
  - Positions required (and when)
  - Skills
  - Training
  - Wage policies/employee benefits
  - Administrative structure
- Legal requirements
  - Structure
  - Taxes
  - Contract/patents/licences
  - Insurance
- Risks
  - Discuss anticipated risks to be encountered (be honest!)
- Growth
  - Identify any future ideas for expansion
- Supporting documentation
  - Includes anything that a potential investor might want to know as well as more detailed information than is contained in the body of the report
    - Résumé(s) of management team
    - Financial documents
    - Primary and secondary market research
    - Information on direct and indirect competitors
    - Organizational charts
    - Job descriptions
    - Interior/exterior layout and design
    - Operation schedules
    - Promotional materials
    - Inventory
    - Insurance policy
    - Leasing agreements
    - Patents, trademarks, copyrights
    - Details on contracts/suppliers
    - Short- and long-term objectives for each strategic area
    - Customer service policies
    - Financial information and ratios
    - Sample partnership or corporation documents
    - Other supporting materials as required

Sources: Kao (1992); Industry Canada (n.d.); Schilit (1987).
PESTE analysis: Political, Economic, Social, Technological and Environmental analysis.
SWOT analysis: Strengths, Weaknesses, Opportunities and Threats analysis.

Strengths	*Opportunities*
Access to full range of trained guides. Close to international airport. Permits to operate in three national parks. Partnership with hotel for cross promotion.	No new permits being offered in national parks, so existing ones will have more value. Creation of social clubs in neighbouring cities provides a potential market. Currency fluctuations are attracting more international visitors.
Weaknesses	*Threats*
Short of working capital. Lack of four-season camping equipment. No Internet site.	May not be able to access other national parks. Areas allowing mountain biking are shrinking.

Fig. 5.7. SWOT analysis – soft adventure company. (Source: Patterson, 1997.)

impacts on the tourism industry include the economic effects of the stock market, social upheavals, technological innovations or setbacks, and extensive environmental catastrophes, such as volcanoes or hurricanes. Even though such events are outside the realm of the operator's control, an understanding of such phenomena, and an action plan to address these issues, is a very important planning consideration. A plan of action may involve continual scanning – much like environmental scanning which aims to anticipate major ecological disturbances – which would enable the service provider to survey the broad social, political, economic and environmental events that affect us regionally, nationally and internationally. The emerging literature on crisis management in tourism will involve an examination of these external forces, and appears to be an area where PESTE analysis has applicability.

SWOT analysis (an acronym for Strengths, Weaknesses, Opportunities and Threats) is a planning approach that allows the service provider to outline a number of pros and cons associated with the ecotourism business itself. In SWOT, the identification of these various catalysts and constraints is important in allowing the organization to anticipate those things or events that are positive and negative, and thus plan accordingly. An example of some of the potential strengths of small tourism firms, include a good knowledge of local and regional markets; innovation, and the ability of the tourism firm to serve the needs of the community (Friel, 1998). At the same time, however, these same firms tend to lack marketing expertise, they have a greater uncertainty of the small business environment, and their knowledge of customers may be distorted by subjectivity and selectivity (Weaknesses). Patterson (1997) writes that Opportunities and Threats to tourism businesses are usually external. In the former case, this may involve the removal of restrictions on travel, and in the latter, the closure of an area to travellers or natural disasters. She has developed the following SWOT analysis for a soft adventure travel firm (Fig. 5.7), which provides a good illustration of the many variables that may or may not lie within the control of the service provider, but yet have a direct effect on his or her ability to operate.

Conclusion

This chapter was focused principally on: (i) participant needs, and (ii) attractions and resources, which are both important in the process of programme planning. In the case of needs, the chapter examined the psychological basis of motives through a series of models that help researchers to understand travel behaviour. It was suggested that needs, in the way they are communicated by tourists, are often superficial as they fail to accurately explain perceived freedom and competence, and the need for optimal arousal and incongruity. Despite what seems a major drawback to a full understanding of why people travel, the needs assessment was introduced as a vehicle that has been used extensively in the field of recreation to evaluate the various motivations of participants. It was argued that the needs assessment is critical to ecotourism programming for the same reasons it is used in recreation. As such, a large part of the chapter was devoted to needs assessment and, in particular, the various aspects that comprise a needs assessment tool. The chapter further explored attractions and resources as key pull factors which influence ecotourism demand. Natural, social and recreational attractions were discussed, among others, as well as the need for ecotourism service providers to undertake inventories of such resources. It was argued that a thorough understanding of the network of attractions and resources is critical to the development of the ecotourism programme, in the same way that needs of the market are important to know. Finally, although the focus of this book is less business oriented, the business plan was briefly examined because of its importance in the process of business and programme development. A business plan outline was included to demonstrate the numerous variables that need to be considered in successfully launching and maintaining a business.

6

Programme Design A: Structure

The woods were made for the hunters of dreams,
The brooks for the fishers of song;
To the hunters who hunt for the gunless game,
The streams and the woods belong.

(Sam Walter Foss)

Introduction

This chapter introduces a number of specific considerations which will be central to the emerging programme. Conventional recreation programming often considers what are referred to as programme areas (activities) and programme formats (patterns or designs) as the basis of the programme. Here, it is argued that in addition to these, ecotourism programming will also need to consider programme settings, mobility (transportation) and lodging. The chapter further discusses the means by which to make programme design decisions, on the basis of the construction of a number of alternatives, through significant brainstorming, and the resources inherent within these various options. This is seen as a critical stage, because it forces programmers to examine a number of options objectively, which they might otherwise not attempt to do. Interpretation is also discussed in this chapter in the context of both outdoor education and environmental education. Effective interpreters, it is argued, are those who are able to incorporate both science (e.g. explanations of natural history) as well as art (e.g. the use of dramatics, visual design, and music) into their programmes, for the purpose of satisfying the educational component that is one of the central tenets of ecotourism.

Designing the Structure of the Programme

The literature on recreation programming, at the programme planning stage, almost universally discusses two main planning themes: programme areas (e.g. the types of outdoor recreation activities that we will build our programme around); and programme formats, which include the constitution through which the activity is organized and presented (e.g. instructional, drop-in, or trips and outings). For example, the service provider may decide that a white-water canoe programme (programme area) is best suited as a competitive event, over a classroom setting (programme format). By connecting activities to such programme formats, we greatly increase the number of options available to the programmer in terms of the organization of a single type of activity. This process which has worked for so long in the field of recreation is just as applicable for ecotourism programming. However, developing the structure of the ecotourism programme should not stop with areas and formats for the simple reason that ecotourism occurs in many remote areas, involves many different types of activities, relies on a number of different forms of transportation, and often requires an overnight stay component.

The result is that, not only do we have to consider the activity itself, and the format, but also these other elements of the broad ecotourism experience. The following section will consider programme areas, formats, settings, mobility and lodging, in establishing a template for further programme planning.

Programme areas

Activity participation has been examined in the literature on recreation chiefly from two main perspectives: (i) the types of activities that we engage in; and (ii) the experiences or outcomes derived from this participation. In the case of the latter, the recreation programmer is in the business of generating positive experiences through their programmes and related services. Not surprisingly however, programmers avoid organizing themselves around experiences because of the limitless number of these, based on the vast number of clients, and their participation in any number of different settings and situations. As such, classification of these experiences would be rather arbitrary and therefore difficult to measure. Although it is important to ensure that participants continue to have rewarding experiences, it is more practical for the service provider to programme on the basis of individual activities or groups of activities. Ecotourism programme areas are thus activities that form the basis of the actual programme. From a purely recreational perspective, and at a broader level, these activities include hobbies, music, outdoor recreation, mental and literary recreation, social recreation, arts and crafts, dance, drama, and sports and games. Table 6.1 is an example of some common outdoor recreation programme areas. Although the list is not exhaustive, it does provide an appreciation of the magnitude of outdoor recreation. To this list of activities we can add environmental studies, which may include volunteer clean-up projects and other leisure activities which allow participants to experience and appreciate nature (e.g. zoos, botanical gardens and wildlife

Table 6.1. Types of outdoor recreation activities.

Camping/outdoor living	*Outdoor sports*
Fire building	Bicycling
Map and compass	Hunting
Outdoor cooking	Fishing
Picnicking	Boating/sailing
Tool craft and care	Skiing
Backpacking	Orienteering
Hosteling	Horse-riding
	Scuba diving
Conservation	Hiking
Landscaping	Surfing
Conservation study	Snowshoeing
Planting for wildlife	Dog sledding
Building wildlife sanctuaries	
Weather observation	*Nature oriented*
Bird census	Nature games
Erosion control projects	Nature crafts
	Animal husbandry
High risk	Tree/plant identification
White-water kayaking and canoeing	Stargazing
Mountain climbing	Nature walks
Rock climbing	Wildlife viewing
Hang gliding	Outdoor photography
Glider soaring	Weather prediction
Caving	Snorkelling

Adapted from Russell, 1982.

sanctuaries), and travel and tourism, which frequently involves travel to scenic or exotic environments (Kraus, 1997).

Although categorized as outdoor recreation, many of the foregoing activities have been also identified as ecotourism, including for example activities like canoeing, bicycling and wildlife viewing. In Canada there are a number of canoe trip operators who presently refer to themselves as ecotourist operators, when not too long ago they did not (the same can be said about some of the angling operators who want to capitalize on the ecotourism market, especially given the decrease in interest in consumptive outdoor recreation activities like fishing). Canoeing is also an activity that some would suggest is adventure based. In many cases these operators combine canoeing with wildlife viewing in justifying the ecotourism label. An important question is which of the two activities – canoeing or wildlife viewing – is the main activity? In some cases the canoe is merely a vehicle by which to take ecotourists to specific regions in order to view wildlife. In other cases, canoeing remains the principal activity, with very little environmental education and interpretation. In this latter case, operators have often been schooled in canoe tripping but not in environmental education. It is thus the former activity that overwhelms the experience, with little regard for effectively interpreting the natural history of the region. The example illustrates that we have two activities built into one ecotour, but vastly different experiences depending on the approach taken by the operator.

How best to categorize activities as ecotourism is subject to some debate. While it is quite easy to categorize ecotourism broadly, the foregoing example illustrates that ecotourism in name and content may be quite different. Nevertheless, as the number of ecotourism programmes continue to increase concurrent with certification, accreditation and other professional programmes, there will be the need to examine the content of some of these more marginal ecotourism programmes to create some consistency in and between regions. For those intent on categorizing ecotourism broadly, programme areas might include: (i) hard-path vs. soft-path ecotourism activities; (ii) associated forms of tourism, such as eco-adventure activities (kayaking), eco-cultural activities (visiting remote tribes), and strict ecotourism activities (wildlife viewing); and (iii) activities that occur in different seasons, such as rainy and dry season activities in tropical environments; or perhaps winter-based ecotourism activities (snowshoeing), spring-based ecotourism activities (bird migrations), summer-based ecotourism activities (viewing whales), and autumn-based ecotourism activities (bear viewing) in more temperate and northern climes. While only speculative, this example illustrates that the classification of activities as ecotourism on the basis of activity alone is not always easily accomplished.

Programme formats

Depending on the motivations of the service provider and the goals and objectives of the programme itself, the ecotourism experience will be structured on one of many different programme formats (see below). Obviously a programme format such as 'trips and outings' will be most often used by ecotourism service providers. This is not to say, however, that it is the only one, or that more than one format could not be used for a programme. Ecotourism providers may develop 'instructional' classes for their potential clients, including, for example, slide shows on penguins for ecotourists who are travelling in Antarctica. This breakdown of different formats suggests that while certain activities are best presented using one type of structure over another, it also important to realize that the appropriate format will enable guides and staff to perform at the levels to which they feel most comfortable. This has the added benefit of not only satisfying staff, but also clients through the implementation of quality instruction or facilitation (Busser, 1990). While a number of authors have developed a range of programme formats, the following are formats identified by DeGraaf *et al.* (1999), and include those that have been developed from a conventional recreation context, and adapted according to the needs of this book.

- *Club and special interest groups.* This format includes groups of individuals organized around a particular activity or purpose, who have an association by common purpose. In ecotourism, this may include a collective body which is concerned about the ecological impacts of tourism, such as Tourism Concern.
- *Competitions.* The focus is on the exhibition of skill, abilities of participants, challenge, where a participant's performance is compared either with other participants or a pre-established standard. Once largely the domain of sports, the competition format can easily apply to other activities such as drama, music and camping. The Texas Birding Classic is an example of a programme where ecotourists compete against each other in time and space.
- *Trips and outings.* Although closely aligned to the travel programme area, it may also be viewed as a specific format for delivering desired benefits in a wide range of programme areas, such as a painting class that decides to take a trip to a nearby wood. This format is effective in changing the pace of a conventional classroom format, by getting people out on brief excursions or extended trips. A natural history society meeting at a local or regional zoo is an example of such a format.
- *Special events.* These usually have a theme around which the activity is organized, and may include trips or involvements on the basis of seasons, festivals, Christmas or craft shows. This format falls outside the bounds of other more conventional formats, and may include a group going to visit an outdoor show or the organization of an ecotourism conference.
- *Instructional.* Usually takes the form of a class, seminar, workshop or clinic, where the focus is to teach and further develop the skill levels (e.g. physical, cognitive or spiritual) of participants. Conventional recreational programmes include, for example, a swimming programme, and for ecotourism, such a format would include a class in animal behaviour organized by a city museum.
- *Drop-in.* As a more informal approach, the drop-in usually takes place at a recreation facility such as a gym, or perhaps a library, and encourages spontaneous involvement of community members. This format allows for much more freedom and is therefore less rigid in its orientation. An ecotourism drop-in may include a programme that enables individuals from the community, who have recently travelled as ecotourists, to show slides, pictures, etc., of their trip to other members of the community.
- *Service opportunities.* Service through volunteering is a fundamental aspect of the recreation field, where individuals perform any number of valuable tasks. The extent to which volunteering is a part of tourism is subject to debate (e.g. the volunteers needed to present an Olympic Games). However, the Earth Watch programme is an example of ecotourism volunteering, where ecotourists give of their time and money to aid a researcher or community with a particular research problem.
- *Outreach.* Refers to the extension of a leisure organization to those groups who do not have the means to participate in conventionally run programmes. Examples include zoos who take animals out on the road for the purpose of education, as well as organizations such as Conservation International who are actively involved in setting up programmes to aid in the development of ecotourism projects in less developed countries.

As was the case with programme areas, programme formats may also be further subdivided into different categories. The 'Instructional' format is a case in point. This format may be further divided into a number of different types of instructional environments, ranging from classrooms to trails, the latter of which may be either self-guided (e.g. interpretive) or guided. Table 6.2 provides an illustration of how the various programme formats apply to ecotourism.

Programme setting

In some publications there appears to be the feeling that setting alone is enough to warrant ecotourism or adventure tourism status. In fact definitions of ecotourism and adventure travel often suggest that the experience is contingent upon a certain type of setting. Based on the foregoing discussion on activities, claims of this nature are often fraught with difficulty. Some remote wild or peripheral settings may be exceptional for ecotourism, but the conduct of operators or government officials charged with managing these sites may be universally inconsistent with the tenets of ecotourism. By same token, urban or rural areas, and the manner in which the environment is used to develop ecotourism, may be more appropriate than in some wilderness settings (Fennell and Weaver, 1997; Weaver and Fennell, 1997). With this in mind, the following typology (Table 6.3) has been established to illustrate that: (i) setting is an important aspect of the ecotourism experience; and (ii) that the setting should not be restricted to only a few idealized wilderness places. In the context of this chapter, however, setting is only one criterion from which to plan and develop ecotourism programmes effectively.

Such an approach enables the service provider to plan actively where the ecotourism event is to take place, with the built in flexibility of combining settings within one main area (e.g. backcountry), or between areas (e.g. backcountry and urban). Programmers should not always be restricted to certain settings, but rather should use various environments in achieving their goals. One of the clearest examples of this type of thinking is the utilization of hot air ballooning over the game parks in Africa, which enables the service provider to access a broad spatial region while minimizing environmental impacts.

Programme mobility

The examples used in the previous section bring to mind the importance of on-site transportation as a means by which to further categorize and explain the various types of ecotourism experiences that occur. While some have been identified as ecotourism activities by virtue of what they are and rep-

Table 6.2. Programme formats and areas.

Programme	Programme formats							
areas	Club	Competition	Trips	Special events	Instructional	Drop-in	Service	Outreach
Ecotourism	NGO concerned with ecological impacts of tourism	Texas Birding Classic	Natural history society trip to zoo	Outdoor show	Class in animal behaviour	Community slide show	Earth Watch in Borneo	CI in developing country
Adventure tourism Farm tourism etc.								

Table 6.3. Programme setting options.

Main setting	Sub-settings
Backcountry	Marine Freshwater Terrestrial Aerial Combined settings
Rural	Marine Freshwater Terrestrial Aerial Combined settings
Urban	Marine Freshwater Terrestrial Aerial Combined settings

resent (e.g. see the discussion on canoeing, above), it is not fully clear the legitimacy of this claim, simply because the ecotourism experience is founded on more than just forms of transportation. In the case of canoeing, it may be for competition, or for enjoyment, which may have little to do with environmental education and nature appreciation, as discussed above. On the other hand, the canoe may be the vehicle that enables the ecotourist to penetrate into many backwater regions for the specific purpose of viewing wildlife. Accordingly, the following typology (Table 6.4) is illustrative of the vast numbers of forms of motorized and non-motorized transportation as they may apply to ecotourism.

Programme lodging

At least part of the overall ecotourism experience is hinged upon the setting in which it occurs and the associated accommodation options that have been made available to the ecotoursts. Lodging options may range from five-star hotels to the most rudimentary camp with few amenities or facilities. Those ecotourists in search of more rudimentary ecotourism experiences will have no problems feeling comfortable in rustic dwellings, while the opposite is true for those who are perhaps in search of more comfort. Wight (1993) made some inferences on the differences between hard- and soft-path ecotourism, on the basis of a number of different lodging options ranging from resource situated ones (e.g. country, forest, mountain) to those which are non-resource situated (e.g. village, city, resort). These were further subdivided, at least in the resource situated case, on the basis of fixed-roof and non-fixed-roof

Fig. 6.1. Ecotourism may also take place in urban settings.

Table 6.4. Programme mobility options.

Main form of mobility	Sub-forms
Motorized	Cruise liners Powerboats Aeroplanes Helicopters Trains Cars and buses Off-road vehicles Snow vehicles
Non-motorized	Canoes and dugouts Kayaks Rafts Yachts Gliders Balloons Mountain bikes and traditional bikes Skis and snowshoes Stock (horses, elephants, camels) Foot travel (hiking and backpacking) Wheelchairs (applies to both non-motorized and motorized categories)

Fig. 6.2. Transport can be arranged to alleviate tourist pressure on sensitive environments.

dwellings. This subdivision is helpful in categorizing the following lodging options (Table 6.5) for ecotourism programme planning purposes.

Among the important decisions which operators and ecotourists must make regarding the ecotour, are the environmental practices and procedures followed by the accommodation sector. This issue has generated a great deal of interest of late, particularly when larger hotel chains move in to places where local people are desperately trying to preserve their lifestyle and biodiversity (Matthews, 1997). In Germany, certain stakeholders have become actively involved in developing concrete solutions to

Table 6.5. Programme lodging options.

Main form of lodging	Sub-forms
Non-fixed-roof	Hammocks Tents Caves Open shelters (lean-to) Teepees Igloos and snow caves Bivouacs
Fixed-roof	Hotels and motels Resorts Bed and breakfasts, and Inns Recreational vehicle (RV) and caravan parks Ecolodges Ranches Huts and cabins Houseboats Tree houses (huts in trees)

Fig. 6.3. Non-fixed-roof dwellings offer an interesting alternative to more conventional forms of accommodation.

a number of environmental issues brought on by the tourism industry. As outlined by EMS Environmental Management Services (n.d.), tour operators have adopted the following measures to help alleviate stress on the environment:

- using recycled paper for catalogues and travel documents;
- carrying out environmental checks in contractual hotels;
- offering ecological training courses for employees and travel guides;
- developing ecotourism into a new market segment.

Although there are a number of such schemes which have evolved over the past decade, one of note is the initiatives of the German tourism operator Touristik Union International (TUI). TUI's ecological concept for the accommodation sector involves reduction of environmental impact, quality assurance and sustainable tourism development through various systematic actions. These include education, hotel management procedures, ecolabelling, environmental quality controls, ecological product controls and environmental impact assessments. In general, while the reduction of pollution may be a short-term goal of a hotel, medium- and long-term goals would include the promotion of conservation and the improvement of the environment (Iwand, n.d.).

Another notable example is the Maho Bay project in the US Virgin Islands, which has implemented a number of environmental measures to coincide with the tenets of sustainable tourism and ecotourism (Maho Bay Camps and Harmony Studios, n.d.). The waste management programme undertaken by this group includes efforts to *reduce* garbage and use of resources; *reuse* items, energy and water; *recycle* rubbish and various resources; and *buy* recycled products. To the hard-path ecotourist this represents a truly committed approach to the delivery of ecotourism services. It not only provides ecotourists with an environmental option, but also stands up as an example for other tour operators and hoteliers to follow. Furthermore, the incorporation of a number of new technologies at Maho Bay will provide a benchmark for further development in a number of ecolodge sites in the future. A sample of some of the environmental measures incorporated may be found in Table 6.6.

Once programme areas, formats, setting, mobility and lodging options are considered, these may be fit within a broader programme matrix (see Table 6.7). This allows for an appreciation of the magnitude of different options available to the service provider, while at the same time providing the basic structure of the programme (essentially the shell of the programme). Programmers select the most appropriate options for the programme within each of the five categories, on the basis of availability, what might appeal most to clients, or other issues including the goals and objectives of the programme, and the strength and weaknesses of staff and available resources. This process of the selection of the most appropriate programme structure is elaborated upon in the following section.

Making Programme Design Decisions

It is an unfortunate fact that many programmes fail, and these failures in part may be attributed to the lack of attention paid at

Table 6.6. A selection of environmental measures used at Maho Bay.

Reduction at Maho Bay
Elimination of polystyrene (styrofoam) food and beverage containers
Low water-use toilets are utilized
Spring-loaded taps are installed in sinks and showers
Reuse at Maho Bay
Old newspaper is the only protective wrapping used in the shop
Water from the sewage treatment plant is used for garden irrigation
Guests are given a 25% discount on every refill of beer if they reuse the same glass
Recycling at Maho
Purchase of a glass crusher is planned to recycle glass into cement
Food and grass clippings used for compost in the organic garden
Used paper is recycled into scribbling pads and photocopying is double-sided
Buying recycled at Maho
Recycled note paper and paper towels are for sale in the shop
Paper towels and napkins are made from recycled paper
Most building materials were made of recycled glass, wood, rubber tyres, etc.

Source: Maho Bay Camps and Harmony Studios (n.d.) Ecotourism at Mayo Bay.

Table 6.7. Programme design matrix.

Programme area	Programme format	Programme setting	Programme mobility	Programme lodging
Nature oriented	Club	Backcountry:	Motorized:	Non-fixed roof:
Bird census	Competition	Marine	Cruise liners	Hammocks
Nature games	Trips	Freshwater	Powerboats	Tents
Erosion-control projects	Special events	Terrestrial	Aeroplanes	Caves
Nature crafts	Instructional	Aerial	Helicopters	Open shelters (lean-to)
Animal husbandry	Drop-in service	Combined settings	Trains	Teepees
Tree/plant identification	Outreach	Rural:	Cars and buses	Igloos/snow caves
Stargazing		Marine	Off-road vehicles	Bivouacs
Nature walks		Freshwater	Snow vehicles	Fixed-roof:
Wildlife viewing		Terrestrial	Non-motorized:	Hotels and motels
Hang gliding		Aerial	Canoes and dugouts	Resorts
Outdoor photography		Combined settings	Kayaks	Bed and breakfasts/Inns
Caving		Urban:	Rafts	Recreational vehicles/caravan parks
Snorkelling		Marine	Yachts	Ecolodges
Risk		Freshwater	Gliders	Ranches
		Terrestrial	Balloons	Huts and cabins
		Aerial	Mountain bikes	Houseboats
		Combined settings	Skis and snowshoes	
			Horse, elephant, camel	
			Hiking/backpacking	
			Wheelchairs	

this stage of the programme development cycle. Simply developing a programme template and implementing it without due regard for a number of key programme criteria will eventually end up in wasted time and resources. In an effort to design the best and most efficient programme possible, programmers will need to consider a process to select the best programme from a number of alternatives. This will entail identifying a number of key criteria and brainstorming to identify a few realistic programme alternatives. The identification of criteria is therefore a critical step, which ultimately depends on what is important to the programmer and to the organization, as outlined by Russell (1982, p. 190):

> This list of criteria represents the dimensions of plausible alternatives. Although the list can be diverse, it should be specific to the unique planning situation of the agency. For example, one agency may specify that 'time to implement' is an important decision criterion, whereas to another agency it is not; 'degree of favorable community image' may be an important criterion to neither or both.

The following represent some programme criteria that may need to be considered at this stage of programme planning, as identified by Russell:

- *Activity analysis.* Begs the questions: 'What activity forms are best?'. These activities can be grouped according to behaviour domains (cognitive, physical and emotional); interaction patterns (social relationship groupings); leadership (does the activity require minimum or maximum leadership?); equipment and facilities (special equipment, and how much); duration (what is the time frame for the activity?); participants (does the activity require a fixed number of participants?); and age (is the activity age specific?).
- *Available resources.* Is there adequate staff, money and space for the idea? Is it a wise use of resources?
- *Alternative resources.* If not, are other configurations possible?
- *Politics.* Do the broader community and other support systems consider the programme desirable? Does it have a favourable public image?
- *Risk management.* Is the programme safe and healthy and does it allow for legally prudent programme conduct?
- *Appropriateness.* Is it suitable for the constituency?
- *Effectiveness.* Will the programme accomplish what is intended?
- *Timeliness.* Is it timely and easy to implement?
- *Accessibility.* Will it be accessible to the target constituency?
- *Integration.* Can it be integrated with other programmes in the agency?

The aforementioned criteria are suitable because they are general enough to apply to just about any programme situation. However, this list may be easily extended in order to make it more specific to the field of ecotourism, to include: (i) degree of environmental impact of the activity; (ii) community involvement; (iii) carrying capacity; (iv) complementarity to other resource uses; (v) reliance on the built environment; (vi) reliance on transportation; (vii) degree of environmental education; (viii) certification/education of guides; (ix) guide/participant ratios; (x) language proficiency of guides; and (xi) following local codes or customs.

In reality, it is probably not feasible or practical to incorporate all of the critical information available into this decision-making process (i.e. the criteria selected to make the decision). Also, the decisions that are made on each alternative may be rather subjective, because the programmer may only be able to estimate the relationships that exist between the alternatives and the criteria selected. However, in order to transcend some of this subjectivity, the programmer may wish to implement a system which enables him or her to provide weighting to each of the alternatives and criteria which have been identified. For example, each criterion may be assigned a weight from 1 to 5, to reflect its importance relative to the others; while alternatives may be scored along a continuum from −2 to +2. The result is a score which would provide a numerical justification for the resultant choice. Table 6.8 is

Table 6.8. Programme criteria and alternatives.

Criteria (and weight)	Alternative 1	Alternative 2	Alternative 3
Degree of environmental impact (5)	High (−2)	Moderate (0)	Low (+1)
Safety of activity (4)	Good (+1)	Fair (0)	Excellent (+2)
Equipment, which must not exceed $1000 per year (3)	$750 annually (+1)	$1500 annually (−2)	$500 annually (+2)
Image in community (2)	Poor (−2)	Excellent (+2)	Good (+1)
Nature of activity (3)	Consumptive/ high density (−2)	Non-consumptive/ mid density (+1)	Non-consumptive low density (+2)
Total score	−13	+1	+27

an example of the process of selecting criteria, alternatives, and the weighting of each to accomplish this end.

Scores in the table are determined by multiplying the weight of a criterion by the associated score for each alternative. For example, the total score of −13 for Alternative 1 was determined by:

- Multiplying 5 (the weight given to 'degree of environmental impact') by −2 (the score assigned to the variable 'high'), resulting in a score of −10.
- 'Safety of activity' with a weight of 4, is multiplied by +1 (the score assigned to the variable 'good') for a score of +4.
- 'Equipment' was given a score of +3, by multiplying 3 (the weight for equipment) by +1 which is the score assigned to '$750 annually'.
- 'Image in community' a score of −4 (2 × −2), and
- 'Nature of activity', a score of −6 (3 × −2).

Clearly Alternative 3 is the most positive based on these scores, and the programmer would thus strongly consider this alternative as the basis for the programme. Consequently, what emerges from this process is a solution that suits the circumstances (situation and setting) and the particular time. In this process, the expertise of the programmer comes into play in assimilating a great deal of information through a process of brainstorming, which is critical in enabling the programmer to develop the perceived best criteria and alternatives in generating the most logical programme decision.

At this stage of the planning process, the programmer will have made some critical decisions about the complexion of the programme. The structure that has emerged is both a reflection of the contents of this chapter, but also the mission, goals and objectives of the agency. It may be helpful to articulate and organize the various aspects of the emerging programme in the form of a worksheet, such as the hypothetical example which follows (Table 6.9). A logical extension of what has been determined to this stage, is the inclusion of some pertinent information that may have resulted from the programme design decisions stage. This may include temporal aspects or other resource issues which might naturally round out the basic structure of the programme.

Designing the Programme's Interpretive Component

Outdoor and environmental education

Blamey (2001) notes that one of the fundamental defining principles of ecotourism is that it is learning or education-oriented. Indeed, this is a perspective discussed in many publications on ecotourism, which suggest that the interest in learning is a key aspect of the ecotourist's experience. Whether the ecotourist sets out on his or her own, or with a group and a leader, there is an implicit assumption that learning will be a product of the ecotourism episode. The focus of learning in ecotourism is on a region's natural history, in the same way it would be on agriculture for farm and ranch tourism; or art for cultural tourism. Although it is beyond the scope of this book to look intensively at the process of

Table 6.9. Programme planning form.

Programme name	Eco-Experiences
Participants	Members of conservation group 'X', but accessible to other members of the park
Programme goal	To expose participants to a number of ecotourism issues in the park/region
Objectives	(i) To enable participants to learn about ecotourism through the experiences of others; (ii) to understand how ecotourism should contribute to conservation and community development; and (iii) to suggest ways on how ecotourism ought to be developed in the region of the programme
Programme area	Ecotourism, with various topics
Programme format	Trip/outing
Programme setting	Temagami Wilderness Area
Programme mobility	Car pool to site, hiking while on-site
Programme lodging	Cabins
Time	15–17 May. Instructional sessions run from 9:00 to 11:30 and 14:00 to 16:30 each day

Topics		
	15 May a.m. Birding and bird banding	15 May p.m. Community development
	16 May a.m. Wildflower identification	16 May p.m. Guiding and interpretation
	17 May a.m. Trail development/upkeep	17 May p.m. Park conflicts

education, it is worthwhile to look at two forms of education that are basal to ecotourism: outdoor education and environmental education. The most widely written on of the two concepts appears to be outdoor education, which, at its root, has involved the basic teachings and knowledge of the outdoors passed down through generations. It is a form of education that has enabled humans to get back in touch with living in and as part of the natural environment, where our roots were once firm and deep (Smith *et al.*, 1972).

Outdoor education was formalized through its links to school camping in the USA, where instructors during the 1820s started to take their students on regular camping trips. (The first outdoor excursions were apparently undertaken by students of The Round Hill School in Massachusetts, during the 1820s.) However, according to Ford (1981), the practice of taking students on such trips fell out of favour when taxpayers questioned the wisdom of spending resources on such trips, without a clear understanding of the value of these trips especially as regards the education of children. Ford writes that the school camping terminology ultimately gave way to outdoor education to depict the educational process in an outdoor setting. The educational label may thus be assumed to have been an important part of the concept in its later incarnation, as the term stuck and instructors went about formulating ideas to incorporate what they wished to pass on to their students in a conventional indoor setting, in an outdoor one.

Ford (1981, pp. 2–7) reviewed a number of definitions of outdoor education in concluding that outdoor education is a form of education which occurs *in* the outdoors, *about* the outdoors (outdoor resources/skills or as a location and a process with the aim of enhancing existing curricula), and *for* the outdoors, implying use (leisure pursuits) and understanding (stewardship and natural processes). In his classic interpretation of the term, Sharpe wrote that 'outdoor education begins when the teacher and pupils close the classroom door behind them' (cited in Mand, 1967, p. 27). Others, such as Priest (1990), write that outdoor education takes place primarily in outdoor settings, but not exclusively. Furthermore, outdoor education is not seen as a separate discipline with its own set of prescribed objectives, but rather a vehicle from which to study a whole range of disciplines including science, social science, health, recreation, physical education, language arts, arts and crafts, and music. The application of an outdoor education approach, in regards to these various disciplines, to learning *in* and *for* the outdoors may be seen in Fig. 6.4. The figure also illustrates the vast number of different environments in which outdoor recreation takes place. These settings, often referred to as outdoor laboratories, include the school yard, local parks, museums, historical monuments or other heritage attractions, provincial/state parks, national parks, wildlife areas, zoos and so on, which may be instrumental in satisfying the following main objectives (see Mand, 1967, pp. 29–31):

- an appreciation of natural resources,
- improved instruction in science and other disciplines,
- development of recreational skills in the outdoors,
- the opportunity for new social experiences,
- involvement in community service projects,
- aesthetic awareness of the settings/resources experienced.

While outdoor education uses the outdoors as a resource for enriching curriculum (use of the environment as a teaching resource), environmental education focuses on natural and social sciences, particularly ecology, political science and economics, as a way to study the environment. Greenall Gough (1990) writes

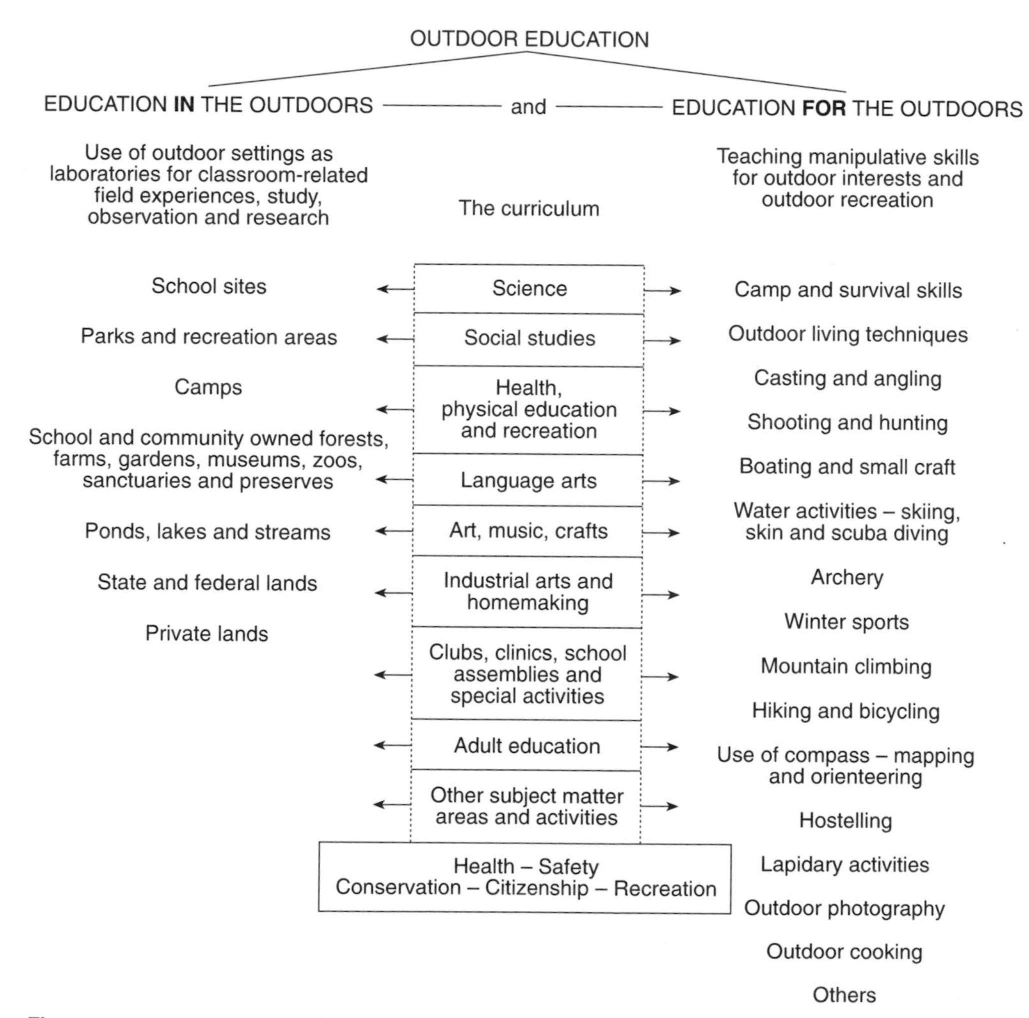

Fig. 6.4. Outdoor education in the curriculum. (Source: Smith *et al.*, 1972.)

that the environmental education movement grew out of the increased awareness of environmental issues which developed during the late 1960s and early 1970s. The work of Leopold (1991), Carson (1987), White (1967) and Hardin (1968) were instrumental in bringing the realities of the day to the fore of research and society at large. Environmental education's focus on ecology, economics and politics is evident in the following goals and objectives endorsed at the 1987 UNESCO–UNEP International Congress on Environmental Education and Training in Moscow (UNESCO–UNEP, 1988). (See also Roseland, 1992, for a list of aims of environmental education which are similar to the ones listed below.) The goals of environmental education are:

1. To foster clear awareness of, and concern about, economic, social, political and ecological interdependence in urban and rural areas;
2. To provide every person with opportunities to acquire the knowledge, values, attitudes, commitment and skills needed to protect and improve the environment.
3. To create new patterns of behaviour of individuals, groups and society as a whole towards the environment.

The categories of environmental education objectives are:

- *Awareness*. To help social groups and individuals acquire an awareness of and sensitivity to the total environment and allied problems.
- *Knowledge*. To help social groups and individuals gain a variety of experiences in, and acquire a basic understanding of, the environment and its associated problems.
- *Attitudes*. To help social groups and individuals acquire a set of values and feelings of concern for the environment, and the motivation for actively participating in environmental improvement and protection.
- *Skills*. To help social groups and individuals acquire the skills for identifying and solving environmental problems.

Environmental education in practice can take many forms, which appear to be a direct function of the theoretical approaches developed on environmentalism. Greenall Gough (1990), for example, writes that perspectives can range from shallow environmentalism to deep ecology. In the former case, environmental education is viewed only as an appendage of the conventional political processes which it seeks to oppose (Bookchin, cited in Devall and Sessions, 1985). Deep ecology, as the opposite end of the spectrum (although theoretical perspectives such as ecofeminism are thought to be 'deeper' than deep ecology), and as an alternative to the present dominant world view, focuses on the development of a new balance and harmony between people, communities and nature. Greenall Gough (1990) writes that certain types of curriculum, such as 'earth education' (see Van Matre, 1979), outwardly reject the shallow environmentalism perspective of much conventional 'nature study', which are characterized by the adoption of instrumental values, and instead seek to develop a sense of identification of humans with and as part of nature. This perspective is illustrated in Fig. 6.5.

The link between environment and education is well developed in the work of Orr (1992), who suggests that not only are we failing to teach the basics about the planet and how it works, but also that much of what we are teaching is simply wrong. We are a society, indeed a planet, of ecological illiterates who are unwilling or unable to observe nature with insight. He stresses that ecological literacy has been difficult to achieve in the Western world because: (i) we have lost the ability to think broadly in an age of specialization; (ii) there is the belief that education is solely an indoor activity; (iii) there is a decline in aesthetic appreciation (we are too comfortable with the ugliness of urbanization); and (iv) there is less opportunity for the direct experience of nature. When we move away from nature to complete urbanization, we lose the sense of place which is often determined by association with soils, landscape and wildlife. Ecological literacy, Orr writes, can be achieved through attempts to comprehend the interrelatedness of life through the disciplines of natural history and ecology (among others) and through the following initiatives (pp. 90–92):

- We must recognize that all education is environmental education.
- Environmental issues are complex and

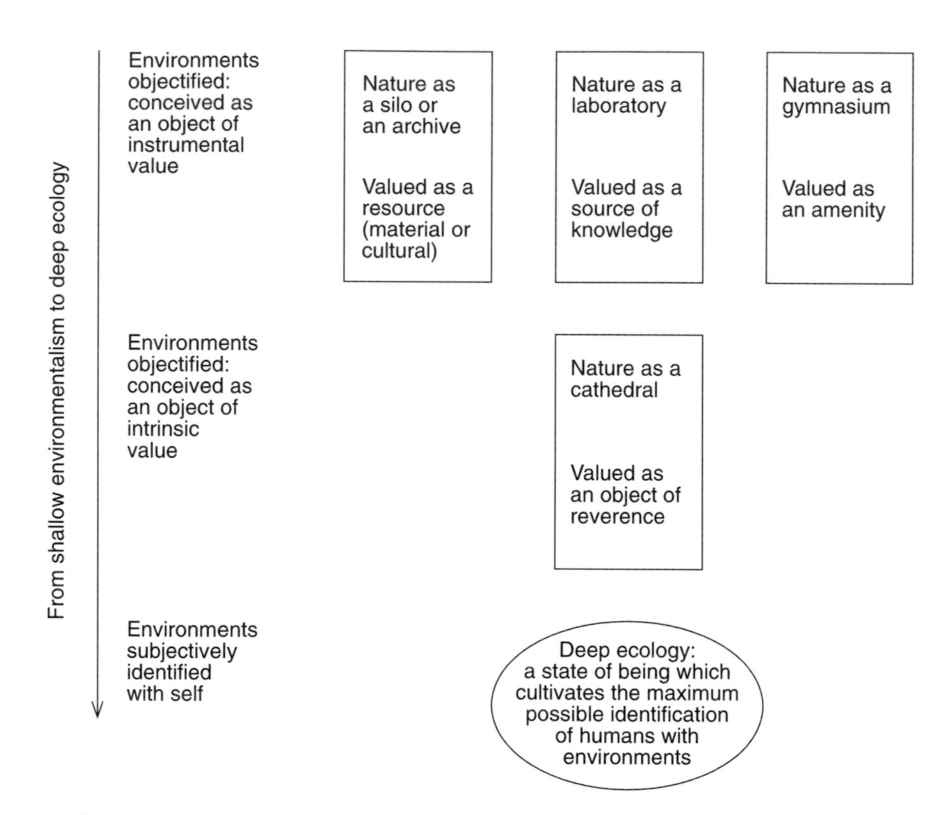

Fig. 6.5. From shallow environmentalism to deep ecology. (Source: Greenall Gough, 1990.)

cannot be understood through a single discipline or department.

- For inhabitants, education occurs in part as a dialogue with a place and has the characteristics of good conversation.
- The way education occurs is as important as its content.
- Experience in the natural world is both an essential part of understanding the environment, and conducive to good thinking.
- Education relevant to the challenge of building a sustainable society will enhance the learner's competence with natural systems.

This attachment to place, as described above, has been explored by many authors including Evernden (1985), who suggests that humans have lost their sense of ecological place everywhere on earth. He writes:

> [W]e are an exotic organism even in our place of origin, wherever that may be. We are exotic in any environment, for in a sense we did not evolve *in* any existing habitat. I say 'in a sense,' for of course we were a part of local ecosystems during most of our history. But in our minds we may have fallen out of context a long time ago, or at least begun the fall.
>
> (p. 109)

In the manner it is discussed above, environmental education holds a significant amount of potential to connect people with environments. Although ecotourists travel for the purpose of learning about the natural world, there appears to be a critical lack of research on the degrees of connection between the ecotourist and the natural world (a point which is supported by Kimmel, 1999). Some of the most recent studies on ecotourism, however, have taken this matter to heart. For example, Weaver (2002) has shown that hard-path ecotourists are much more biocentric in their attitudes than soft-path ecotourists. Furthermore, a stronger link between environmental education – especially from the curriculum standpoint – and ecotourism may be forged in view of the work of Stapp *et al.* (1996), who has attempted to synthesize teaching and environmental learning from two different perspectives: Action Research and Community Problem Solving. Action Research is a process of democratic decision making and equitable distribution of power, where affected parties are empowered to act on a problem. Through analysis, conceptualization, fact-finding, planning, implementation and evaluation, participants are involved in a process of task resolution. On the other hand, Community Problem Solving is 'an approach to learning that focuses on resolving or improving local issues through a problem-solving process' (Stapp *et al.*, 1996, p. 3). Stapp *et al.* combined both methods in the development of the Action Research and Community Problem Solving process. It is defined as follows:

> a process that enables students and teachers to participate more fully in the planning, implementation and evaluation of educational activities aimed at resolving an issue that the learners have identified. Technically this could be any issue, depending on the group engaged in the process.
>
> (p. 3)

Students become the practitioners of their own learning and are catalysed into the role of explorer, researcher, theorist, planner and actor (as outlined by the authors). As such, they assume more responsibility for their own learning, with the teacher taking on the role of facilitator, who may also learn as a result of the task and the information that may be generated. The application of this mode of thinking has potential for ecotourism operators who are intent on making a clear distinction between what they offer in their programmes compared with their competitors. The Action Research and Community Problem Solving process would enable ecotourists to undertake worthwhile projects on ecological rehabilitation or community development which, for example, would have a positive tangible impact on the ecology and culture of the region visited. In cases like this, however, the ecotour operator must take a much more active role in developing a plan of action that could address certain problems in the community. Although unpalatable to some who lack the motivation to go to such lengths, others who wish to adopt such an approach would be rewarded and, in the process, carve out a certain niche, which would be very attractive to a certain type of ecotourist, who has very specific needs, attitudes and values.

Programming Tip 6.1.

Environmental education.

The responsibility towards the environment goes well beyond reducing, reusing and recycling at Maho Bay and Harmony Studios, US Virgin Islands, to an environmental education programme designed to educate guests on a variety of on-site initiatives. Some of these include:

- Evening programme with speakers and slide shows on flora and fauna of the region.
- Carrying water to tents induces guests to be more thoughtful of their consumption.
- Cookies baked in the solar oven are sold, with information labels.
- Guests are asked to separate food scraps for composting when clearing plates.
- Plants identified with small signs and an 'Adopt-a-plant' programme has been developed.
- Placards posted around the resort advise guests of water consumption tips.
- Guests can monitor their own energy and water consumption via computer.
- A 'Four-hour worker programme' offers people the opportunity to obtain training in responsible tourism.

Ecotourism service providers need to be as innovative as possible in providing the best programmes possible to clients who are travelling specifically to be educated on many different environmental trends and practices.

Source: The Ecotourism Society (n.d.).

Interpretation

Knudson *et al.* (1995) state that interpretation and education can be distinguished by the former being a means by which to convey the meaning of something through explanation or exposition, while the latter is the knowledge derived from the experience. Successful interpretation has the ability to reach into people's minds in creating interest, understanding, delight, and perhaps even a life-long interest (Edwards, 1980, p. 12). In generating this affect, the communicator relies on science (the knowledge base of botany, geology, animal behaviour, physics, for example, and its relationship to a broader environment), but also art. Art is the catalyst, according to Edwards, that makes the interpretation event happen. The value of art in this equation is underscored by J.R. Butler (1993), who writes that the interpreter will rely on elements such as dramatics, visual design and music to improve the process of communication. This, according to Butler, enables the interpreter to deal with complex relationships and thus deal with many holistic concepts beyond an approach that focuses on isolated facts. The movement away from a reductionist approach to information dissemination is echoed in one of the most oft-quoted definitions of interpretation by Tilden (1957, cited in Knudson *et al.*, 1995, p. 4), who said that 'interpretation is an educational activity which aims to reveal meanings and relationships through the use of original objects, firsthand experience, and by illustrative media, rather than simply to communicate factual information'.

In general, there are two main forms of information service which fall under the interpretation label. The first includes non-personal services such as visitor centres, exhibits, signs, interpretive trails and publications (Butler, J.R., 1993). The second includes the personal services, which involve direct contact between tourists and the interpreter(s). Ecotourism service providers must be open to using a range of different personal and non-personal services in their programmes, including those provided by the service provider him/herself, but also those which are established by the parks and protected areas (or other areas) they visit. In the case of the former, service providers (e.g. guides) must hold their professional duties in the highest regard. This includes the dissemination of appropriate and accurate information for ecotourists, but also accountability to many different stakeholders, including hosts, operators, and land managers, and their values systems (see Weiler and Ham, 2001a). The following set of guidelines will help ecotourism guides and operators more effectively accomplish their interpretive goals (Northwest Arkansas Resource Conservation and Development Council Inc., 1997):

- The ecotour guide is physically present with the ecotourists.
- Interpretation can occur at ecotour headquarters, during transport, at the ecotour destination site, during ecotreks, around campfires, and while overnighting on the trail.
- The ideal presentation is *live* and *spontaneous*, not canned or rote.
- Ecotourists interact with the ecotour guide(s).
- The ecotour guide remains flexible enough to adapt his or her comments to the audience's interests and actual conditions encountered during the ecotour, and always looks for teachable moments.
- The ecotour guide should be prepared to refer ecotourists to additional information related to their area(s) of interest.
- The ecotour guide should study and prepare for activity or skill demonstrations, which are interpretive demonstrations that feature an individual who actually performs a cultural act, such as a person demonstrating the craft of split white-oak basketry or playing a traditional musical instrument.

Models of interpretation

Successful interpretive programmes are the result of time, effort and planning. Consequently, the interpretive planning process may be viewed as a microcosm of the overall programme planning process, including the consideration of objectives,

Fig. 6.6. Interpreters must often rely on props to deliver the message.

budgets, promotion and so on. In Chapter 4 there was a discussion of the who, why, when and how of programming. The same holds true in the case of interpretive planning for the purpose of developing the best programme possible for the intended audience. Furthermore, it is no good for the interpreter to simply show up, pass on some facts, ask for questions and terminate the programme. To be truly effective the interpretive programme will be played out in the mind of the interpreter well before the event, especially as it pertains to the audience, the theme and message, scripts, props to be used, the site and other programme-related elements. The comprehensiveness of this interpretive planning process can be appreciated through models developed by the Bureau of Land Management in the USA, as cited in Knudson *et al.* (1995), and the Department of Conservation (2000) in Australia (see Bucy, 1991, for another model of interpretation), which are illustrated in Table 6.10 and Fig. 6.7.

Table 6.10. Bureau of Land Management (USA) interpretive planning model.

A. Formation of planning team:
  1. From the public sector
  2. From the private sector
B. Interpretive inventory:
  1. Audience analysis
  2. Cultural and natural resources
  3. Current interpretive services
C. Preparation of the plan:
  1. Area overview
  2. Goals, objectives and themes
  3. Programme components
D. Implementation requirements:
  1. Personnel
  2. Equipment
  3. Site development
  4. Maintenance
  5. Time schedule
  6. Programme budget
E. Monitoring and evaluation:
  1. Plan monitoring
  2. Effectiveness evaluation
  3. Plan revision

Source: Knudson *et al.*, 1995.

The most comprehensive of the two models appears to be the one developed specifically for ecotourism by the Department of Conservation in Western Australia (see Fig. 6.7). Although lengthy it is worthwhile showing it in its entirety because of the template that has been developed. The benefits from a programme-planning standpoint are numerous, including the ability to document a number of different interpretive themes and programmes in any number of areas. Cataloguing these would enable the programmer to gain an understanding of potential gaps in programme offerings as well as overlap and redundancy which may also occur, especially between interpreters within the same agency.

A note on trails

In most cases ecotourism operators will be using public and/or private natural areas in their programming. These will have established trail systems which will be used on an ongoing basis, which will allow groups to

Interpretive Activity Planner

TITLE

TOPIC

- Natural Values
  - Landscape/Seascape/Processes
  - Plants
  - Animals
  - Ecology
- Cultural Values
  - Aboriginal/History/Heritage
    - Lifestyle
      - Customs
      - Religion
      - Law/Lore
      - Entertainment
    - Resource use
      - Land/Water/Sea
      - Food
      - Tools
      - Shelter
      - Clothing
- Management Issues
  - Visitors
  - Fire
  - Ferals
  - Weeds
  - Wise use

THEME/MESSAGE

The theme is your intended message for clients to take away with them.

CAPTION/SLOGAN FOR ACTIVITY (OPTIONAL)

DESIGN TECHNIQUE

Talk
Walk/drive
Wildlife observations
Concept exploring
Arts and craft
Sensory
Problem solving

THE EXPERIENCE

Use as many learning styles as possible. What will participants:

See?

Hear?

Do?

Make?

Feel? (emotionally)

AUDIENCE

Consider 'visitor profile' (country of origin; cultural preferences; special interests)

☐ Children 6 - 10
☐ Children 11 - 12
☐ Teens 13 - 17
☐ Young Adults 18 - 25
☐ Adults
☐ Family
☐ Groups

Maximum

Minimum

OBJECTIVES

These are your desired outcomes.

What *knowledge* will participants gain?

What *skills* will the participants gain?

What *attitudes/values* will be discussed/integrated into the activity that the clients can take away with them?

What *actions* will participants be able to take as a result of the activity?

INTERPRETIVE ACTIVITY PLANNER

OUTLINE

The major points that hold the activity together - Include introduction, activity steps or points, and conclusion/theme.

Introduction

Body

Conclusion

SITE

Describe site essentials.

☐ Map attached (optional)

TIME LENGTH

TIME OF DAY

☐ Morning ☐ Evening

☐ Afternoon ☐ Anytime

PROPS

For leader to supply

For participants to bring

PROMOTION

What? (write the advertisement)

When/where? (state time and place)

How? (the media)

INTERPRETIVE ACTIVITY PLANNER

BUDGET

PARTICIPANT COST

THE SCRIPT AND TIPS FOR GUIDE

☐ Completed and attached

EVALUATION

Number attending the activity

Age groups

☐ 6 - 10 ☐ 18 - 25

☐ 11 - 12 ☐ 26 - 60

☐ 13 - 17 ☐ 61+

Overall audience response

Self Evaluation

Which objectives did you achieve?

How could you improve your activity?

Participant Evaluation Survey

☐ attached

REVISED ACTIVITY PLANNER

☐ Revised planner attached

Fig. 6.7. Interpretive activity planner. (Source: Department of Conservation, 2000.)

reach any number of intended attractions. While it is often the case that the trail itself is not the focus of the programme, there are times and instances when the trail can be the focus of discussion (e.g. such as overuse of trails leading to erosion). Trails are also noticed if they are divided, obstructed, difficult to traverse or if they suddenly disappear. Trails come in all shapes and sizes, with different purposes for different users. For example, Hultsman *et al.* (1998) suggest that in understanding what constitutes a good trail, one must first consider the type of trail and who is to use it. They further note that what is good design for a hiking trail may not be good for an interpretive trail. This suggests that trail systems are created with different user groups in mind, including those who hike, cycle, ride horses or who require the use of a wheelchair. In the case of hiking trails, Warren (1998) writes that there are six categories, each having similarities, yet attracting different users for different reasons. These include:

- *Wilderness trails.* Long-distance routes for experienced campers who seek challenge for at least two nights. Focus is on nature, with few facilities and few other users.
- *Backcountry trails.* At least one night's camping in a natural environment. More reliance on built structures (e.g. bridges) to make movement easier for the moderately experienced hiker.
- *Frontcountry trails.* Located in semi-wilderness or rural areas, there is less concern for being away from civilization. More amenities offered, with easier trail conditions.
- *Day-use trails.* Accommodates many users with elaborate structures and signage often with a number of short trails from which to choose. Frequently adapted for people with disabilities.
- *Urban trails.* Located in developed settings, these trails are used by many people with a variety of interests. Extensive amenities are offered such as toilets, along with parking and information.
- *Interpretive trails.* Characterized by signage, interest points and viewing areas, this type of trail does not have separate development specifications. Such trails highlight natural, heritage and historical attractions, and connect people with the best parts of an area.

Ecotourism service providers should be aware of the different types of trail in protected areas, in association with the requirements of the programme and participants. They should also be sensitive to the fact that trails have different capacities to absorb use. Urban and frontcountry trails will be more resilient than those in backcountry regions. In addition, trail systems will be subject to intense periods of use at times (spatially and temporally), and the guide must be prepared to adapt accordingly (see Chapter 3). For example, the lure of photographing small and elusive beasts is often too much for even seasoned ecotourists who see no reason why they cannot venture off trails, even when instructed not to do so. This is the case in Point Pelee National Park (about 10 km^2 in size) in Canada, where the annual spring migration of birds brings hundreds of different species. Unfortunately the competition to secure a sighting often means that trails are full of tourists in specific areas and that the overuse results in spillover on to areas that contain many rare plants.

Programming Tip 6.2.

Focus on leading.

The groups will look to guides for just about everything, especially in places which are culturally and ecologically different. The service provider needs to free the guide to focus on what he or she might do best: lead the group. All hotels and restaurants need to be pre-booked and paid for in advance of the trip so that the leader can focus on the needs of the group.

Conclusion

Although the dynamism of the programming process will prevent the programmer from completely isolating certain aspects of the programme cycle in time, the culmination of planning to this point will result in the ability to structure a number of key aspects of the emerging programme. Initially these will include an understanding of the types of programme areas and formats available to the programmer, as well as setting, mobility and lodging considerations. In some cases, service providers will have a good grasp of the complexion of their programmes, as regards the above variables. In many other cases, however, they will not. A consideration of the needs of participants, along with the agencies goals and objectives, will help in determining the direction of the programme. Additionally, the programmer should endeavour to construct a matrix of different programme alternatives in order to decide objectively on a programme that is truly reflective of the resources and competencies of the service provider. This will entail a significant degree of brainstorming on the part of the programmer, perhaps in association with other members of the agency. This chapter further discussed the importance of outdoor and environmental education, and the differences between the two, in the context of ecotourism. While the former is viewed as a resource for education, the latter focuses on the natural and social sciences as topics or disciplines that allow for the understanding of environmental issues. Interpretation, as a form of communication, was also touched upon in the latter stages of the chapter. Subsequent to the discussion of the value of science and art to interpretation, several models were discussed as well as a number of interpretation guidelines which apply specifically to ecotourism.

7

Programme Design B: Gearing Up to Go

Never for me the lowered banner, never the last endeavour.
(Sir Ernest Shackleton)

Introduction

In this chapter the focus switches from the structural aspects of planning, to more of an emphasis on the practical and logistical components which are so important to the programme. Many of these practical elements include briefing of the client, physical and mental preparation, transportation, persons with disabilities, health and hygiene, first aid and survival, trip planning, clothing and equipment, permits, and environmental conditions. Consequently, this chapter discusses many of the pragmatic elements of ecotourism and, in doing so, reinforces the notion that the transportation of ecotourists to remote regions means that many precautions must be taken in order to ensure a successful venture. Missing from this chapter is a discussion of leadership, safety and risk management. These topics will be considered in the following chapter.

Pre-trip Information: Briefing

The distribution of pre-trip information to the client, well in advance to the start of the programme, is essential given: (i) the educational focus of ecotourism; and (ii) that ecotours often involve travel to places that are ecologically and culturally different from the places they reside. In their classic model of the total recreation experience, Clawson and Knetsch (1978) suggested that the first of five phases of the travel experience involved the anticipation and planning of the event. These authors note that pleasurable anticipation is almost a necessity, and that travel agents cannot underestimate the importance or attractiveness of the areas they advertise. The ecotourism service provider does him/herself a great favour by generating a good deal of enthusiasm in the client through the dissemination of articles, lists, pamphlets and other bits of information on the natural and cultural attractions of the areas to be visited. The operator may even go so far as to recommend readings and other sources (videos, web sites, and so on), in allowing the client to prepare to an acceptable level. In addition to the specific information on the natural and social attractions themselves, the service provider will also want to pass on the customary information on the various practices of the local people at the destination. This may include information on taxes and tipping, postal services, shop hours and information on restaurants, while other pieces of information may be absolutely essential, as illustrated in Table 7.1.

When to send the information is also of considerable importance, both to operators and ecotourists alike. In most cases information will be disseminated once the ecotourist

Table 7.1. Travel checklist.

Passports and visas	Anyone in need of a passport, visa, or other travel documents should put that item at the top of the list. No amount of pleading will get you into a foreign country if you don't have the required documents. Submit your application as early as possible
Health certificates or vaccinations	Find out from your service provider whether you require special health certificates or vaccinations. Do not wait until the last minute in case you have a negative reaction to the vaccination
International drivers' licences	If planning on driving in another country, ask your travel agent if you need an international drivers' licence or any other documents
Health insurance	A short hospital stay in another country can cost thousands of dollars. Depending on one's health plan, it is a good idea to purchase additional health coverage through a private plan
Dates, times, flight numbers	Always confirm flights before travelling. Telephone before you leave for the airport to make sure the flight is on time
Return trips	In some countries, and with some airlines, it is essential that you confirm your return flight in order to reserve a spot. Seats are often lost by neglecting to confirm
Importing	Before buying anything with the intention of bringing it home, check Customs regulations. The item may be prohibited or you may be required to pay duty
Money	Travellers' cheques are the safest form to carry, but it is also wise to have some cash, especially small denominations, in the currency of the country in which you will be travelling. Don't forget to carry a small amount of cash of your home country to use on your return home
Luggage	Clearly label all baggage with your name and home address. Be sure bags are locked, or bound shut (a common approach to safety these days). Some items will not be covered by the insurance company of the airline (e.g. cameras). Be sure to carry all valuables in your hand luggage, including overnight kit
Prohibited items	As a result of the terrorist attacks in the USA and other places, airports now require passengers to take care not to include certain items in their hand luggage. Such items include scissors, nail clippers, pocket knives and anything else that may be used to injure others
Home security	Stop delivery of newspapers and inform neighbours/friends of vacation plans. Ask someone to empty the mailbox, shovel snow from paths, and generally make the house look occupied
Personal care	Carry spare medication, eye glasses or contact lenses with you. Carry all prescription medication in original container with your name and physician's name on it

Source: Adapted from Ontario Ministry of Consumer and Commercial Relations (1995).

has financially committed to the trip, which usually means a down payment or payment in full. Payment may be requested, for example, 2 months before the actual trip. Information could be sent to ecotourists a month later to sustain the interest between the time of the booking and the trip itself. Also, 1 month is a sufficient amount of time to educate oneself about the attractions of the destination. Understandably, the information sent must accurately reflect the types of attractions and social and ecological conditions of the site itself. It must also be consistent with the brochures and other promotions which ecotourists used to select the trip, including the chances for viewing certain species, climate and weather patterns, and so on. In their *Ecotourism Guidelines for Nature Tour Operators*, The Ecotourism Society (1993) proposed that pre-departure programmes should use a number of techniques, including:

- Provide pre-departure packages containing introductory information on the people and ecosystems to be visited. Stress the importance of reading pre-departure information, such as selected bibliographies, and review additional resources for each destination.
- Keep information objective and accurate, using examples of phenomena visitors might encounter.
- Provide general travel ethics addressing standards for behaviour in natural areas and with local cultures.
- Provide information on the equipment, clothing and personal supplies suitable to the regions being visited.
- Warn against bringing disposable goods that contribute to the solid waste burden in the region.
- Provide information on products that are illegally traded.
- Provide information, as required, on avoiding the accidental transport of foreign, exotic species into isolated ecosystems being visited.

The Ecotourism Society publication states that ecotourists will benefit, as will the destination region, by having visitors attuned to the full range of opportunities to experience culture and wildlife; they will become aware of their personal responsibility to minimize impacts at the destination; and they will further understand the value of proper gear and clothing required for the natural and cultural environments to be visited.

Mental and physical preparation

For trips that are more physically demanding (e.g. kayaking), ecotourists will need to gauge their level of fitness accordingly. This is important in view of the fact that many ecotourism trips are quite costly, and the cost–benefit consideration is put at ease when tourists are able to participate at a level that they deem as being worthwhile. Poor mental and physical preparation also puts the ecotourism operator in an uncomfortable position. He or she may not wish to exclude people from the trip who are not prepared. However, the operator needs to protect his or her right to refuse a client on the basis of physical ability, not only for the safety of the client, but also for the safety of others. This is especially important in activities where there is high level of risk involved in the activity, but also where there is a known danger because of the physical conditions of the site, which may ultimately constrain involvement (e.g. terrain is too steep, rivers are too high). Ultimately the responsibility for appropriate levels of conditioning should lie in the hands of the ecotourist; whereas, the conditions of the programme setting are more the responsibility of the service provider.

In their study on ecotourism development, the Northwest Arkansas Resource Conservation and Development Council (1997) writes that Ozark Ecotours ranks their tours on the basis of terrain type, ecotour length and physical exertion required, for the purposes of programme efficiency, as well as for the satisfaction of clients. The tours are ranked as *easy*, *moderate*, *moderately strenuous*, *strenuous* and *very strenuous*. Such categorization, and the associated descriptions of such trips, give the ecotourists the opportunity of placing themeselves in the correct category or programme. Despite the fact that these rankings are published so consumers can make informed choices, Ozark Ecotours also conducts careful screenings of the physical abilities of clients over the phone. This phone call also gives the operator the chance to talk about clothing, footwear and other issues related to the trip. As suggested above, this step is critical in minimizing, as much as possible, the problems (time delays, exhaustion, irritability of clients and leaders, and so on) that go along with having inexperienced clients in situations that they cannot handle.

Programming Tip 7.1.

The Guide and the Setting.

Guides should be well acquainted with the wildlife and physical conditions of the setting in which the programme occurs. This includes dangers and distances, and a continual assessment of weather conditions. This may well mean the development of a contingency plan in the event that conditions do no permit entry into a particular setting. Consideration should be made regarding the sites that do and do not have facilities for the needs of participants. Finally, it is important for the guide to project an 'ecology-friendly' image in his or her day-to-day practices. In the minds of participants, the guide *is* the organization, who must uphold the mission and image of the company at every stage of the way.

Health Precautions

People have travelled for health-related reasons for hundreds of years. Foremost among attractions were health spas, hot springs and climates, which were often prescribed for numerous different ailments (e.g. tuberculosis). More recently, the increase in interest in health and well-being have stimulated the creation of a very specialized travel market, particularly in Europe (Hall, 1992), but also in a number of other developed countries such as the USA and Australia. Such tourists are in search of a whole host of physical, cognitive or spiritual tonics, including relaxation, wilderness therapy and philosophical enlightenment. Health tourism, however, should not be confused with travelling healthy or travellers' health, which is a fast emerging area of research concerned with the physical and mental health of travellers, before they leave on trip, during and after. Preparing for travel is an ominous task considering the arrangements for transportation, passports, visas, language problems, accommodation, foreign currency, insurance coverage, fears of flying and terrorism, travel timetables and concerns about home (Forgey, 1990). Often these are compounded when visiting far off regions for extended periods of time. Unfortunately, the most important consideration when travelling abroad – one's health – is often neglected.

As Dawood (1992) suggests, staying healthy abroad is not a question of luck (although luck may certainly play a part), but rather a result of a process of prevention. Travellers must take an active role in taking care of their health while abroad. This usually takes the form of becoming educated on the proper immunizations, prophylaxis (drugs for malaria, as an example), food hygiene, water, facilities, insects, sex, sun and other aspects of the broad environment. While health rightfully should be the responsibility of the traveller, there is no reason why the ecotourism service provider should not play an active role in the dissemination of health-related information to ensure the overall safety of clients. After all, word of mouth often plays a significant role in the promotion of ecotourism programmes. Travellers who return from such trips and who have been unlucky enough to contract a preventable disease will unquestionably pass this information on to friends and colleagues.

Efforts to stay healthy while travelling become especially important in view of the research on traveller's health. For example, in a study of over 14,000 travellers to various European and Mediterranean destinations between 1973 and 1985 (see Dawood, 1992), 37% of travellers reported being unwell. Twenty-eight per cent reported having alimentary problems, 6% were 'other ailments (not mentioned)', while 3% were of a respiratory nature. Dawood further illustrates that all destinations are not the same when it comes to the chances of becoming ill. In the aforementioned study, illness attack rates were much higher in North African and Eastern European countries than other regions. These statistics corroborate the results of other studies, including one by

Cossar *et al.* (1990), who found that 36% of travellers reported illness, and another by the World Health Organization, which reported the following statistics for every 100,000 individuals travelling overseas to tropical destinations:

- 50,000 have health problems,
- 30,000 travellers get diarrhoea,
- 3000–4000 will contract malaria,
- 8 will die,
- 2 will contract typhoid,
- 0.1 will contract cholera.

In their introduction to tourism and health, Clift and Page (1996) provide many documented cases of illnesses that have been contracted by those visiting tropical countries. In one case a man from Britain contracted three tropical illnesses (typhoid, malaria and filariasis) during a trip to Cameroon, as outlined in the following quote:

> Symptoms to accompany these three illnesses did not appear until after my return to Britain. Because my GP diagnosed influenza – despite me informing her that I'd just been to West Africa, had regular fevers of 105° and was violently shivery – treatment for malaria did not begin for a week which led to the development of a fourth tropical disease, blackwater fever.
> (p. 7)

As travellers continue to seek the furthest reaches of the planet, there will always be concern over whether proper precautions have been taken before travel to these regions, the facilities available to the traveller while there, and the level of expertise of the provider in the event of an accident (see Wilks and Atherton, 1994). As a result of the emergence of the travel medicine field, there appear to be many more education- and prevention-based services available to the tourist. Figure 7.1 illustrates some of the sources of, and settings for, health information and services for tourists (Clift and Page, 1996). Sources and settings are organized along a continuum based on the decision-making phase, the pre-travel stage, the travel-holiday phase and finally the post-travel phase. The bottom of the figure illustrates the issues, illnesses and individuals who may play a part in the process along this continuum, while the top portion pertains to the information and services which are available to the tourist at various stages of travel.

There would appear to be an opportunity for the ecotourism operator to play a role at just about all phases of this continuum in Fig. 7.1. In the decision-making stage, ecotourists may opt for one service provider over another on the basis of one's reputation for safety (as alluded to earlier). Once this decision is made, the pre-travel phase would involve having the operator to help facilitate or encourage clients to seek immunizations and other preventative measures. The travel or on-site phase might involve a number of efforts to maintain the safety of clients (e.g. food and water precautions); while the post-travel phase may find the service provider encouraging clients to follow-up with their physician when they arrive home, especially if the tourist has experienced any signs or symptoms during the course of the trip, or upon returning home.

Two conclusions from the work of Page *et al.* (1994) are important in the context of the responsibility of ecotourism operators. First, accurate and up-to-date information from the service provider is a necessity if the traveller is to be able to understand potential health risks which may be encountered at the destination. This entails having the ecotourism operator communicate as much information as possible about any health-related issues (see Appendix 2 for an example of a medical form that every ecotourist should fill out before the trip). Second, and closely linked to the first, is that holidays are often advertised and sold on the basis of their health benefits. However, if over a third of travellers are experiencing health problems, as suggested earlier, than the holiday itself may not be fulfilling its psychological and physiological role, which may ultimately be a significant cause of customer dissatisfaction.

As hinted earlier, there are now a number of resources that the tourist may utilize in his or her attempts to stay healthy abroad. The obvious ones include tropical disease centres and government regional health programmes, which routinely educate tourists on destinations the world over. An example of other professional programmes which

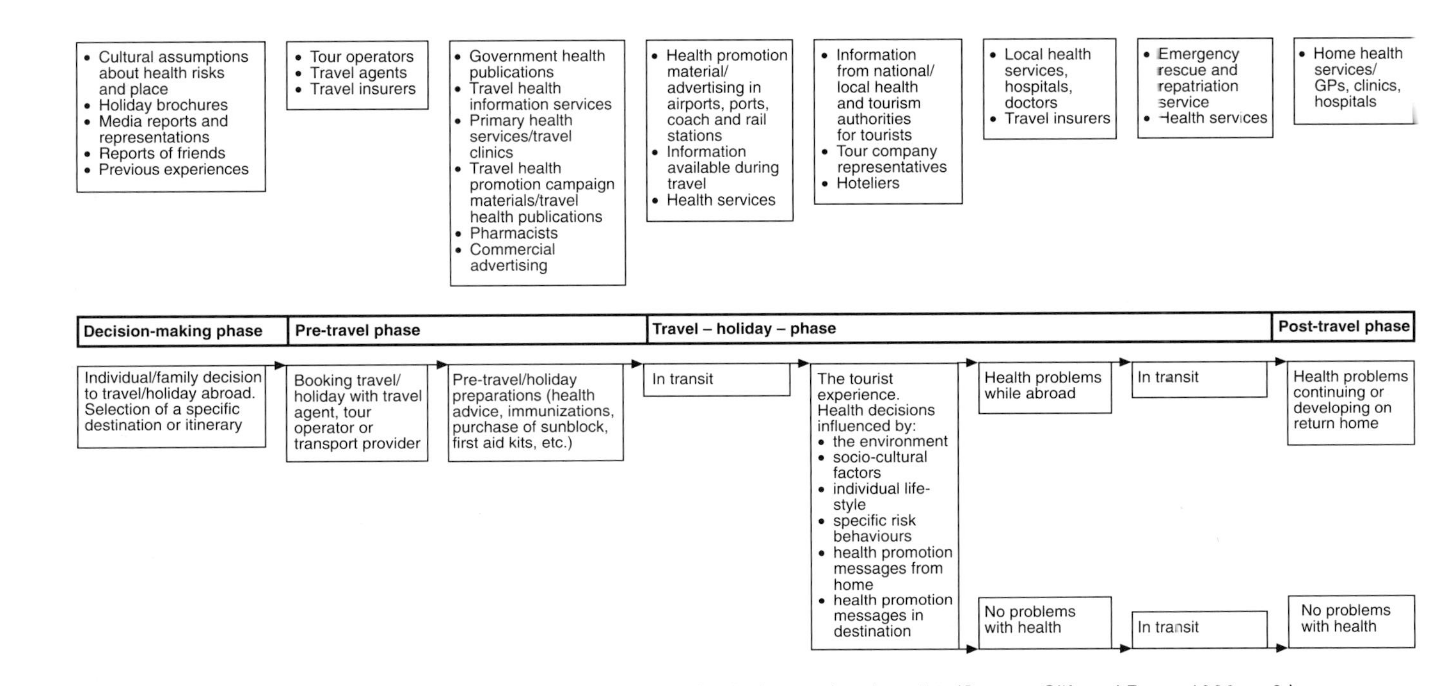

Fig. 7.1. Sources of, and settings for, health information and services for the international tourist. (Source: Clift and Page 1996, p. 9.)

have developed to assist travellers include the International Association for Medical Assistance to Travelers (IAMAT). This organization provides a directory of physicians in various countries who have fixed fees, who speak English, who have passed credentials set out by IAMAT, and also sends out publications on various tropical diseases, including useful world maps on the geographical distribution of diseases.

Transportation

It is widely recognized among those who programme in the field of recreation, that the most hazardous part of a programme is not the activity participation phase, but rather the transportation of participants from one site to another (Gilchrist, 1998). Unfortunately, this is not widely recognized by participants who place a good deal of trust in the people charged with moving clients from place to place. To ensure that accidents do not occur, service providers must be cautious in who they enlist to drive vehicles, the vehicles themselves which are chosen to transport, and the length of time that drivers are asked to transport participants. The following list compiled by Hanna (1986) is representative of the number of concerns related to this critical part of the ecotour programme (Table 7.2). In following these guidelines, the service provider may at least minimize the chances of something going wrong.

In addition to the points raised above, the service provider should avoid the following: never let participants drive vehicles; transportation needs should be organized well in advance, and confirmed a week in advance; and an emergency vehicle should be identified and accessed in the event of an accident or mis-communication. These points suggest that transportation is a critical part of the tour – and although participants and operators alike may not regard this aspect of the trip highly, they must at least appreciate many of the inherent and statistically documented dangers involved in the process of moving people on a day-to-day basis.

Persons with Disabilities

A person with a disability is one who has a mental or physical impairment which substantially restricts a major life activity such as walking, talking, breathing, sitting or learning (Cordes and Ibrahim, 1996). The disability, however, is the expression of a mental or physical limitation in a social and environmental context (PLAE Inc., 1993). The disability is thus the interaction that exists between the mental or physical limitation in relation to the social and environmental conditions encountered. The word disability is different

Table 7.2. Checklist of transportation planning considerations.

1. If renting a vehicle, review the terms of the contact carefully.
2. Make sure that all drivers are covered with an insurance policy to protect the company, but also potential accidents involving others.
3. Be sure drivers have the appropriate licence for the vehicle and number of passengers.
4. Check the condition of the vehicle and ensure it has a spare key, tyre, jack, and fire extinguisher (as well as other emergency items).
5. Participants must be properly seated in the vehicle, and told to use seatbelts.
6. Trip gear should be carried outside of the passenger compartment or lashed down under seats.
7. Employ safety procedures such as safety education, and loading and unloading.
8. Drivers should avoid driving continuously for extended periods of time.
9. Be sure the vehicle will be reliable in cold weather.
10. Check ahead that winter travel routes are cleared of snow.
11. Park in parking lots to reduce the chances of getting trapped by other vehicles.
12. If more than one vehicle is to be used, make sure each driver has a route map and time schedule for rendezvous.

Source: Adapted from Hanna (1986).

from handicap, which was commonly used to describe or label people with disabilities. According to Kennedy *et al.* (1991) 'disability refers to a specific impairment or disorder, whereas a handicap results from actions of the person with the disability or by society' (p. 29). It is therefore the attitudes that exist within society which lead to unjust labelling and treatment of individuals as 'handicapped'.

Disability has been described not simply as a state of being, but rather as a process which has at its root an interruption of normal bodily functioning and ending with the disability itself, such as the inability to hike at an advanced level. The stages of this process have been identified through the Nagi classification, which has been used as a method to classify a number of different disabilities, as outlined in Fig. 7.2.

Another framework, referred to as the Enabler Model (see PLAE Inc., 1993), consists of 15 different conditions which are based on mental, sensory and motor impairment. These include difficulty interpreting information; severe loss of sight; complete loss of sight; severe loss of hearing; prevalence of poor balance; incoordination; difficulty in moving head; difficulty reaching with arms; difficulty in handling and fingering; limitations of stamina; loss of upper extremity skills; difficulty bending, kneeling, etc.; reliance on walking aids; inability to use lower extremities; and extreme size and weight. These general categories can be further examined on the basis of particular conditions, but also on the basis of degrees of severity. Difficulty in interpreting information may relate to a communication limitation. At a severe level, a condition like aphasia would prevent someone from communicating on many levels. At a more moderate level, persons may have difficulty in communicating owing to a propensity to stutter, or because of dyslexia. Furthermore, limitations of stamina may result from breathing difficulties. At a severe level, such breathing could be a result of cystic fibrosis or asthma; whereas at a moderate level of affliction the individual may have episodic difficulty with breathing as a result of sinus conditions, allergic reactions, or hay fever (see Vinton and Farley, 1979).

The tremendous range of different mental, sensory and motor impairments means that there are a number of challenges which must be addressed in being inclusive and accessible to the entire population. If the operator is one

Pathology ➢	Impairment ➢	Functional limitation ➢	Disability
Interruption or interference of normal bodily processes or structures	Loss and/or abnormality of mental, emotional, physiological, or anatomical structure or function; includes all losses or abnormalities, not just those attributable to active pathology; also includes pain	Restriction or lack of ability to perform an action or activity in the manner or within the range considered normal that results from impairment	Inability or limitation in performing socially defined activities and roles expected of individuals within a social and physical environment
Level of reference: Cells/tissues	*Level of reference*: Organs and organ systems	*Level of reference*: Organism – action or activity performance (consistent with the purpose or function of the organ or organ system)	*Level of reference*: Society – task performance within the social and cultural context
Example: Denervated muscle in arm due to trauma	*Example*: Atrophy of muscle	*Example*: Cannot pull with arm	*Example*: Change of job; can no longer swim recreationally

Fig. 7.2. The Nagi model. (From Institute of Medicine, 1991.)

who is keen on the development of an 'accessible' operation, he or she will actively strive to overcome some of the attitudinal barriers which may exist to accessibility. These may include a move away from: (i) the belief that people with a disability are all the same; (ii) there are too many communication barriers; (iii) transportation logistics would be too difficult to manage; and (iv) disabled persons could not handle the rough or uneven terrain found in most wilderness areas. In regards to this last and very important point, there is an age-old dilemma which exists in outdoor recreation on the degree to which an agency modifies the natural world to cater to the needs of disabled persons (e.g. do all trails have to be modified to make them wheelchair accessible?). To these agencies the answer is probably no, simply because it is not practical to modify trails as such, and often not consistent with the planning and management of natural areas. While some of the more front-country environments have been modified to suit the needs of persons with disabilities, there does not appear to be the expectation for more extensive modifications. It is also important to consider, as outlined above, that many disabled persons are more 'able-bodied' than those who are not disabled. In such cases either minor modifications or no modifications need to made in order to facilitate the experience.

Several strategies in recreation have been used as interventions to induce persons with disabilities to participate more frequently. While counselling and education are well established conventional methods which are quite easily facilitated, other more ambitious programmes have persons with disabilities participating in adventure pursuits in wilderness settings. Wilderness Inquiry Inc. in the USA, is one such programme which focuses on integration, group dynamics, and self-discovery, as essential to the overall experience; with the ability to handle any logistical, medical and legal concerns for this population. Their staff procedures have been honed over a period of 20 years, and include a curriculum based on disability awareness, social integration, risk management, group processing, specific disability information, technical disability skills such as dispensing medications, and adapted equipment (McAvoy and Lais, 1999). These authors state that their insurance costs to run such a programme depend not on the fact that they cater to a disabled population, but rather on the organization's record of safety, credentials of leaders, and other safety policies (see also Wilderness Inquiry, 1999, for an overview of this company's philosophies regarding integration, and staff policies and practices). While there appears to be a growing number of such programmes for adventure recreation, the same cannot be said for ecotourism where there seems to be an unoccupied niche. Those intent on developing ecotourism programmes for persons with disabilities would be wise to follow the lead of a company such as Wilderness Inquiry, which has shown that the setting should not be viewed as a constraint to participation in wilderness pursuits. It is a model worth emulating.

Personal Hygiene

Food

No matter how civilized or primitive the programme setting, there must be the utmost concern for the manner in which food is disposed of, stored and served. It also follows that to keep food safe, is to know the environment in which one is travelling. Urban environments will be different from wilderness ones; while hot humid environments will be different from cold and dry ones. Furthermore, it is not just bacteria and protozoans which need to be considered in keeping food safe, but also other beasts including, for example, ants, mice, raccoons and bears, all of which are very resourceful in gaining access to packs and other storage containers. In wilderness settings, operators may need to use airtight containers for food storage, or they may opt to keep food safe by hanging it from trees and thus out of the reach of nocturnal animals. It should also be noted that guides must be open about cleanliness, by making it a practice to wash hands and dishes often and in full view of participants, to ensure that both parties feel comfortable with the food-handling process.

Programming Tip 7.2.

Selecting tea times.

Besides the natural mealtime occurrence, tea times might be selected on the basis of weather, a lull in activity, or signs of group fatigue or boredom.

Source: Mitchell (1992).

For ecotours that occur in backcountry settings, it is important to follow some fairly standard procedures for the disposal of food. The burn-bash-and-bury philosophy used so frequently in the past has fallen out of favour. For example, it used to be appropriate that after consuming a can of food, the paper off the can was burned, the can was bashed with a boot or axe, and subsequently buried somewhere on the campsite. In addition, plastics were often thrown in fires, along with just about everything else, including glass. Organizations such as the National Outdoor Leadership School in the USA are very clear about what is appropriate to take into remote locations. Glass is now frowned upon because it is heavy and difficult to handle. Plastics, especially disposable plastics, need to be minimized through repackaging and/or packing out (taking home) after use in the wilderness. And because garbage is always a problem, especially with big groups, the amount of work before the trip in minimizing things such as plastics and cans, will help in the latter stages of the trip when one person or another will be charged with the responsibility of carrying it out of the wilderness. The following are a few tips on how to handle garbage on trip (Curtis, 1998, pp. 111–112):

1. Minimize garbage by repacking food before your trip.
2. Avoid leftover food by carefully planning your meals.
3. When leftovers do occur, they should be carried out in plastic bags.
4. Do not try to burn food unless you have a hot fire and can completely incinerate the food. Partial burning leaves a charred food mess in the fire site.
5. Food particles (like noodles) that invariably occur in dish washing should be treated like bulk leftovers and carried out. Fish viscera are a natural part of the ecosystem. In high-use areas, your goal is to minimize other people seeing or smelling them, so consider burying them in a cathole. In remote areas with few visitors, you can scatter them widely, away from camp and trails, to reduce the chance that other people will come across them.
6. If you are in bear country, garbage should be hung just like food.

One of the delights of wilderness travel is the building and maintenance of fires for cooking, warmth and aesthetics. As much as possible, operators should encourage the use of stoves for cooking (obviously for suitable trips in wilderness settings) rather than wood. This is a controversial item in minimum impact camping, and different groups subscribe to different approaches. Many of the most environmentally minded researchers and practitioners suggest that cooking should only take place on gas, butane or alcohol stoves, and that fires should be built in situations where they are required (e.g. cold), or periodically for aesthetic purposes. In this latter case, smaller fires should be built using existing fire pits. However, fires for aesthetic purposes should not be lit in cases where there is not an abundance of wood or when there are fire warnings because of dry conditions.

Water

As suggested earlier in the chapter, stomach ailments appear to be one of the most common health problems for travellers. The usual culprits are food and the consumption of contaminated water. In the case of the latter, contamination comes in the form of bacteria (e.g. *Escherichia coli, Salmonella typhi, Vibrio cholerae*), protozoans (e.g. *Giardia lamblia, Cryptosporidium*) or viruses (e.g. hepatitis), which appear to be much more prevalent in tropical destinations (Brock, 1979). The largest of these, the protozoans, have been the topic of great interest, especially in relation to the devices which have been designed to eliminate these from drinking water supplies. Because of their size (10–100 μm in size, compared with bacteria which are generally 2 μm in size, and viruses which are 0.1–0.2 μm), protozoans are more easily captured through the various water filtration devices that are on the market.

One of the most troublesome protozoans, at least from the standpoint of many water filter manufacturers, is *G. lamblia*, which appears to be transmitted through the waste of humans and other animals. This microorganism destroys the intestines' villi, which are responsible for the absorption of food nutrients. Incubation usually occurs over a period of 2 weeks, with the host eventually experiencing symptoms such as diarrhoea, vomiting and cramps. Owing to its longer incubation period, the infected individual does not often equate the illness with their trip, in situations where people have been home for as long as 10 days. Those who fail to seek proper treatment of *Giardia*, become chronic carriers of this protozoan and may experience periodic episodes over time. The protozoan *Cryptosporidium*, also affects the small intestine of an infected individual by stimulating an outward flow of water from the body over a period of 10 days. One further example, schistosomiasis, or safari fever, (caused by *Schistosoma* spp.) is problematic to travellers who visit many of the less developed countries, especially those with tropical climates. This protozoan lays its eggs in freshwater, which later infect snails that reside in the water. After a period of development in the snail, a further incarnation appears in the water supply, where humans become exposed to this microorganism through either drinking or swimming. Schistosomiasis is a contributor to fever, lethergy, cough, rash, nausea and abdominal pain, with more severe cases resulting in a variety of different liver complications (Forgey, 1990).

In general, there are three ways in which to kill the pathogens that inhabit water: boiling, filtration and chemical decontamination. The first, and most efficient, is boiling. Although most organisms are killed instantly at a boil, it is recommended that water is boiled for a period of 5 minutes. The drawback to boiling is that it takes both time and fuel. And for the guide who is responsible for providing water for the entire group, the inconvenience is often not a welcome one. Depending on the unit, filtration can be very effective. There are literally dozens of filtration units on the market, all claiming to ensure a safe quality of water. Unfortunately, these are not all created equal. The best filtration systems are those which are dependable, easily maintained, and give the user some indication as to when the life of the filter cartridge (whether it be carbon, ceramic or other material) is over. Users should be wary of those which claim to filter thousands of litres. Also, some are just filters, while others will filter and purify. In the case of the latter, the filter removes suspended solids by straining, while the purifier actually disinfects or kills the pathogens via chlorine or some other disinfectant. One of the main problems with filters is that they are burdensome to take on trips for extended periods of time, unless the trip is a wilderness excursion. For those conventional tourism trips to far off destinations, Gifford-Jones (1988) recommends taking the Water Tech Purifier cup which purifies water as it passes through the cup. It releases iodine in trace amounts which kills bacteria, protozoans and viruses on contact. At less than 250 g and no more than 10 cm high, it is very easily packed for those areas where clean water is debatable and bottled water is scarce. The third method of decontamination is chemicals alone. Iodine tablets, for example, may be placed in a litre of water for a period of time rendering the water safe. While this is highly affective, some people are allergic to iodine, and the protozoan *Cryptosporidium* is resistant to both iodine and chlorine.

Recreational impacts on water

The attractiveness of water to a whole range of different outdoor enthusiasts has meant that these systems, especially those adjacent to large populations, have had to endure a significant amount of pressure. Much of the finger-pointing has been at the motorized recreational vehicle crowd (e.g. motor boats), which have contributed literally millions of litres of oil and petrol in water systems around the world. Only recently have many of the major manufacturers engineered four-stroke engines which are significantly more efficient than their two-stroke counterparts. Some resource management authorities have gone so far as to ban two-stroke engines, while others have banned motors altogether. Liddle (1997) suggests that there is little in the way of quantitative data describing what substances actually appear in the aquatic environment from motor boats. Interestingly, he cites a series of studies (from the 1950s and 1970s) which suggest that the impacts from outboard motors are negligible – oil and petrol are probably discharged out of the lake or river system – and that the effects of crude sewage and litter are more serious. On the other hand, canoes and other non-motorized means of transportation have been praised for the much reduced impact on water systems. This may certainly be the case, but these modes of travel also enable the traveller to penetrate far more deeply into wilderness settings. The behavioural patterns of these excursionists, in these settings, have also been the topic of much debate.

The principal issue surrounding recreational impacts on water seems to be in regards to the frequency of use of areas and the associated pathogens introduced to the water supply from human waste. In the case of the latter, the concern is that human waste is introduced into the system at close proximity to water supplies, and leaches into the water supply before it has the time to break down. Facilities, such as pit toilets, are notorious for high levels of contamination, especially if they are infrequently emptied (if designed to do so). Depending on the physical conditions of the water supply (e.g. inflow and outflow, depth, and so on), and the frequency of use, contamination will vary. Scientists measure the level of faecal pollution through the presence of coliform bacteria, which are easily detected and not pathogenic. The following are some observations based on the work of Hammitt and Cole (1987), and drawn from a rich base of data on water and outdoor recreation:

1. Human faeces contain over 100 viruses, bacteria and protozoans that are known to cause disease.
2. Bacteria in the faeces of wildlife are just as much a problem as those from recreationists.
3. Light recreational use of an area is said to improve the bacteriological quality of water as it scares away wildlife, who would otherwise continually contaminate the water.
4. Most studies of contaminated water supplies examine only the surface waters. The benthic regions of the same water source however have been found to contain significantly higher concentrations of organisms. For example, surface water concentrations of faecal coliforms ranged from 2.1 to 8.0 ml^{-1}, while concentrations have reached as high as 48,000 ml^{-1} on some river bottoms.

Water contamination has been addressed by quite a number of wilderness organizations, who regularly publish on topics related to backcountry ethics. Following the example of the US-based NOLS (National Outdoor Leadership School), those who set up camps in wilderness settings should do so at least 60 m from any water source (see Simer and Sullivan, 1985). Larger groups in wilderness areas are encouraged to dig latrines to deposit human waste, while smaller groups should dig what are referred to as catholes – a 15 cm deep hole which is filled in by the traveller after use. These too must be at least 60 m from any water source. In addition, some regions or authorities will require ecotourists to pack out (remove) their human waste (e.g. the Colorado River), where human waste has created health concerns and detracted from the participants' experience. Similar concerns have been reported in the Himalayas, where the 'toilet paper trail' has had many effects on the water systems of the region (Robinson and Twynam, 1995; Bilodeau and Trembley, 1996).

Decisions on where to set up camp in a

Fig. 7.3. Although a more ecologically sound form of transportation than other means, canoes allow ecotourists to penetrate into many remote pristine environments

wilderness environment are subject to the region in which the group is travelling. In many national and regional (state or provincial) parks, hardened camp sites have long been created adjacent to water supplies. These sites should be used by ecotourists so that impacts are not created on pristine areas adjacent to these sites. In cases where hardened sites cannot be used, ecotourists must be careful in creating as little disturbance to an area as possible. Where and how to cook, camp, and set up the latrine are important considerations in these situations. Wilderness purists suggest setting up camp at least 60 m from a water source, and separating tents, latrines, washing areas and cooking areas by predetermined distances (Wiard, 1996, p. 136). Abiding by these fairly rigorous rules is difficult in situations where group numbers increase in size. Still, special attention must be paid to minimizing, as much as possible, the compounded effects such use may have on land and water.

First-aid and Survival

Few wilderness guides neglect to mention the importance of wilderness first-aid and survival. While an in-depth treatment of this topic is beyond the scope of this book, it is worthwhile mentioning a few conditions that may be encountered on trip. At the very least, operators should insist that their guides carry up-to-date backcountry first-aid guides in the event that first-aid is required. There should be the recognition that first-aid in wilderness settings is different from first-aid in urban settings. In the latter case, responsibilities surround breathing, bleeding, splinting and safety. In the wilderness, care goes beyond what is presumed to be 'first-aid', to other concerns related to exposure, shelter, travel, sanitation, innovation, and the administering of proper fluids and food (Merry, 1994). Furthermore, guides should be aware of access and egress points while in the field, know where the nearest health-care facilities are located and how to get there, carry a mobile phone, as well as a list of a number of pertinent contact numbers in the event of an emergency. In addition, guides should be encouraged to seek advanced certification in first-aid in order to be as current, and as prepared, as possible.

Cold and heat

Hypothermia or exposure occurs in individuals when the core body temperature dips

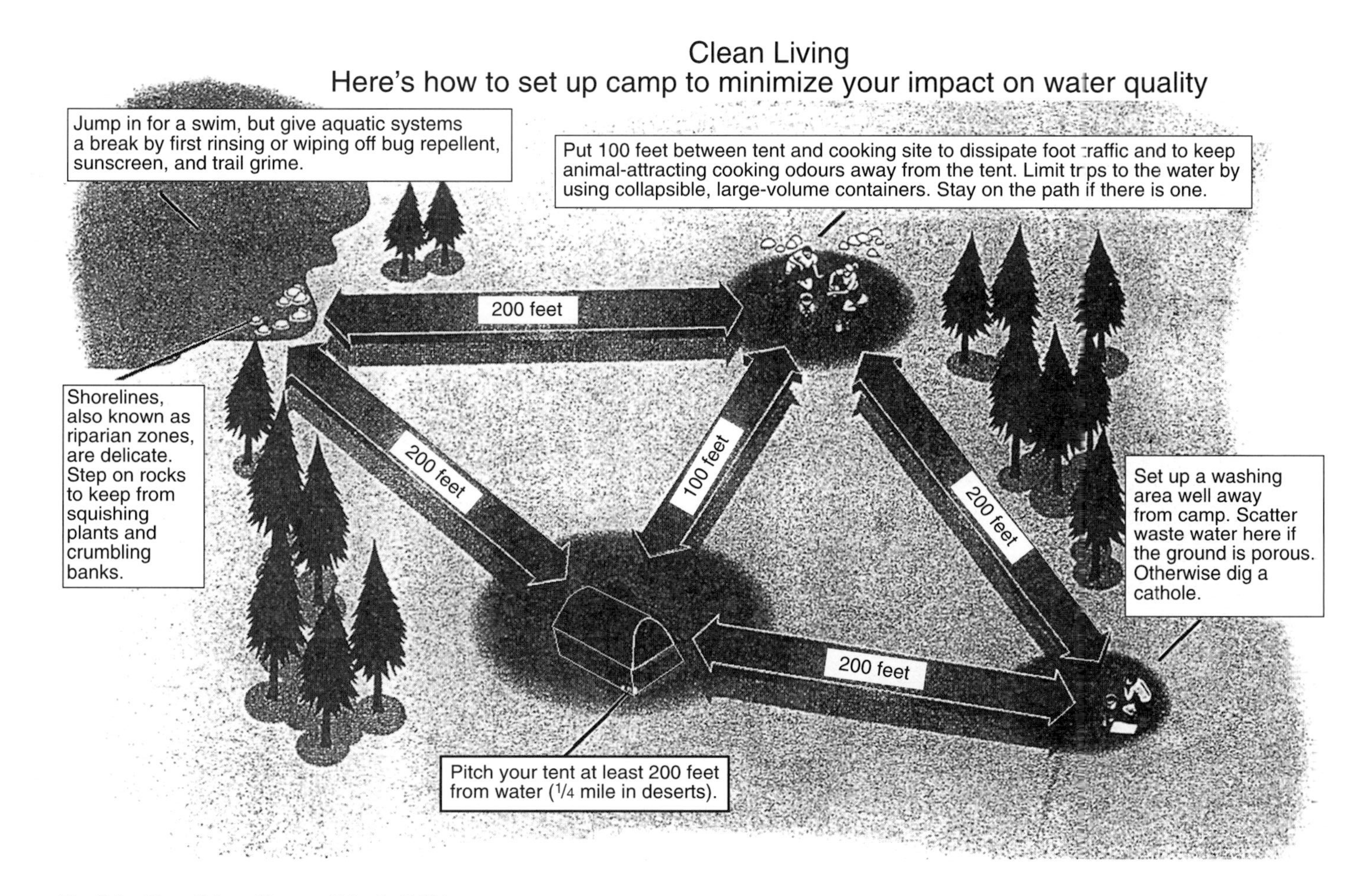

Fig. 7.4. Clean living. (Source: Wiard, 1996.)

below 37°C. It can develop in almost any region with temperatures below 10°C, but especially those with colder temperatures, where the condition can accelerate. Travellers in mountain regions are especially prone to hypothermia because of the transition, for example, from sunny mountain ridges to very cold and damp valleys. Unfortunately travellers often forget to make appropriate clothing changes (Province of British Columbia, 1981). In general, the body loses heat in five different ways, as outlined by St John Ambulance (1995):

- *Radiation.* Heat radiates from the body into the air around it, especially from the head.
- *Breathing.* Cold air is inhaled, warmed by the body and exhaled, causing heat loss.
- *Evaporation.* Body heat is used to evaporate liquid on the skin.
- *Conduction.* Heat moves directly from the body to a cold object that the body is touching, such as a rock.
- *Convection or wind chill.* The thin layer of warm air around the body is replaced by cooler air, which the body must now heat.

The signs of hypothermia are much different depending on the magnitude of the condition. In general, there are three states to consider: mild, moderate and severe. Although it is not difficult to know that someone is experiencing hypothermia, a greater challenge rests in determining the onset of more serious stages, and what to do at each stage. This is important to detect, however, because the more advanced conditions can be life threatening. Table 7.3 illustrates the signs of the various stages of hypothermia, as well as what to do at each stage.

Probably more dangerous, because of the number of people who travel to warm destinations, are the effects of heat on human functioning. Physiologically, the body removes excessive levels of heat through the dilation of blood vessels which allows blood to reach the surface of the body. This in turn allows heat to escape, and the body to cool (The Canadian Red Cross Society, 1994). In hot weather conditions, particularly if the ecotourist has been involved in physically demanding activities, the body may lose excessive amounts of water and salts through sweating. If too much water is lost, the body may respond by cramping, especially in the legs and abdomen. The condition may be offset both by stretching and ingesting liquids, which helps to restore the body's optimal levels of sugar and salt. If exposure to heat continues, along with continued participation in the activity, the condition may progress to heat exhaustion and perhaps further to heat stroke. The signs of these two conditions, and what to do, are outlined in Table 7.4.

Table 7.3. Signs of hypothermia.

Mild exposure: The casualty is awake, can answer questions, complains of being cold, shivering, slurs words, lost interest in what s/he is doing
What to do: Stop travelling, prevent any further loss of heat, get casualty to shelter, replace wet clothing, allow shivering to continue (for heat), give food and/or sweet hot drinks

Moderate exposure: The casualty is confused and illogical, doesn't want to move much, is clumsy, stops shivering, shows signs of muscle stiffness, slow breath and pulse rates, has a fruity odour to breath, dilated pupils
What to do: Treat as for mild, except avoid rough handling and do not let casualty walk, do not give fluids unless wide awake and understands what is going on

Severe exposure: The casualty is barely conscious or is unconscious; s/he has slow, shallow breathing and a weak, slow, irregular or absent pulse; has a very pale, cold, bluish skin
What to do: Handle the casualty very gently, prevent further heat loss, move gently to medical care if possible. If not possible, move to warm shelter; apply heavily wrapped warm water bottles to chest, neck, groin; keep warm and let recover gently

Source: Merry (1994).

Table 7.4. Heat-related conditions.

Heat exhaustion: Rapid heart rate, dizziness or faintness, nausea, pale or red skin, headache, exhaustion
What to do: Place casualty at rest, give water or salty fluids as soon as can be taken, loosen tight clothing, cool the body by any means (ice)

Heat stroke: Marked by confusion, poor coordination, delirium, or unconsciousness; hot skin; no sweating, irritability, rapid pulse and breathing
What to do: Generally the same as above, but much more of a need for this, including immersing the casualty in water. *The longer the delay, the more danger of brain damage or death*

Sources: Merry (1994); The Canadian Red Cross Society (1994).

Clothing and equipment

In efforts to avoid hypothermia and heat-related disorders, the responsible ecotour service provider should ensure that their clients are outfitted in appropriate clothing. This means passing on appropriate information before the trip either by phone or through information packets to ensure that they bring along clothing that will keep them warm, dry, or otherwise safe. In cases where clothing gets ruined or lost, operators should make a practice of bringing along spare clothes such as jackets, gloves or hats. This is particularly important in regions of extreme cold and heat. In rare cases, service providers may give a water- and wind-resistant jacket to participants as part of the trip fee. This is the case with Quest Nature Tours, who provides a waterproof jacket to each and every participant who signs up for their Antarctic expedition. In doing this Quest knows that at least all participants will have the right jacket for the conditions of the far south.

In all cases, service providers should provide detailed information about the region in which the tour is to take place, especially for purposes of packing (i.e. which clothes to take). Recommendations on the weight of clothes, the number of clothes, style of footwear and hats, toiletry items, as well as other miscellaneous items will be greatly appreciated. Jeans, for example, are not recommended in the heat nor do they do well when they get wet. The ecotourist is better off with an assortment of nylon shirts and trousers, comfortable fitting shoes or walking boots for the field, a wide-brimmed hat, and a knapsack to carry jackets, waterproof trousers, or other items during the day. Long trousers and shirts are especially important for keeping the sun off the skin, also long trousers can be tucked into socks to keep small biting and sucking insects, ticks and mites at bay. Appendix 3 contains a list of items to consider for an ecotour (subject to change based on the length of the trip, the activity, and the setting).

There is a saying among outdoor leaders that 'there is no such thing as poor conditions, just poor clothes'. The concept of layering has evolved to enable the ecotourist to feel comfortable in almost any environmental condition. For those colder climate regions, for example, or those with changing conditions, three layers have been recommended. The first, or underwear layer, includes close-fitting garments which have the ability to transport moisture away from the skin, while at the same time being breathable. Synthetic fibres such as polypropylene have been widely used, however there are many fibres available, which generally accomplish the same end.

The second, or clothing layer, usually includes fibres which are: (i) quite durable, (ii) they move with you, (iii) they transport perspiration away from the skin, and (iv) dry quickly. Nylon is a good choice over cotton because of its durability but also because it dries quickly and does not shrink. Many of today's high performance nylon-based garments are even more useful because of more rugged materials, the use of a variety of coatings and seam-sealing options for different weather conditions, and through thoughtful design. Fleece, which is highly versatile, light, and warm (even while wet) is a good mid-layer choice because it provides the insulation needed, but is also involved in the moisture management process begun by synthetic underwear. Fleece picks up moisture from the surface of the underwear and allows it to move further outward, where it can disperse either directly from the fleece itself or through to another layer. It also comes in a number of different weights depending on the type of activity, as well as the setting.

The outer or third layer garment should insulate while at the same time be durable and reliable. By weight, goose down is the best option because it provides a significant amount of insulation per unit area. However, although heavier, and bulkier, synthetic fibres absorb less water, dry much more quickly, and maintain their insulating value when wet. For comfort in severe rain conditions, there is little that can compare with Gore-Tex® fibre, which is waterproof, light and breathable. Because of the demands of the various environments in which they recreate, ecotourists need to be prepared for all conditions. Layering provides the traveller with the opportunity to add and subtract clothes as needed.

Just as the right clothes are important for the variable conditions of the field, the right equipment is also essential for conducting safe and fun trips. Briefly, equipment must be in good working order, it must be appropriate for the type of outdoor pursuit, and there must be enough of it to ensure participation at a comfortable level. It is therefore a wise practice to continually monitor equipment, and to replace it before it wears out.

Medical supplies

Not surprisingly, one of the most important items to bring along on all outings, even those which are short in length and time, is the first-aid kit. Knowing what is inside the kit and how to use each item is critical in any emergency situation. While many guides bring larger kits, the guide may also function properly with smaller kits, but which are stocked with what are viewed as the most appropriate items for the nature of the trip. It also helps if the kit is sturdy and waterproof. Guides should have a list of all of the items in the first-aid kit, and these items should be organized and easy to find. Furthermore, the kit may also hold medications for group members, but only as a backup to the medications which should be carried by ecotourists themselves. First-aid kits will range in size depending on the length of the trip, the number of people involved, the setting and climate in which it takes place, and the distance from a medical facility. Appendix 4 includes some suggestions for a basic extended trip for four people (Murdoch and Trygstad, 1993). To this list, one should add a wilderness first-aid book or guide which would help leaders and participants with procedures to follow for many of the possible problems that may occur along the way. The kit may also include a cold pack, tweezers, feminine hygiene items, trauma scissors, latex gloves (take care that some may be allergic to latex), antacids, alcohol swabs and sun screen (see Skaros, 1998, for suggestions on rounding out a group first-aid kit). The first-aid kit should also contain forms which will enable the guide to record some valuable information in the event of an emergency. In general, this may include an injury checklist an illness checklist and a consciousness record (see Appendices 5a, 5b and 5c). This information may be vital to a health care worker who may need to work quickly to evaluate and care for an individual.

Guides who are travelling in very remote regions should consider bringing along a survival kit. This kit consists of a number of items which will enable the group to sustain itself under very adverse conditions. Survival kits should be small, and packed in a lightweight, watertight metal container. The kit should also be accessible at all times, as should the first-aid kit, which means locating it on an outside part of a pack, especially when the pack may need to be discarded because of bears, water, or other environmental hazards. Some of the items to include in a survival kit include firelighting items (such as fire sticks), various food items such as salt and dextrose, fishing line and hooks, insect protection, and a space blanket for warmth. A more comprehensive list of items is located in Appendix 6.

Programming Tip 7.3.

Keeping a journal

Participants should be encouraged to keep a journal of the events of the trip as they unfold. Ecotrips are special occasions which generate wonderful experiences and which should be cherished for future reference. The journal is an important record of events in their purist form. Entries need not be lengthy, but may include bullet points of important facts, occurrences, or perceptions. Entries may deal with the senses (smells, sightings, tastes of unique foods, textures), aspects of space and place, temporal events, problems encountered, successes, goals, people, distances, things learned, and so on. The journal can be carried during the day for immediate entries, or filled out at night. If the former, ensure that it is kept dry in sealable plastic bags. Some may prefer to use a microrecorder on trips, but this will entail transcribing the tapes at a later date. There is also the problem of batteries or other malfunctions that may occur along the way.

Permits

Those who organize and conduct ecotours in natural areas will need to be aware of any permits that may have to be purchased or attained in order to access an area. In general, one of the principal reasons for a management agency to develop a permit system for their park or protected area is to limit use. This was the case in Algonquin Provincial Park in Ontario, Canada, where visitor use in many of the most accessible lakes had to be controlled, especially at peak periods during the summer months. However, there are a number of other very good reasons for developing a permit system. These include providing valuable information on the user, such as group size, method of travel, entry and exit dates and points of entry/exit (Hammitt and Cole, 1987). Otherwise, the permit system allows the management agency to come into contact with the user, with the intent of providing opportunities for education and the mutual exchange of important information on a day-to-day basis (Hendee *et al.*, 1990). Despite the cost of adopting a permit system within a park or park system, its advantages are quickly realized through the ability to document the itineraries of each and every group, and thus provide an element of safety to wilderness travel. Licences too may need to be purchased by the ecotourism service provider. These are contingent upon where the operator wants to go, the method of travel, and the types of activities that form the basis of the ecotour. Table 7.5 is an example of the possible ecotourism permits and licences that may be required in Australia.

Table 7.5. Possible ecotourism permits and licences.

Permits and licences	Cost and fees
Transit and entry permits (aboriginal reserves)	No cost
Permit to travel along state barrier fence	No cost
Commercial tourist activity licence	A$50 application fee; A$250 licence charge
Permit to conduct tours in national parks	Variable depending on statutory body
Permit to conduct commercial activities in national parks	Variable according to activity and location
Omnibus licence and permit/temporary licence	A$4.25 per passenger seat per annum; permit A$10 per journey
Approval for temporary caravan park	No fee
Business name registration	A$90 for the first 3-year period; renewal $75

Adapted from Department of Conservation (2000).

Environmental Conditions

Wind and weather

Having had the opportunity to lead trips in many of Canada's more northerly realms, the author has a healthy respect for the elements and the affect they have on the success of such journeys. In fact, the number one dimension that dampens spirits on trip is poor weather. Students often take it for granted that the summer months are warm, and that even if it does rain, it will not last long. Despite my warnings to bring along appropriate rain gear (which does not have to be expensive), there are many who do not listen. The times when it has snowed in June or July have been difficult ones. So too have been the times when it has rained for 3 or 4 days solidly, or when the wind has been so severe that canoeing 3 or 4 km has taken the better part of an afternoon (often times it is just sound judgement to wait it out). Some things are controllable in the field, such as shelters and appropriate gear, others like wind and rain are not, and the latter are tolerable only if one has adequately prepared (see discussion above on clothes). The adage 'Pack for the worst, hope for the best', is one that ecotourism service providers must take seriously any time they are taking clients out into the field. Scouting daily and long-range forecasts should be part of every ecotour operator's daily regime. This can be done via the Internet, television, newspaper, Coastguard, or weather band radio. Even so, remember that weather systems are inherently chaotic.

Failing the ability to gauge the weather through one of the aforementioned means, an understanding of cloud patterns has been helpful to wilderness travellers. Ecotour guides who will be in the field for 2 or more days will be warned of an approaching warm front (the lead edge of an air mass is warmer than the one it is overtaking) by a succession of different cloud formations that decrease in height over time. The warm front is characterized by very high wispy cirrus clouds which precede the surface position of the front by as much as 800–1000 km (about 2 days). These clouds are than replaced first by altostratus (dense veils or sheets of grey) and then nimbostratus (low, dark grey rain clouds) with associated precipitation. Curtis (1998, pp. 176–77) provides a very useful list of natural indicators which may be used to gauge weather.

- *'Red sky at morning, sailors take warning; red sky a night, sailor's delight'.* In North America, red in the sky in the morning (west), means moisture is beginning to move over you. If red is seen in the evening (east), it means that moisture has moved past you.
- *'When leaves show their underside, be very sure that rain betide'.* A period of damp air softens leaf stalks and bends them more easily in the wind as a storm approaches, especially true for aspen and maple leaves.
- *'Rain before the wind, topsail halyards you must mind. Wind before the rain, soon will make plain sail again'.* Rain falling ahead of a strengthening wind indicates a gale-force storm approaching. Wind ahead of a rain will die out as the rain arrives.
- *'No dew at night, rain by morning. No dew at morning, rain by the next day'.* Dew is the sign of a fair tomorrow. On a clear night, without cloud particles, moisture is released from the air and collects as dew.
- *Animal sounds.* Crows, woodpeckers and blue jays are typically very noisy before a storm, whereas insects stop making noise.
- *Rhododendron act as thermometers.* At 16°C (60°F), the leaves are spread out. At 4°C (40°F), the leaves are drooped. At −1°C (30°F), the leaves are curled. At −7°C (20°F), the leaves are tightly curled.
- *Crickets.* This insect can tell you the temperature. Count the number of chirps per minute. Subtract 40, then divide by 4. Add 50 for the approximate temperature in degrees Fahrenheit.

A final weather note concerns lightning. There is a rule of thumb that every 5-second interval between a lightning strike and thunder means 1 mile of distance (see San Diego Chapter of the Sierra Club, 1993). Experts on the subject warn that, while in the field, travellers should avoid exposed areas such as ridges and taller trees which may receive direct hits. Furthermore, ditches and caves which often hold water are also not good places to take shelter. All metal objects should be discarded (glasses, canteens, and so on) and the traveller should crouch as low as possible to the ground, but not directly on the ground. Instead, he or she should sit on boulders or his or her own pack to avoid ground currents. Also, the group should move apart, approximately 8 m (25 feet). While sitting directly under a large tree is unwise, crouching close to the tree (half the distance of its height) is wise. This moves the traveller far enough away from the tree to dissipate ground currents, but close enough to the tree to avoid being a target.

Plant and animal considerations

The range of health risks that travellers are exposed to is extensive considering the number of plants and animals that have the ability to inflict pain or discomfort. Concern usually exists over the largest and most toxic animals which inhabit a particular region (e.g. crocodiles, spiders and snakes in Australia; bears in Canada; snakes in India). In reality, it is the microscopic organisms, as we have seen in the section on water, which usually cause most of the problems. The effects of plants and animals may be indirect by, for example, dissuading people from travelling to a region, or direct, such as in the case of a first-hand encounter with a plant or animal.

The dinoflagellates (a form of unicellular marine and freshwater plankton), the cause of red tide, is an example of how a microorganism can have direct and indirect effects on tourists but also the tourism industry as a whole. The infrequent outbreaks of this organism, which colour the water red or brown, have been known to kill hundreds of thousands of fish, and infect shellfish with a toxin. Humans who consume shellfish are at risk of becoming sick from the toxin, as suggested by Raven *et al.* (1986) who reports that shellfish consumption caused 26 people to become sick in Cape Cod (direct effect), crippling the shellfish industry for years after (indirect effect).

Plants and trees throughout tropical and temperate regions of the planet are known to produce toxins in various concentrations. While some people have mild reactions to these toxins (e.g. rashes), other people have more severe reactions such as anaphylaxis, which is detectable by signs that include difficulty breathing, nausea, swelling of the tongue, hives, and later no pulse and loss of consciousness. Poison ivy, poison oak and poison sumac are three examples of plants containing contact poisons, which provide a great deal of discomfort to travellers in North America. The poison ivy (*Toxicodendron radicans*), for example, includes vine-like and groundcover varieties, making its detection more difficult. The plant may be identified by its three-part variable leaves, green flowers and white berries. Poison ivy contains a substance that causes a blistering, itchy rash that is very red and often raised. Because the sap stays active for months, people who have come into contact with it unknowingly spread it by leaving an invisible 'trail' of toxicodendrol where they sit, touch, sleep or bathe (www.quickcare.org/skin/poison.html). Oak-n-Ivy Brand Techno® is reported to be quite effective in removing plant oils from the skin, and should thus be included in a kit for those areas where these plants are indigenous. An awareness of plants such as poison ivy is obviously of great importance to the ecotourism operator who must be responsible for the safety and comfort of clients.

In just about every region of the world one can find plants that are poisonous if ingested. In more temperate zones, examples include the yew family (*Taxaceae*), foxgloves (*Digitalis*), monkshood (*Aconitum*), hemlock (*Conium maculatum*), deadly nightshade (*Atropa belladonna*), baneberries (*Actaea*) and death camas (*Zigadenus venosus*). This latter plant is often confused with wild onions, which makes its proper identification that much more critical. The same can be said for the berry-bearing plants, such as baneberries, which are frequently mistaken for other similar looking plants (see Wiseman, 1986 for a description of edible and non-edible plants, or many of the existing web sites under poisonous or toxic plants for descriptions of plants in various world regions).

As regards the Kingdom Animalia, ecotourists must be wary of a huge number of beasts. Some of the smallest offenders, but most troublesome, include the ticks and mites which comprise the Order Acarina, of the Class Arachnida. This group of chelicerates (two segments, no antennae, with the first pair of appendages – chelicerae – as feeding structures) is said to be the most important of the arachnids from the standpoint of human economics. Barnes (1980) suggests that numerous species (there are about 25,000 species of mites alone) are parasitic on man, domesticated animals and foods. In general, mites and ticks feed by tearing or piercing tissue and ingesting the fluid of the host. One of the most familiar parasites to ecotourists of this Order is the chigger mite found throughout many tropical regions. The larva of the species *Trombicula* emerges from an egg, which has been deposited in the soil, to attack almost any terrestrial vertebrate. The larva feeds off the epidermal tissue which is broken down by enzymes. Feeding, according to Barnes (1980), takes place for up to 10 days, after which time the larva simply drops off and undergoes further development into an adult. The resultant itch and red irritation remains after the mite has been removed as a result of the various secretions of the larva. The common chigger, found for example in Africa, is a flea, *Tunga penetrans*. The gravid female embeds itself under the skin of the foot, causing a painful lump or ulcer.

It is these secretions that are most problematic to humans. Mites and ticks are carriers of pathogens which are the cause of many human diseases such as Lyme disease, Rocky Mountain spotted fever and tularaemia. Lyme disease, which is especially well known in North America, is a case in point, where the secretions from the deer tick, *Ixodes dammini*, infect human hosts with the bacterium *Borrelia burgdorferi* through its feeding behaviour. The deer tick has highly specialized hooked mouth parts that enable the invertebrate to maintain a solid hold on the host. To aid in its feeding, the tick will also flatten itself on the host to avoid being brushed off. Early symptoms of the disease include a skin rash at the point of contact, and later a doughnut-shaped rash. This is accompanied by flu-like symptoms, swollen glands, stiff neck, backache, nausea and vomiting. Later stages of the illness may include problems with joints, inflammation of the heart, neurological problems (temporary paralysis of face muscles), with many of these symptoms being highly variable between patients. Because it is a bacterium, Lyme disease may be treated with a regimen of antibiotics, typically over a 10-day period of time.

The first line of defence against mites and ticks is to wear lightweight, long trousers and shirts. Trousers should be tucked into socks, and the socks should be lighter in colour which may allow the traveller to detect these invertebrates. Some travellers may consider the use of repellants, including those that have concentrations of DEET (*N*,*N*-diethyl-*m*-toluamide). These products are very effective, especially those which are 95% DEET. However, with such protection comes certain risk. Kodis (1996) writes that DEET is an insecticide and must be treated with caution. Suggestions are made to use lower concentrations (many come in concentrations of 15% or 17%), spray it on clothes instead of skin, and place it on cuffs and other places where clothes are adjacent to exposed skin. There are also many other less toxic alternatives, such as citronella, which are not as invasive but are less effective than products containing higher concentrations of DEET.

The invertebrate which appears to have the most profound effect on people is the mosquito, particularly the female from the genus *Anopheles*, which infects more than 200,000,000 people annually with one of four protozoans from the genus *Plasmodium* carried in the gut of the mosquito (Bezruchka, 1988). One of these in particular, *Plasmodium falciparum*, is deadly. Consequently, it is not the mosquito itself that causes malaria, but rather the protozoan harboured within the gut of the animal introduced into humans through the mosquito's sting that does. The female mosquito requires blood for the production of eggs, and she may lay up to 200 eggs at a time, in standing water, more than once per night.

Fig. 7.5. Ticks.

Over the course of her life (anywhere from 3 to 6 months) she may lay upwards of 3000 eggs (IAMAT, 1995). Unfortunately, many less developed countries are without the resources to educate and cure the innumerable cases that arise each year. Travellers from more developed countries visiting the less developed countries, are fortunate to have health-care systems which recognize the global extent of malaria, as well as the geographical variability of the disease (different drugs are prescribed for different strains in different areas, as some protozoans have become immune to certain anti-malarial drugs). In a study of 376 travellers departing out of Gatwick Airport, Coole *et al.* (1989) reported that 70% had sought advice on malaria prophylaxis. The study also found that only 63% of these travellers visiting malaria zones were taking prescribed drugs, while 22% travelling to malaria-free regions were taking drugs unnecessarily. As inferred above, malaria is both preventable and treatable through prescribed drugs. Prevention, as in the case with ticks and mites, is through the use of long trousers and shirts and repellants. In those areas that are reported to be high risk, ecotourism operators may insist on the inclusion of mosquito netting for the beds of their clients, as it is between dusk and dawn that the *Anopheles* mosquito is most active.

Outdoor enthusiasts need to have a good deal of respect for some of the larger dangerous animals such as bears, tigers, cougars and water buffalo. While such animals do not generally stalk people, they may do so if they are chronically hungry, mentally unwell (as in the case of rabies), attracted by a particular scent (e.g. menstruating women), or if the individual has taken the animal by surprise. It is important to realize that healthy animals such as bears or the big cats will almost always flee an encounter with humans. Notwithstanding, it is always safest to view these animals as unpredictable and the responsible ecotourism service provider should take the proper precautions when or if an encounter occurs. Bears, for example, follow established trails, so camping close to trails is not advisable. Ecotourists should keep food out of sleeping bags and tents, and practise 'clean' camping skills while in the field.

Research on bear attacks illustrates that black bears can be easily frightened and intimidated, and are less likely to attack than their larger and more powerful counterparts (such as the grizzly bear). If need be, the ecotourist must make loud noises, adopt an aggressive stance, throw rocks, show no fear, and fight back if the bear does attack. By all means, however, avoid running away as this will trigger the instinctive need to pursue. Degrees of success have been achieved through the use of devices which are designed to discourage bears from pursuit. These include bear bangers or fog horns which discharge at high decibels, and pepper sprays which are effective up to about 3 metres (Morton, 1997). The problem with bears is that they often become habituated (learning how not to respond to a stimulus), especially in parks where bear feeding was common. Habituation has also been noted in zoos as well as in African game parks, where lions and other large carnivores show little concern towards ecotourism vehicles. Again, the issue appears to be the unpredictability of the animals and the safety of ecotourists. Other cases include deer who have gored tourists, monkeys that bite, as well as koalas that are less than hospitable. Respect must be shown for the space that these animals require. Responsible organizations will develop codes of conduct for tourists and tour operators, and dutiful individuals will adhere to these codes.

Sun

In the last decade, the depletion of the ozone layer, and the factors which contribute to ozone depletion, have generated a great deal of concern among academics and governments. The ozone layer is a band of particles about 10 km thick, reaching a distance of about 20 km up in the earth's atmosphere. This gas is essential as it protects plants and animals from the harmful effects of UV radiation. Scientists in the 1970s first discovered that chlorofluorocarbons, used in aerosol cans, refrigerators, air conditioners, and so on, played a part in the destruction of the ozone layer (massive holes were detected in the ozone layer over the Antarctic, Australia and Chile). It was further discovered that ozone depletion was linked to global warming through increased concentrations of CO_2 and other gases, along with the disruption of planktonic photosynthesis, which is the base of oceanic food chains (Goudie, 1990a; Simmons, 1990). The reduced protection from UV radiation that has come with ozone depletion has meant that tourists and outdoor recreationists have had to be more conscientious of prolonged sun exposure, especially in regions that fall under ozone holes. Studies have indicated, however, that even though tourists recognize the link between sun exposure and skin cancer, many (23.3%) still spend a significant time in the sun without the use of sun screen protection (Ross and Sanchez, 1990). Ecotourism service providers must be aware of the effects of prolonged exposure to the sun (as should tourists). Although guides need to respect people's decisions to sun tan, it may be helpful to set a good example by using sun screen often and by bringing along extra for those who want it.

The Trip Information Sheet

In preparing for the field, the service provider is well advised to use a trip information sheet for the purpose of recording a number of critical details related to the on-site aspects of the trip. Information that will need to be considered may include where the group is going, for how long, at what elevation, the distance expected to be travelled, the people to be contacted for safety reasons, and so on. It may also be a requirement for service providers to notify the management agency responsible for the area (e.g. Ministry of Natural Resources) about the trip, including information on the number of participants, the itinerary, level of experience and the number of days. This becomes very important, especially in the case of remote trips, as part of the risk-management plan. If the park authority is contacted in the event of an injury, they will immediately have a sense of where the party is, the number of participants, and other critical pieces of information. It is also critical that a group de-register after coming out of the wilderness. This lets the management agency know that the group is out and safe. Failing to do so places the agency in a very difficult position. Some agencies in some settings have regulations that if they have not heard from a group 24 hours after the time they are supposed to de-register (e.g. the 8th day of a 7 day trip), they will send out a search-and-rescue team. In general, the information contained within Table 7.6 should be included on the trip planning sheet. One copy should be carried with the group at all times, another copy should be left with the management agency, and one more should be left with an administrator at home.

Conclusion

This chapter has sought to illustrate that trip planning involves a tremendous number of considerations. Ecotours often take place in very remote settings and for long periods of time. For this reason the responsible ecotourism service provider will very thoroughly examine the ways in which to brief clients, consider mental and physical preparation, take adequate health precautions, insist on first-aid and cardiopulmonary resuscitation training for trip leaders, pass on information on appropriate clothes, obtain the appropriate permits, make an effort to understand environmental conditions, and draw up an effective trip information sheet. It should be noted that many of the aforementioned details will no doubt be omitted in the programming process because of a lack of knowledge, resources or motivation. In an era of accountability and greater demand for professionalism the incorporation of these important components will hopefully enable the service provider to develop effective programmes that help ecotourists to enjoy safe, rewarding experiences.

Table 7.6. The trip information sheet.

Area visited
Dates
Personal contacts at home and at the site
Road and trail maps needed
Guidebooks needed or consulted
Management agency at a ranger station or other park services station
Permit requirements
Maximum group size
Resource information such as fire usage or water availability
Long-range weather forecasts
A breakdown of each day on the basis of elevation, kilometres travelled, campsites to be used and their coordinates
An understanding of sunset and sunrise times
etc.

8

Programme Design C: Leadership and Risk

> The boss drives his men; the leader coaches them.
> The boss depends upon authority; the leader on goodwill.
> The boss inspires fear; the leader inspires enthusiasm.
> The boss says, 'I'; the leader, 'We'.
> The boss assigns the tasks; the leader sets the pace.
> The boss says, 'Get here on time'; the leader gets there ahead of time.
> The boss fixes blame for the breakdown; the leader fixes the breakdown.
> The boss knows how it is done; the leader shows how.
> The boss makes work a drudgery; the leader makes it a game.
> The boss says, 'Go'; the leader says, 'Let's go'.
> The world needs leaders; but nobody wants a boss.
>
> (D. Dodge)

Introduction

The discussion in this chapter follows on from that in the previous chapter, but with a focus on the importance of leadership and risk management as essential elements in the success of the programme. The importance is sufficient enough that they warrant a chapter to themselves. While tour organizers often become concerned with the logistical aspects of the experience exclusively, it should be remembered that although logistical issues are important, it is often the efforts of the leader that makes or breaks ecotours. Basic aspects of leadership are discussed in addition to the concept of followership. Just as it is important to have a good leader, good followers are also essential to the success of a trip. Leadership theory is included as a means by which to underscore the important relationship that exists between the leader, the group and the various situations that occur on trip. The chapter also contains a discussion on professional development, with an overview of a number of skills and guidelines which should be required in ensuring that ecotours are conducted in the most professional manner possible. Considerable attention is paid to safety and risk management. These were not treated seriously in the tourism industry until rather recently, but the nature of the ecotour experience compels the operator to understand better the inherent risks in this form of tourism.

Leadership

Leadership has been likened to beauty in the sense that it is hard to define, but you know it when you see it. In general, it has been described as the process whereby one person influences other members toward a goal (Yukl, 1989). People, companies and nations can be transformed through the efforts of one individual who has the insight and vision to see clearly the way to the future. Accordingly, we can never underestimate its power. At one time it was thought that leaders emerged as a product of time and circumstance. Individuals were induced to greatness in response to an event which consequently acted as a springboard for their actions. This 'Great Man' Theory is representative of, among others, Winston Churchill, Martin Luther King and Gandhi. An adaptation of this view of leadership posited that leaders were born with inherent characteristics and traits that enabled them to emerge as great leaders.

A significant amount of effort has been directed towards an understanding of the types of traits which are thought to characterize effective leaders. Those commonly associated with good leadership include intelligence, strength, conviction, confidence, maturity, initiative and understanding (see Chemers and Ayman, 1993). In a study involving over 2600 managers in the USA, Kouzes and Posner (1987) reported that the most important leadership characteristic was 'honesty' (83% of managers selecting) followed by 'competent' (67%), 'forward-looking' (62%), 'inspiring' (58%), and 'intelligent' (43%). Bennis and Nanus (1985) write that from an organizational perspective, successful leaders often share some common skills, including: (i) creating and conveying a vision; (ii) providing meaning through communication; (iii) empowerment; and (iv) self-understanding. This last skill involves the ability to understand his or her own strengths and weaknesses and use the human resources available to compensate and spillover into areas that the leader may not be able to satisfy (see also van der Wagen and Davies, 1998). Good leaders are said to be eager to receive feedback, and continually take stock of themselves. Brandwein (reported in Edginton and Edginton, 1994) writes that leaders must consider the following eight great 'ates' as essential in being effective at their jobs:

1. *Motivates.* Guides and leaders must maintain a good, positive attitude which will activate participation by ecotourists. One inexperienced leader with a positive attitude is worth far more than an experienced leader with an unhealthy attitude.
2. *Communicates.* To be a good communicator, leaders and guides must be good listeners. They must be in tune with the mood of participants and act accordingly to include all in the day-to-day events of the programme.
3. *Accommodates.* Leaders who are willing to accommodate the needs of others, often show the ability to compromise. Leaders must be team oriented and flexible.
4. *Anticipates.* Problems and situations that may arise. By using a proactive approach a problem may often be averted without the group being aware of it.
5. *Facilitates.* Key to facilitation is the ability to allow the participant to think freely, explore for him/herself, and make some of their own decisions. The leader is thus a catalyst or conduit to the experience itself.
6. *Creates.* To create a caring environment conducive to learning and sharing of information.
7. *Educates.* Leaders must be role models, leading by example, in regards to environmental management, recycling, stewardship, and so on.
8. *Celebrates.* Leaders and guides must demonstrate genuine enthusiasm and excitement about their role. Openly participating in a number of activities that celebrate the complexity and diversity of the natural world.

To this list of 'ates' we could add 'validates' as an important characteristic which leaders must bring to the fore. Participants will often have concern over conditions or situations (e.g. biting insects, water conditions, temperature extremes) which are far removed from the ones they are used to in their daily lives. In such cases, it is important for the leader to validate or affirm these concerns instead of dismissing them as nonsense or improbability. It is only through the recognition of such fears that the participant is able to recognize that his or her perceptions are indeed real, and that there may be others who are also experiencing the same fears.

Followers

No matter what the circumstance, setting, population or conditions, a conscientious leader will go to bed each night while on trip

Programming Tip 8.1.

Guide–participant relations.

The guide should make it his or her business to understand the abilities of the participants. It is also the responsibility of the guide not to leave the group, under normal circumstances, for any reason. Furthermore, guides should be careful about spending too much time with certain members of the tour. It is good practice to empower members of the tour to support each other on longer walks, inclement weather, or other difficult or adverse conditions.

with a great deal on his or her mind. Questions related to group health, group dynamics, participant enjoyment, leadership effectiveness, and so on, are a natural consequence of involvement with a group of participants over a period of time. Good followers make the leader's job easier by requiring a great deal less effort, which will thus allow the leader to get on with his or her job. This means that a clear distinction must be made between who leads and who follows. There is therefore the acceptance of a leader on behalf of the group, and the recognition that each have much different roles. Failing to accept these roles, perhaps by refusing to commit to the decisions made by the leader, disrupts the balance within the ecotour, and seriously questions the integrity of the leader, the company and the trip as a whole. Sessoms and Stevenson (1981) write that followers should never lose sight of their initial reasons for deciding to follow, which may include: (i) efficiency: goal fulfilment is easiest and most efficient by giving responsibility to some else; (ii) satisfaction: some are more satisfied with being a follower – it puts them at ease; and (iii) experience: people simply do not have the experience to be a leader or to venture off into the wilderness without a guide.

Approaches to Leadership

Any comprehensive survey of the literature on leadership will reveal a rich variety of different leadership approaches and theories. Constraints on space preclude the inclusion of an in-depth discussion of these various approaches. Instead, a brief description of three different styles of leadership will be discussed here (autocratic, democratic and laissez-faire), along with one theory which examines the relationship between the leader, group and the various situations that arise from events like ecotourism trips.

Autocratic leadership

The most authoritarian of the three approaches discussed here, is based on the premise that leadership comes from the leader alone. The autocratic approach seldom considers the opinions of the group (group responsibility is low), and puts the leader in the position of single-handedly deciding on the goals and activities of the group. This 'big stick' method of leadership forces participants to become dependent and submissive to the upper hand of the 'boss', by relinquishing their ability to make decisions for themselves. Ecotours are often characterized by tightly controlled itineraries which force the group into a number of prearranged engagements. This is not to say, however, that there may not be pockets of time where the group can make decisions about which trails to take, whether to skip a stop on the itinerary (if this is not problematic to others who are anticipating the groups arrival), or changing a social activity which may be planned at night. Autocratic leadership seems to be more useful in emergency situations when a leader must organize individuals for evacuation, injury, or other similar matters. At times this may require the trip leader to relinquish control of the group to someone else, e.g. a physician, who has the ability to address the emergency adequately.

Democratic leadership

In this shared or participatory form of leadership, the leader works with, for, or in the group without losing control. The democratic leader actively seeks the advice of participants for the purpose of making informed decisions. As such, he or she encourages all members of the group to share their views. Decisions are often made by having the group leader share information or options about a situation, and, as a group, deciding on the most appropriate alternative. There is a high level of responsibility afforded to the group, and thus a great deal of trust and respect may be common between group members and the leader. The democratic style of leadership should not be employed in situations where group members are not knowledgeable or skilled to make such decisions. Allowing participants, for example, to vote on a critical element of the trip (e.g. to continue canoeing in the dark), is subject to many problems such as risk and safety. Although it is probably not the form of leadership which is most prevalent for the essential in-the-field aspects of ecotours, leaders will win points by providing an element of flexibility during certain activities whenever he or she can.

Laissez-faire leadership

The laissez-faire style of leadership is the exact opposite of autocratic, and is characterized by offering the group complete freedom to control and direct decision making. Leaders will at times provide information to the group, but only when asked. It is felt that complete harmony can be achieved through this freedom, but this comes at a cost. Group members may sense a lack of direction and may also feel as though the leader is incompetent and lacking in confidence. Ecotour operators must ensure that guides have the appropriate level of training before engaging a tour. Not doing so opens the door to this type of leadership where decisiveness and content are lacking. This is problematic in ecotourism, given that tourists will very often come to expect a level of knowledge and professionalism that is of the highest level, especially for tours that are expensive. This is the least attractive style for ecotour operators to adopt, given the demands of this type of traveller.

Programming Tip 8.2.

The ice breaker.

When the group first comes together at the beginning of the trip for a briefing or information session, it is a good idea to start by introducing people to one another. Instead of doing this informally, have each individual find someone in the group that they do not know. Instruct them to gather a bit of information on that individual such as: where they live, where they work, family-related matters, recreational interests, and so on. Have each individual give a short synopsis to the entire group.

The comprehensive–interaction–expectation theory of leadership

Jordan (1996) has proposed a style of leadership which is based on two different theories of leadership: (i) the comprehensive view of leadership; and (ii) the interaction–expectation theory of leadership. These she has combined into one dynamic theory, referred to as the comprehensive–interaction–expectation (CIE) theory of leadership. The first part of this overall model, the comprehensive view of leadership, suggests that leadership is a combination of leaders, followers, and the special circumstances of a given situation (see Fig. 8.1). The situation includes the many factors that influence a group, including the environment, time, temperature, the physical setting and the cultural milieu in which the event is taking place. Jordan suggests that it is the interaction of these three components which determines the style of leadership that should be adopted, e.g. autocratic, democratic or laissez-faire.

The second of these models, the interaction–expectation theory, is based on the notion that leadership should be a function of the way in which group members and leaders interact, accept, and reinforce each others' role-oriented behaviours (Jordan, 1996). Leader–group interactions, and the expectations which evolve from these interactions given any situation, range from no interaction between leader and group members to overwhelming interaction where there is no differentiation between leader and follower roles (see Fig. 8.2). Jordan explains these two extremes of the model:

> A lack of interaction between leader and group members with recognition and acceptance of role behaviours (or leadership title) will result in blind following. Group members who follow blindly lack the interpersonal relations necessary for fulfilling recreation experience. This type of experience may also diminish the quality and effectiveness of the leadership process ... The fourth representation of the theory illustrates an overwhelming amount of interaction that allows for no recognition of roles. This might occur when roles are confused, the leader has a weak personality, or the leader and group members become overly involved with one another's roles. Effective leadership is impossible in this situation and action must be taken to define roles more explicitly.
>
> (pp. 51–52)

The optimal relationship exists in example (c) of Fig. 8.2, where there are optimal levels of interaction and recognition of role behaviours, which continually strengthen leader–group effectiveness.

The dynamic and interactive nature of leadership can be examined through a combination of the two aforementioned theories into the CIE model. Jordan writes that the effectiveness of the CIE theory is measured in the recognition, acceptance and reinforcement of leader role behaviours within a particular situation (which reinforces the dynamism of the model). As illustrated in Fig. 8.3, within each situation some or all group members may interact with the leader. Consequently, each interaction (of those who do) within each situation is unique for each individual. The interaction is said to act as a causal agent which induces the selection of a particular leadership style (e.g. autocratic, democratic or laissez-faire) at a particular moment. For ecotourism guides this would suggest that there should be some acknowledgement of the differences that exist between group members and their reactions to different events. The leader who is experienced will have an advantage in determining the right course of action given the

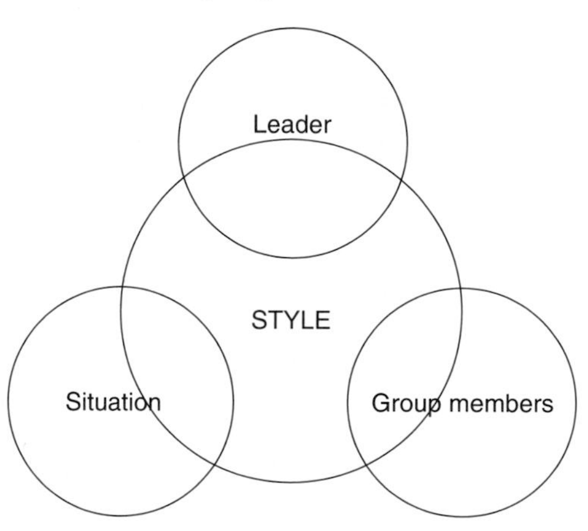

Fig. 8.1. Comprehensive theory of leadership. (Source: Jordan, 1996.)

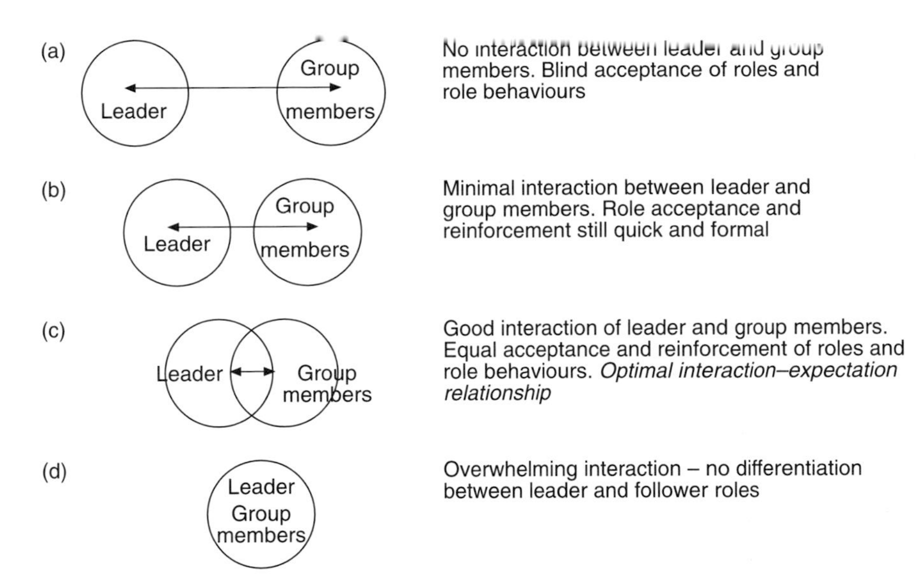

Fig. 8.2. The interaction–expectation theory of leadership and the possible role relationships within the group. (Source: Jordan, 1996.)

situation and the different personalities of the individuals involved in the trip. These decisions on the part of the leader will be important ones in upholding the objectives of the programme and the overall mission of the agency, but also the experiences of the clientele. Acting with the best interests of the participants in mind, through integrity, honesty and responsibility, has immediate benefits, but also long-term benefits in the form of repeat visitation and word of mouth promotion.

The leadership of Robert Greenleaf

One of the most significant paradigmatic swings in leadership philosophy came as a result of the work of Robert Greenleaf who, while at the helm of AT&T and later in life, developed a much different view of leadership from conventional approaches based on policies and economics. His was one that placed high value on vision, beliefs in people, enabling values through a network of research, experience and systems of mentors. To Greenleaf, it was people, not products, that counted, with the ultimate goal of allowing people to become healthier, wiser, freer, more autonomous, and more likely themselves to become good leaders. His concept of the servant-leader is one which characterized the 'leader' as a servant first who, through conscious choice, is inspired to lead (Greenleaf, 1977). This person is different than the conventional leader, Greenleaf explains, who assumes leadership positions for power and material possessions, and who serves as an afterthought – once leadership is secured. Greenleaf writes:

> The natural servant, the person who is *servant first*, is more likely to persevere and refine a particular hypothesis on what serves another's highest priority needs than is the person who is *leader first* and who later serves out of promptings of conscience or in conformity with normative expectations. My hope for the future rests in part on my belief that among the legions of deprived and unsophisticated people are many true servants who will lead, and that most of them can learn to discriminate among those who presume to serve them and identify the true servants whom they will follow.
>
> (p. 14)

More specifically, Greenleaf's philosophy of servant leadership is based on the following ten principles (from Spears, 1995; pp. 4–7), which place employees, customers and community as the number-one priority, encompassing increased service to others, a more holistic approach to the work environment, a sense of community and shared decision-making power:

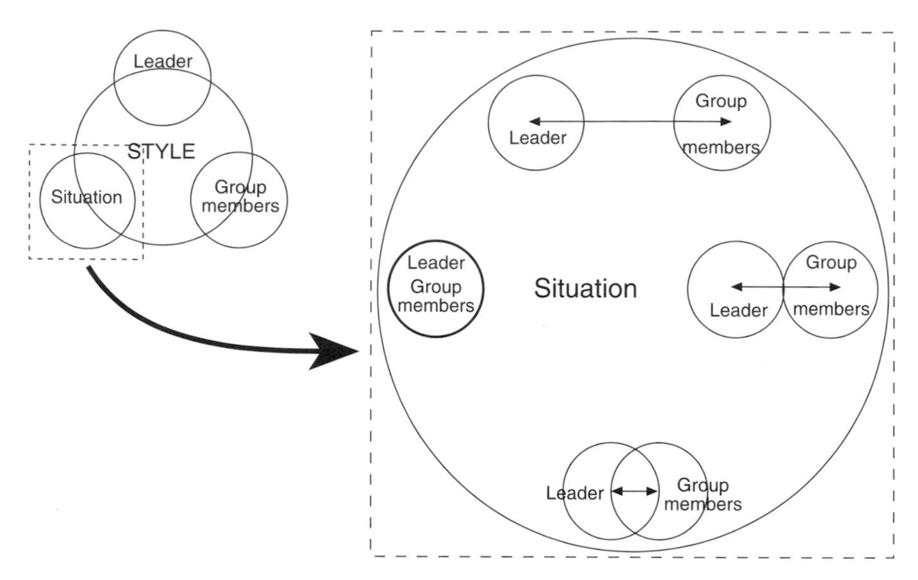

Fig. 8.3. The comprehensive–interaction–expectation model of leadership combines the comprehensive theory with the interaction–expectation theory in examining role relationships within the specific situation. (Source: Jordan, 1996.)

- *Listening.* Servant-leaders make a deep commitment to listening to others and seek to identify the will of the group.
- *Empathy.* The most successful servant-leaders are those who have become skilled empathetic listeners.
- *Healing.* Learning to heal is a powerful force for transformation and integration. One of the great strengths of servant-leadership is the potential for healing one's self and others.
- *Awareness.* General awareness, and especially self-awareness, strengthens the servant-leader. Awareness also aids in understanding issues involving ethics and values.
- *Persuasion.* Servant-leaders seek to convince others, rather than coerce compliance. The servant-leader is effective at building consensus within groups.
- *Conceptualization.* The ability to look at a problem (or an organization) from a conceptualizing perspective means that one must think beyond day-to-day realities.
- *Foresight.* Foresight is a characteristic that enables servant-leaders to understand the lessons from the past, the realities of the present, and the likely consequences of a decision for the future.
- *Stewardship.* Chief executive officers, staff, directors, and trustees all play significant roles in holding their institutions in trust for the greater good of society, which assumes, first and foremost, a commitment to serving the needs of others.
- *Commitment to the growth of people.* People have an intrinsic value beyond their tangible contributions as workers. Servant-leaders are deeply committed to the personal, professional and spiritual growth of all people in the institution.
- *Building community.* Servant-leaders seek to identify a means for building community among those who work within a given institution. The shift from local communities to large institutions as the primary shaper of human lives has changed our perceptions and caused a sense of loss.

In particular, the awareness, foresight, stewardship and community principles outlined above are those which should be especially relevant to the ecotourism service provider, who must try to 'serve' by example. Responsibility rests with the service provider as the tangible link between local communities, governmental policies and ecotourists. However, responsibility for the development of an ethically sound ecotourism industry must not fall on the service provider alone, but rather on those who are charged with the task of serving many different stakeholder groups. This includes government and industry representatives who must be sensitive to the needs of local people and the environment, while at the same time attempting to realize certain economic benefits.

Professional Development

Guiding: skills and guidelines

In citing the CEN Working Group, a European committee for standardization, Collins (2000) writes that a tourist guide is 'the person who guides visitors in the language of their choice and interprets the cultural and natural heritage of an area' (p. 22). Collins further alludes to the fact that guides may possess area-specific qualifications, which provide an element of expertise to the tourist experience. The issue regarding qualifications is an important one to ecotourism, and tourism in general, because it addresses the concept of professionalism in tourism. The guiding qualifications that exist are not standardized, and may be region- or sector-specific. For example in Canada, some provinces and territories such as Ontario and the Northwest Territories will make certifications available for freshwater angling guides and outdoor guides (core skills), while other regions will not (although there may be national certification programmes for these) (Canadian Tourism Human Resource Council, 1996). Furthermore, food and beverage and accommodation sectors appear to be much further along in terms of the requirement for employees to hold certifications than other sectors, such as the ecotourism and recreation sector. In Scotland, a tour-guiding course takes 2 academic years of part-time study, and involves: (i) core knowledge including geology, environment, and so on; (ii) professional guiding skills, such as practical days and assessments; and (iii) detailed area studies for practical training (Collins, 2000).

According to Knudson *et al.* (1995) guiding has been with us for some time, although decidedly focused on national parks, historic homes, archaeological sites and environmental education centres. Examples include the guides who accompanied tourists through many of the rugged regions of the national parks of North America during the late 1880s and throughout the 1900s. This guiding legacy has quite naturally spilled over into the field of ecotourism, where the interpretation of the natural history of a region is often a central part of the experience. However, despite what has been a central aspect of the ecotourist experience, little has been done in the way of research on the role and value of the ecotourism guide (see Weiler and Ham, 2001b), or the skills required of guides.

Guide skills were the focus of a project in Costa Rica which sought to examine the reasons why local people were motivated to sign up for a local naturalist course (Paaby *et al.*, 1991). This project was designed to train rural residents as naturalist guides through a programme of 40 hours of lectures and 103 hours of field walks. The programme, facilitated by the Organization of Tropical Studies, included an introduction to ecological principles, conservation education, plants, birds, mammals, reptiles and amphibians, aquatic ecology, ecology, ecotourism and field guiding techniques, among other aspects. The organizers attribute the success of the programme, and the persistence of the students, to the motivation to earn an income, which appears to have been the main benefit for involvement.

Research by Black *et al.* (2000) suggests that to be effective, the ecotourist guide, especially those in the less developed countries, needs specialized knowledge and skills to ensure visitor satisfaction, assist natural resource management, educate and interpret, as well as act as a cultural mediator between clients and the local community. This last element is essential as ecotourists often grapple with the demands of a language and culture that places both community members and tourists in uncomfortable positions (Gurung *et al.*, 1996). The comprehensive work on ecotourism and guide training put forth by Ham and Weiler, 2000 suggests that there are six main principles which should be considered in the development, design and delivery of guide training programmes, including (pp. 3–8):

- The initiative for training should come from the host country and ownership should remain within the host country.
- Training content and methods should be

informed by the literature on what constitutes good or best practice ecotour guiding and the adult training literature, with appropriate customization to meet local needs and the target trainee group.
- Training efforts must be systematically evaluated, and lessons learned from these evaluations must be documented and disseminated widely.
- Training can and should be made accessible, both logistically (location, time and cost) and intellectually to residents in peripheral and rural areas.
- Delivering cost-effective guide training by mobilizing resources for multiple 'ends' is an important sustainable development strategy.
- Training and supporting in-country trainers is essential for ensuring that ecotourism is of benefit to host economies.

While there appears to be a dearth of research on ecotourism guiding, there is also a lack of programmes which have been designed to train, qualify or register guides in the field (Weiler, 1999). Such has been the case in Australia, by many accounts the world's leader in ecotourism research and development. More recently, however, Australia has responded to this shortfall by establishing the Ecoguide Certification Program, which is an initiative developed by the Ecotourism Association of Australia. This voluntary, industry-led certification programme provides a qualification that is designed to reward guides that achieve various competencies. The programme is not designed to train guides, but rather to assess their skills, knowledge, actions and attitudes (Crabtree and Black, n.d.), with an overall purpose of instilling a philosophy of best practice among practitioners. Table 8.1, provides a comprehensive overview of some of the benefits which may accrue to a number of different tourism stakeholders, as outlined in the EcoGuide Program (pp. 8–9).

Programmes, such as the one identified above, address the need for guides to have many diverse skill sets. These skill sets have been examined in the research on business management (Mintzberg, 1973; Katz, 1974), but also in the context of outdoor recreation and adventure recreation. In general, three broad skills sets have been identified, including: (i) conceptual skills, which include those related to decision making and problem

Table 8.1. EcoGuide programme benefits.

Benefits to nature and ecotour guides
- A recognized industry qualification
- An opportunity to promote guiding services as genuine nature/ecotourism
- A defined competitive edge rewarded through factors such as better job opportunities
- Access to relevant, appropriate and reduced cost training materials/networking options

Benefits to nature and ecotourism operators
- A simple method of recognizing and recruiting quality guides
- Identification of training gaps/needs within operations
- Improved guiding practices leading to fewer negative environmental and cultural impacts and increased client satisfaction

Benefits to nature and ecotourism consumers
- An assurance of guides that are committed to providing quality nature or ecotourism experiences in a safe, culturally sensitive and environmentally sustainable manner

Benefits to protected area managers
- Improved guiding practices that lead to fewer negative environmental impacts
- Guides who role model and ensure good environmental/cultural behaviour
- A framework of standards applicable to interpretive rangers

Benefits to the environment
- Guides providing relevant and appropriate interpretation that inspires clients and workplace employees and encourages minimal impact actions and a conservation ethic

Source: Crabtree and Black (n.d.)

solving, (ii) people skills, such as handling conflicts, listening, praising and communicating; and (iii) technical or hard skills, such as climbing, cooking, driving, and computer application, which enable the individual physically to get the job done. This means that the most effective guides will be those who not only have the technical capabilities to get the job done, but also the human skills to understand group dynamics, especially given the tremendous variety of environmental or social situations that may surface on trip. It is therefore crucial for the leader to achieve certain levels of proficiency, either through experience, or education, or preferably both, in the activities that form the basis of the ecotour. Table 8.2 illustrates a number of hard and soft skills which are useful for outdoor leaders. More recently, researchers have discussed other conceptual skills (Phipps and Swiderski, 1990) or meta skills (Priest and Gass, 1997), in the context of adventure education consistent with work in the field of business (see above). These conceptual skills are thought to be instrumental in allowing for the safe and effective

Table 8.2. Hard and soft skills.

Hard skills	Soft skills
Physiological	Social
Understanding the human body	Maintenance of group dynamics
Assessment of physical strengths	Resolution of conflict
Assessment of physical weaknesses	Group support skills
Administering first aid	Empathy
Treatment of backcountry ailments	Recall of names
Environmental	Allowance for personal growth
Group sanitation	Psychological
Interpretation of weather systems	Create a climate of trust
Promotion of environmental ethics	Stimulation of motivation
Understanding ecological principles	Management of psychological stress
Knowledge of local natural history	Promotion of value/positive attitudes
Prevention of negative impacts	Team building
Safety	Assess mental and emotional strength
Awareness of hazards	Development of ecological ethics
Practice of group security	Communication
Search and rescue competencies	Ability to think on one's feet
Treatment of water for drinking	Ability to speak in front of groups
Implementation of risk management	Interpretation of non-verbal body language
Checking the safety of equipment	Ability to listen and respond
Demonstration of safe driving	Persuasiveness
Technical	Ability to transfer information
Proficiency in outdoor pursuits being led	
Maintenance of equipment	
Teaching of activities being led	
Knowledge of knots (if needed)	
Life saving in aquatic environments	
Other land-based competencies	
Administration	
Establishment of programme goals	
Implementation of risk management	
Understanding legal liability	
Trip planning	
Supervision and evaluation skills	
Compliance with resource regulations	
Well versed in safety procedures	
Proficiency in organizational skills	

Source: Adapted from Phipps and Swiderski (1990).

Fig. 8.4. Leader training is critical in both developed and developing countries.

execution of hard and soft skills, and include skills such as experienced-based judgement (analysis of alternatives, anticipation of the unexpected), creativity (perceptions of trends and generation of new ideas), professional ethics (not coercing people into going beyond their limits), and being flexible with leadership styles (as outlined above).

A number of agencies have focused on the development of competencies or guidelines that leaders should follow in *safely* conducting tours in the field. The most accessible of these appear to be ones which have been developed almost exclusively for adventure recreation (e.g. camping, hiking, skiing, rock climbing, and so on). An example is provided in Appendix 7, which illustrates a checklist for effective leadership in hiking and backpacking for daytrips. This activity may be further broken down into overnight tripping and extended tripping, with the understanding that longer and more extensive tripping demands further qualifications. In general, guideline categories include experience, fitness, navigation, environmental factors, emergency training, and loss of participants and food (Hanna, 1986). The guidelines must be quite specific to ensure that leaders have the abilities to conduct the tours safely and efficiently. They may also specify appropriate leader-to-participant ratios, which will be different for different types of activities, settings and levels of experience. The most important criterion for setting effective guide-to-participant ratios should be safety. Too many people in a group reduces the opportunity for all people to assimilate information equally. And for those more remote settings it is important to have at least two leaders, not only for better attention

Fig. 8.5. During a trip, the leader is on the job 24 hours per day.

to individuals and their needs, but also in case one of the leaders becomes incapacitated (see Hanna, 1986; Ford and Blanchard, 1993). As a rule of thumb, acceptable participant-to-leader ratios appear to be around 12:1 for ecotourism, with any higher ratios taking away from the experience of the participant, and making the job more difficult for the leader. (Reference to skill sets and guidelines for ecotourism or other outdoor pursuits quite naturally leads to a discussion on certification and accreditation, especially these days, as such topics appear to be on the minds of many. These programming elements will be considered more fully in Chapter 10.)

Safety and Risk

Much of the allure of ecotravel is based on the movement of tourists to geographically marginal places which are far removed from the comforts of home (the environmental bubble in Cohen's, 1972, terms). Accompanying this movement is the reality that such regions place us in a different institutional setting as regards health and safety. In venturing off, however, travellers are often not aware of the chances of becoming ill or injured, nor are they told of the facilities, precautions taken, or the standards of care which may or may not be available. In the developed world, people are used to a particular standard of care in the event of an illness or injury. Do travellers have the same health care standards in less developed countries? Are guides trained in first aid and cardiopulmonary resuscitation (CPR)? Does the operator have an emergency action plan developed for his or her trips? If ecotourists are injured through negligence in a foreign country, what legal recourse do they have to protect themselves and their family? These and many other questions are worthy of consideration if the ecotourism field is to value the best interest of participants.

Obviously there is no way to anticipate or control all the events which take place in the natural world in ways that might allow for the avoidance of risk in ecotourism and other forms of outdoor recreation (particularly adventure pursuits). This means that the ecotourism service provider needs to be as diligent as possible in recognizing the dynamic nature of the natural environment, as well as providing the best possible equipment and facilities (Johnston and Bruya, 1995). Added to this is the recognition that programme participants differ along a number of dimensions, including age, physical condition, level of experience, training, motivation, occupation, habits, perception of risk (see Johnston, 1992), values, peer group characteristics, familiarization of the resource and acceptance of authority (Gold, 1994).

The element of control is very important in the implementation of safe programmes. As Clark (1998) writes, aspects such as flawed judgement, the physical and mental abilities of the providers/participants to perform the activity, poor equipment and poor planning all appear to be the most likely causes of accidents in programmes. Clark defines safety as 'a conscious and intelligent attempt to minimize hazards we must face to gain the enjoyment, satisfaction, and physical and mental fulfilment of participating in the outdoor recreation activity' (p. 5). Both participants and providers have certain safety responsibilities to consider before becoming involved in an activity. And contrary to what people may assume, most of this responsibility rests with the participant. In general the participant has four general categories of safety controls to consider in participating in an outdoor activity (Clark, 1998):

1. *Planning.* Envisioning tasks you intend to pursue, their hazards, and the measures to control those hazards.
2. *Training.* The effort required to ready your body for the environment and the activity required to meet the tasks.
3. *Practising.* The effort required to gain skill to perform those tasks effectively.
4. *Adapting.* Envisioning 'what if,' considering backup planning, then using the experience you have gained.

The provider, on the other hand, is generally responsible for: (i) equipment, including clothes, tools, and transportation; and (ii) services, including guides, transportation, food and supervision. As suggested by Morgan (2000a) in reference to adventure tourism in New Zealand, 'Appropriate safety levels obligates risk management at a margin appropriate to both individual and to group competencies. Within this safety margin, the operator will attempt to meet their clients' desires for excitement, arousal and feelings of achievement' (p. 86). This means that the provider will have to balance the level of perceived risk (the estimation of risk within a situation) against the level of actual risk (the true potential for loss), with the understanding that ecotourists and outdoor recreationists have a right to expect to participate in an activity that is free from hazards. Kaiser (1986) has proposed a series of general supervisory obligations which, if followed, would significantly increase the standard of care to participants and staff and thus reduce the potential for liability, including:

- Where an activity or area involves a high risk of serious injury, specific supervision is required.
- Safety and operational rules for areas, facilities and programmes must be developed, posted and enforced.
- Supervisors must remain at posts or with clients, and only leave in emergency situations.
- Participants should be warned of any hidden dangers present in the programme or in the maintenance of the area.
- Facilities or vehicles that present a high risk of injury should be made inaccessible and/or regularly inspected. Defects should be reported immediately and dealt with promptly.
- Provide appropriate, properly designed, and well maintained protective and safety equipment, if required.
- Evaluate participants' physical conditioning and skill level before they undertake high risk recreational activities.
- Warn participants of the unique and particular risks of an activity. Never assume that participants know and understand all of the risks that may result in injury from participation.
- Develop written policies that specify emergency procedures for the provision of appropriate medical assistance for participants.
- Institute, practise and follow these emergency procedures.

If an accident does occur, questions will be raised concerning the various practices of the ecotourism service provider, including the use of equipment and its quality and up-keep; the setting in which the activity has taken place; as well as any other concerns related to leadership, facilities or transportation. For example, by not properly fixing equipment, putting too many people in a vehicle, or exposing participants to dangerous situations or animals, the service provider may be deemed negligent. Negligence is 'the unintentional violation of a duty to take into consideration the interests of others that results in injury … [or] to act as a reasonable and prudent person would have acted under the same circumstances' (Russell, 1982, p. 230). In cases where the service provider has been negligent, victims may consider legal action under tort law. In general, tort is the body of law that is concerned with compensating victims who have been injured, or wronged, as a result of the conduct or misconduct of others (Hanna, 1991). This includes assault, battery, false imprisonment, defamation and negligence. Legal references state that there are basically four components which comprise negligence, all of which must be proven for a party to be compensated. These are illustrated below, as outlined by Prosser (1971, in Hanna, 1991, p. 32):

1. A duty of care owed by the defendant to the plaintiff, requiring that the defendant meet a certain standard of care.
2. A breach of the established standard of care or a failure to conform to it.
3. Actual injury(ies) suffered by the plaintiff.
4. A proximate connection between the defendant's conduct and the plaintiff's injury(ies).

The issue of standard of care is an important one, and links back to the discussion on the various standards that have been established for many outdoor recreation activities for the purpose of protecting participants and providers. Examples include standards established by the American Camping Association (1984), the Association for Experiential Education (1993), and the various ecotourism accreditation programmes which are beginning to surface in different regions of the world.

Programming Tip 8.3.

Refund policy.

Document the refund policy in the participant contract for extreme cases where it is unsafe to undertake a tour or when conditions are such that they prevent you from completing a tour.

Risk Management

At its most fundamental level, risk management is the process developed by an agency to avoid costs. Generally, these costs can be as a result of injury, damage to equipment and facilities, as well as wrongful actions. The main sources of risk (i.e. the things that exist in one's programme which contribute to the types of risk outlined previously) include facilities, equipment, the programme itself and people (Corbett and Findlay, 1998). More succinctly, risk management may be defined as the 'formal process of assessing exposure to risk and taking whatever action is necessary to minimize its impact' (National Association of Independent Schools, cited in Ammon, 1997, p. 174). It is thus a sophisticated phrase for a very basic management tool, and the need for an appropriate risk-management plan cannot be understated for all recreation and tourism providers. Naturally, however, in consideration of all these various forms of tourism and recreation, some such as adventure pursuits, require very detailed and specific plans, for as Liddle (1997) suggests, such activities are based upon the perceived potential for loss, with the occurrence actual loss – which may mean the loss of life – being unacceptable. In general, the risk-management plan will involve a series of steps which will enable the operator to identify risks, classify these, develop risk-control measures, implement these measures, and monitor and modify them (Watson, 1996). The identification and classification of risks may be accomplished by outlining a number of steps to enable the programmer to identify problematic areas for each and every programme, as outlined in Table 8.3.

Potential risks are then subject to an evaluation of either 'low, medium or high' or perhaps some other method such as '1' (low) to '10' (high). For example, under the category of 'Location' the site may be evaluated as low, the weather as low, the routes and accommodation as medium, and the facilities as high. Corbett and Findlay (1998) suggest that the next step is to determine the measures that may be employed to control the significant risks which have been identified on the basis of the following four options.

1. *Retain the risk.* Nothing is done because the likelihood of occurrence is low and/or the consequences are slight.
2. *Reduce the risk.* Practical steps are taken to reduce the likelihood of occurrence and/or the consequences, usually by changing human behaviour or actions. This may include: designing and following a regular maintenance programme for equipment and facilities; helping staff to pursue appropriate professional development; or posting appropriately worded warnings and rules within a facility.
3. *Transfer of risk.* The level of risk is accepted, but passed on to others through contracts. This may include: the insistence that all participants sign a waiver of liability; the purchase of insurance which is appropriate in scope and amount for all activities; or the contracting out to other agencies (e.g. transportation).
4. *Avoid the risk.* Steps are taken to restrict, limit, postpone or cancel certain activities. This may include: postponing events in dangerous conditions; not travelling in bad weather; restriction to participants with certain levels of expertise; insistence on main-

Table 8.3. Outline for risk-management plan.

1. General Description
   - Name of programme
   - Type of activity
   - Level
2. Dates and Times
   - Dates
   - Times
3. Goals and Objectives
   - Organizational
   - Activity
4. Location
   - Site/Area
   - Weather
   - Routes/Accommodation
   - Facilities
5. Transportation
   - Mode
   - Routes/Destinations
6. Participants
   - Number
   - Skill level
   - Characteristics
7. Leaders
   - Number/Roles
   - Qualifications
8. Equipment
   - Type and amount
   - Control
9. Conducting the activity
   - Preactivity preparation
   - Group control
   - Teaching strategy
   - Time management
10. Emergency preparedness
    - Policies
    - Health forms
    - Telephone numbers

Adapted from Ford and Blanchard (1993).

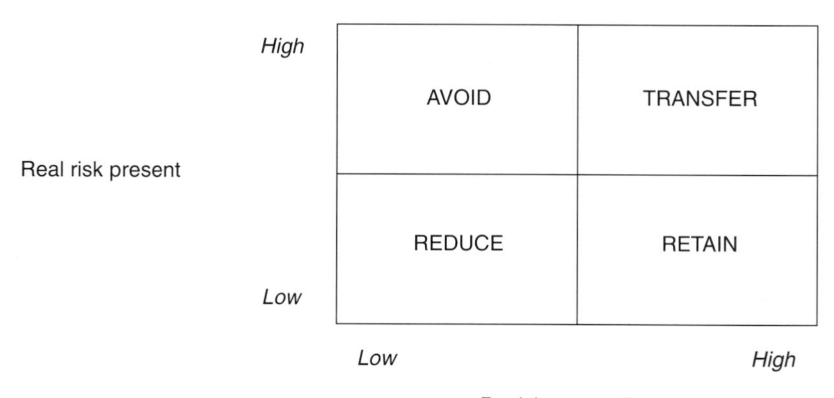

Fig. 8.6. Risk–readiness matrix. (Source: Hanna, 1991.)

Anticipated accident severity	Low frequency	High frequency
High	TRANSFER – INSURANCE – WAIVERS	AVOID
Low	RETAIN	REDUCE

Anticipated accident frequency

Fig. 8.7. Severity–frequency matrix. (Source: Hanna, 1991.)

taining a 'no alcohol' consumption policy for certain activities; or adhering to an organization's or region's policies.

These four options may be graphically charted according to a number of programme-related variables, including the real risk present and participant readiness (Fig. 8.6), as well as the anticipated accident severity and anticipated accident frequency (Fig. 8.7). The illustrations are thus vital in allowing the operator to discern under which conditions to employ one of the four identified options. Inherent in this process is the need to monitor the types of accidents which occur in certain settings or with certain types of equipment. Furthermore, the operator must routinely inspect parks and other settings which may prove to be problematic for ecotourists, and train staff on a regular basis to ensure that they are following proper procedures and employing proper methods in the safe conduct of ecotourists.

The waiver

One of the key aspects of a risk management strategy is the waiver, which is a legal contract between a participant in a programme and the organization offering the programme. More elaborately, Cotten (1997) defines a waiver as 'a contract in which the participant or user of a service agrees to relinquish the right to pursue legal action against the service provider in the event that negligence of the provider results in an injury to the participant' (p. 63). Unfortunately, a vast number of ecotourism providers do not insist that participants sign a waiver. Even for those who use waiver forms there are issues related to the sophistication of these devices, and their ability to hold up in a court of law. Also, it is critical to know that waivers are not deemed to be of equal value across jurisdictions, whether they be state or federal. Courts are thus more willing to uphold a waiver as valid in some states, for example, over others. Morton (1997) writes that good waivers that will stand up in court have the following characteristics (see Appendix 0 for an example of a waiver):

- They are clear that the participant is signing away their right to sue anyone who may be liable (including political bodies).
- The waiver must clarify to the signer the risks to which they will be exposed. Terminology will state: 'risks including but not limited to ... ' because of inconceivable risks which may not be readily apparent.
- The waiver will have the participant initial at several points throughout, e.g. after each paragraph.
- The waiver must be signed and the signature witnessed.
- Participants should know well in advance that they will be required to sign a waiver, instead of arriving and being required to do so.
- It is better if all members of a party sign a waiver together (this is difficult to accomplish in many circumstances), under the same circumstances, after the same briefing.
- A good waiver may even prevent someone from legitimately getting something to which they are entitled (through gross negligence). This makes it important for the tourist to carry his or her own personal insurance coverage that pays up quickly regardless of who is at fault.

In regards to this last point, the ecotourism service provider may insist that their clients have some type of documented medical insurance or extended coverage before being accepted on the trip roster. It will also be helpful if the service provider provides some helpful information about the rules, laws and policies of the country/region in which they are visiting, along with details of the types of activities and levels of fitness or participation required, as suggested in Chapter 7. This should not be viewed as a barrier to participation by operators and clients, but rather as a matter of good business practice, which underscores the best interests of the ecotourist.

The importance of a clearly detailed and implemented risk management plan is emphasized by van der Smissen (1998), who illustrates that it is vitally important to set policy statements and operational procedures to be followed over time. Depending on its size, the agency may wish to appoint a person to act as the risk management officer, so that these policies and procedures – detailed in a risk management manual – may be vetted with some consistency. It follows that a great deal of emphasis should be placed into ensuring that employees receive in-service training so that the policies and procedures established are implemented in a consistent fashion, especially in relation to environmental conditions, equipment, and areas and facilities.

As stated at the outset of the chapter, risk and safety in tourism have only recently started to gain the attention they deserve. The journal *Tourism Economics* developed a special edition on tourism safety and security in one of their year 2000 editions, with the aim of identifying risks associated with travel and tourism; ways in which to reduce such risks; and ways by which to overcome the negative consequences of illnesses and accidents (see for example, Mowby, 2000). Furthermore, the fourth international congress on safety and security in tourism focused specifically on legal aspects of risk management, issues related to liability insurance, risk associated with various segments of the tourism industry, and the effects of perceived risk on destination image. Finally, the University of Queensland has recently developed a Centre for Tourism and Risk Management, which deals with a number if issues, including the development of regional adventure tourism audits.

Conclusion

Leadership is described as the process which enables a person or persons to influence others towards a common goal. Good leaders are those who inspire and create a clearly articulated vision for others to follow, and who are cognisant of a variety of different leadership styles, which are dependent on the setting, situation and group dynamics. Autocratic leadership is more important in situations that demand expertise, while a more democratic style may be employed in situations where group consensus is important. Just as good leaders are essential for successful programmes, so too are good followers who must recognize, accept and reinforce leader role relationships within the group. It was further suggested that good leaders are servants first, by placing the welfare of the group first and by striving to allow people to become healthier, wiser and freer. The chapter followed with a discussion of professional development, and in particular the importance of skill sets, leader guidelines, certification and accreditation. It was suggested that service providers will be able to enhance their offerings through the 'professionalization' of guides and other leaders. While there does not appear to be a substitute for experience, there are a number of individual- and programme-related guidelines that may help the ecotourism industry over time. The chapter finished with a discussion of the importance of safety and risk management, and the realization that a number of elements of the tour setting cannot be controlled. On the other hand the service provider can control the human side of programming through effective planning and management to avoid, for example, flawed judgement, and thus create the basis for safe, enjoyable experiences. This includes a clearly articulated risk-management plan as well as an effective waiver.

9

Programme Implementation

The known is finite, the unknown infinite; intellectually we stand on an islet in the midst of an illimitable ocean of inexplicability. Our business in every generation is to reclaim a little more land, to add something to the extent and solidity of our possessions

(T.H. Huxley)

Introduction

The notion that 'fair' ideas well planned and appropriately implemented are more useful than brilliant ideas that are never acted upon (Mason, 1960), is a good one and is representative of the fact that the link between planning and implementation in programming is important. Consequently, the effort placed into planning will have merit if it leads to programme implementation. In this chapter the focus is on the most prominent features of programme implementation, including the programme life cycle, marketing, quality, staff training, public relations, budgeting, implementation strategies, and schedules and itineraries. The reader should note that, as stated earlier, many programming tasks have been placed within certain sections for purposes of convenience. In fact many of these are ongoing and apply to different stages of programme development, which is the case with some of the topics in this section.

Product Life Cycle

The concept of a product life cycle has been examined in the business, marketing and tourism literature for some time. The basic idea behind the cycle is that the demand for goods and services, over time, may be shown to pass through a number of defined stages. This typically includes a product development stage, followed by an introduction of the product, growth of the product, maturity and finally product decline (see Fig. 9.1). The product development stage – the finding and development of a new product – is marked by investment costs, and zero profits; while the introduction stage is marked by a relatively low level of sales, due to a product's infancy, as well as little or no profits due to the high costs incurred in product development. Growth is characterized by a period of rapid market acceptance and increasing profits. Sales begin to slow down in the maturity stage as a result of a saturation in the marketplace, and profits fall off due to the increase in marketing expenditures which are geared towards defending the product against competition. Finally, decline occurs when sales fall off and profits drop (see also Kotler, 1994).

A classic example of the application of this model to the tourism field is Butler's (1980) tourist area life cycle model, which suggests that tourist destinations experience increases and later decreases in visitation primarily through industry efforts to accommodate more tourists in the face of increased competition. The model has strong implications for the ecological and sociological carrying capacities of these regions, which are

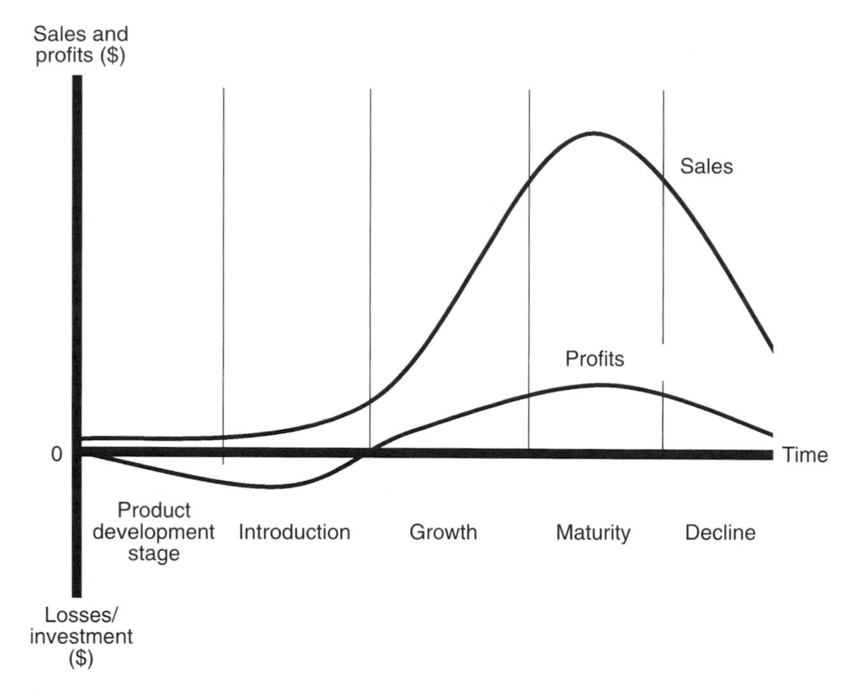

Fig. 9.1. Product life cycles. (Source: Kotler *et al.*, 1996.)

exceeded by the efforts of the industry to sustain economic growth over time. In the latter stages of the tourist area life cycle, multinational investors typically withdraw their financing owing to diminished returns, resulting in serious effects on services and the upkeep of the facility (see also Doxey, 1975).

Some products, like fashions, have relatively short lives. Others, such as tourism destinations, may have longer product lives which may be sustained over many years. The following hypothetical example involving a whale-watching programme provides an example of the long-term viability of a programme. After programme development, the introduction stage is characterized by marketing efforts, including word-of-mouth, which generate increasing support and acceptance of the programme by the market. Growth is characterized by increases in participation in the programme, and an associated need for less marketing. Such growth may take place over 5 or 6 years, for example, or until numbers of participants are maximized, either through external policies restricting growth (e.g. government regulations), or internal factors which limit use (e.g. number of boats kept to a minimum according to the service operator's code of practice). Although profits may be sustained during the maturity stage, there may be a levelling off of sales as a result of waning interest or competition (other operators successfully launching their own whale-watching operations). The programme may rely on repeat visitation (loyalty), but the drop in interest may prompt the programmer to pursue other options in light of the efforts required to stimulate growth. The continued drop in the original programme may mean that the programme will have to be terminated or continued on a limited basis. However, rejuvenation may occur if the operator is able to include other attractions or resources to the programme, such as a shore-line studies programme, seal and sea bird viewing components, and so on.

Up until recently it has been thought that no matter how successful a company is, a decline is inevitable. However, Morrison *et al.* (1999) write that there appears to be an infinite number of ways to offset the decline stage of the product life cycle through initiatives that enable the entrepreneur to reposition and relaunch products, as inferred in the preceding example. One way in which to accomplish this is to decouple the product's life cycle from the life cycle of the aggregate sector, where the good or service is located. While there may be a general decline in the hotel sector on the basis of competition from other lodging sectors, for example, hotels may wish to introduce innovative schemes such as the green hotels concept and thus inject new life into the product.

Marketing

Perhaps as a result of the sustained growth of the ecotourism industry over the past couple of decades (see Canadian Tourism Commission, 1997a; Luhrman, 1997), there appears to be unbridled optimism about the prospects of starting an ecotourism venture and making it successful. However, the author has witnessed first-hand the many problems that small-scale ecotourism service providers encounter (e.g. seasonality, waning demand, defining the market) in their operations on a yearly basis. For the ecotourism industry in the 21st century, the adage 'to be' is 'to survive' no longer holds true as a result of competition, which appears to be getting stronger and stronger, year after year. Statistics in the business literature reveal that approximately seven out of every ten businesses fail in the marketplace, and the same figures seem to apply to ecotourism (S. Edwards, personal communication, 15 May 2000). The other significant force working against the entrepreneur is that time has a way of determining who will be successful and who will not. In just about any industry, it is usually the biggest and most efficient that will survive and prosper. This is not to say that the small ecotour operator is doomed, but rather that there are large dominant players out there, which have an overwhelming share of the market and who are constantly looking for more of it (see Knowles and Grabowski, 1999).

There will always be areas around the world that stand as ecotourism icons (e.g. the Galapagos Islands). For those areas that lack a truly international reputation, however, there is always the possibility that through increased competition, the operator who at one time had to turn away business, may be confronted with challenges to fill seats. At present there are too many exotic destinations that offer good quality programmes at competitive prices. Faced with this prospect, operators must initialize creative, effective strategies in an effort to capture as much of the market as possible. This means taking business away from his fellow competitors, who are equally willing to do the same. This may be accomplished in large part, but not exclusively, through effective marketing which allows the customer to recognize a business; which in turn persuades the potential customer to buy (Davidson, 1985). Marketing is thus a social and managerial process by which individuals and groups obtain what they need and want through creating and exchanging products and value with others (Kotler *et al.*, 1996).

The conventional marketing process is derived from four main aspects which comprise a marketing 'mix'. These include product, price, place (distribution) and promotion. The interrelatedness of these four components represents a systematic process whereby programmes, which satisfy the needs of a target market, are developed, and later delivered in the most efficient way possible. Each of the four aspects of the marketing mix is dealt with briefly, below.

Product

As suggested earlier (see Chapter 1), products are both goods and services. It was noted that tourism products might more appropriately be viewed as services, instead of goods, because tourism is an experiential phenomenon that brings people and places together over defined periods of time. Davis *et al.* (1985) note that the service provider–customer intersection allows the former to control the quality of service better, as well as to adjust service levels as required. While a service-oriented approach to ecotourism is important, it should also be noted that there are a number of different types and levels of service. This is one of the main issues in the ecotourism field, where services range considerably in regards to the levels of education, nature-oriented focus, consumptiveness, impact, ethics and responsibility, benefits to local people, and so on. The product (or programme in the context of this book) is what sets ecotourism apart from other forms of tourism, as well as one ecotourism operator from another.

Price

Establishing the right price for the programme is a decision that causes the most anxiety among managers (Pavesic, 1989). Pricing too high will understandably lead to lower levels of programme interest, while in pricing too low the service provider runs the risk of sacrificing profit. At the heart of the matter rests the need to price according to the ecotourist's ability or willingness to pay. However, there are a range of other factors which must be taken into account when determining price. These include an understanding of who the clients are, the demand for the programme, and an understanding of what competitors are charging (McKenzie and Smeltzer, 1997).

While some destinations (e.g. Antarctica) are simply out of the financial range of some people, more reasonably priced ecotours may become even more reasonable through a system of variable pricing in order to access as many potential buyers as possible. What this means is that a number of programmes may be offered in the same setting and focused on the same activity (e.g. wildlife viewing), but which are differentiated on the basis of time (e.g. 8 day, 5 day or 3 day experiences), conveniences, amenities, involvement of specialists, accommodation and specific guarantees (e.g. the chances of making a siting). In regards to this last point, operators will often give potential clients a percentage chance of viewing certain animals. The author has seen whale-watching operators post the probability of seeing a whale each and every day, based on the movement patterns of a pod. For example, the chances of viewing a humpback whale on Saturday may be estimated at 80%; whereas the percentage may be 70% on Sunday. Although this gives the client more information in making a decision as to whether to join a tour, it does mean that the tourist is subject to the reliability and experience of the operator.

Crompton (1981) writes that price setting may be based on: (i) actual costs; or (ii) not based on cost. In the former case, costs may include the total cost of all fixed elements in the programme/service (e.g. facilities, overhead charges, leadership) and the variable costs (e.g. food), which increase with the number of participants. Variations may include basing the charge on variable costs only (divided by the number of participants) or on variable costs with some portion of overhead costs added. Charges not based on costs are typically based on the going rate, where the price reflects the average charges set by other, similar organizations, for similar services. They may also be based solely on consumer demand, in other words, 'what the traffic will bear'. In ecotourism it may be difficult to establish charges not based on costs, because of the nature of diverse programme offerings which exist in an environment (i.e. operators not offering the same types of experiences, in exactly the same environments, and under the same conditions).

The financial success of an ecotourism venture will be hinged upon a number of cost pressures. The cost of supplies, equipment, utilities and salaries are only the tip of the iceberg. The service provider must be cognisant of the competition (potential dissatisfaction with his or her programmes), as well as the changing expectations of ecotourists over time. Ecotourists will continue to demand a high quality experience consistent with the amount of money they have spent.

Operators should also be prepared to monitor the condition of equipment and facilities regularly. These are often costly to replace, build and/or maintain, especially given the location of many ecotourism operations and the ability to mobilize a portion of the workforce to come and actually do the labour. Adequate short- and long-range planning will enable the entrepreneur to examine the changing needs and the costs of the programme. Although the pay-off from long-range planning may be somewhere in the future, and perhaps require an expenditure of time and money, the result will be a responsive programme which will be financially sound and resilient. Such an approach has a variety of advantages including the flexibility to meet changing interests and perhaps shorter start-up times for new programmes.

Place

Place, or distribution, refers to the activities of the firm, and other entities, that make the service available to the market for purchase. This includes the location and character of booking options (e.g. one's own office, travel agent, direct line, online, inclusive tour operator), and the ease and speed of booking (Bennett, 1996). Becherel (1999) writes that advances in electronic reservation and communication systems have been transforming the manner in which tourism services are distributed. He cites the example of Air Canada's self-service airport check-in kiosk for its customers, considered a more convenient method of checking in at the airport. The service is in English and French, and the customer has the choice of using the new service or the person-to-person services to check-in. Having experienced this new system, there is a great deal of pressure by Air Canada employees to use the new system, thus cutting down on the need for human resource support.

Promotion

Promotion is an area of marketing which is designed to create awareness of the firm's products and values for smaller firms, as well as to reinforce buying habits for larger firms (Friel, 1998). This may be accomplished using a variety of means, including the Internet, personal selling, local and national advertising, price promotions, sponsorship, merchandizing and sales literature (see Thomas *et al.*, 1997). Swarbrooke (1999) writes that promotion is a key element in the marketing mix in creating more sustainable forms of tourism, especially in raising tourist awareness of various social and ecological issues at the destination. He also notes that promotional items, such as brochures, have often lead consumers into believing that aspects of the destination, such as accommodation and transportation, are better than they actually are.

Much has been written of late on the Internet as a key promotional device which may be used to better position tourism businesses in the marketplace. Ecotourism is a perfect example, where service providers in far-off destinations have effectively used the Internet, relatively cheaply, to promote their services. There is the added bonus for the service provider of potentially being able to develop links with other ecotourism sites, such as parks or other nature-based attractions in the region, which would help to draw attention to the ecotourism business. As this medium appears to be a tool that will be with us well into the future, it is advisable that at least one employee in the firm gain expertise on how best to set up web sites and links. (See Appendix 9 for the Reservation Form used by Ozark Ecotours which is included in their web pages of mission statement, programme offerings, and so on.)

In a 1995 study of Canadian tour operators, advertising (the paid, non-personal promotion of goods and services by an identified sponsor) was found to be the second highest business expense at 15%, after salaries, wages and benefits which accounted for 41% (Canadian Tourism Commission, 1997b). It was further noted that most tour operators used a combination of advertising approaches, with brochures being the most often used method (85% of

Fig. 9.2. Effective marketing is critical for placing tourists in the right place, at the right time and at the right price.

operators employing this method), followed by attending tradeshows and the marketplace (71%), newspaper advertisements (58%), sales trips (55%) and magazine advertisements (49%). Only 14% of operators placed TV advertisements, while only 10% used the Internet (Canadian Tourism Commission, 1997b). This latter statistic has probably changed dramatically over the course of the last 7 years.

A focus on people, systems and green

Despite the fact that ecotourism operations are often small in scale and do not possess the resources to develop their own marketing departments, there is no reason why they should not attempt to market their programmes. These efforts may be smaller in scale but may be just as effective as the efforts put forth by marketing entities of much larger firms (Friel, 1998). Service providers must also realize that programmes at times fall out of favour, and that others need to be developed to take their place for the purpose of ensuring a level of growth that allows the firm to stay viable and profitable. Effective marketing also demands that the agency knows the consumer. This means gaining an understanding of his or her attitudes as well as being sensitive to how ecotourists make decisions. Davidson (1985) emphasizes this through the example of a businessman who appeared in one of his studies:

> He is the chief financial executive of a Fortune 500 corporation who, when describing the trip that he took to a company location in Florida, reported that he flew first class, he rented a Lincoln on arrival, and stayed at the VIP floor of one of the major up-scale hotels. He said, in fact, in an interview, 'I worked to get to this position and I deserve it!' But later on we talked of a vacation trip that he took to Florida with his wife. They flew People Express, their car was a Rent-A-Wreck, and they stayed at the Days Inn … The gentleman was the same, the psychographics were the same, his demographics were the same, he was the same person. These trips were in fact three weeks apart. But they were entirely different occasions, different events, and different decisions.
>
> (p. 106)

While products will sell much more easily by developing services and prices which are of good value, and promoting and distributing them in an effective manner, the importance of knowing the consumer and satisfying their needs strikes to the heart of marketing (Kotler *et al.*, 1996). This entails combining the above factors in a strategy that enables the organization to adopt and implement a set of systematic procedures, including adherence to a series of established objectives, in efforts to secure financial success and participant satisfaction. The importance of strategy is clearly emphasized in the following systematic marketing process, as outlined by Bécherel (1999: 37; see also Faulkner, 1999), which is further outlined in Fig. 9.3:

1. Gaining a firm understanding of the organization's business and its reasons for operating.
2. The use of research to understand how the industry is expected to perform in the future.
3. Possible future scenarios are identified which consider different strategies, where the most appropriate are selected.
4. Operational plans to implement the strategies that are developed.
5. The strategies are implemented.
6. The plan is constantly monitored to ensure that it is on target and, and if not, to take the appropriate actions to bring it back on course.

Working from the principles set forth in the mission statement, the organization will undertake an external and internal analysis of the business environment through PESTE and SWOT analyses. The strategic analysis and choice step in the process addresses the question of 'Where do we want to go?'; while short-term operating strategies address 'How the agency needs to get there?' Finally, the control and evaluation element examines 'Did we get there?' This planning approach will thus include the philosophy and mission of the organization, the goals and objectives which reflect the overall direction of the firm, and the analytical procedures and data-gathering techniques

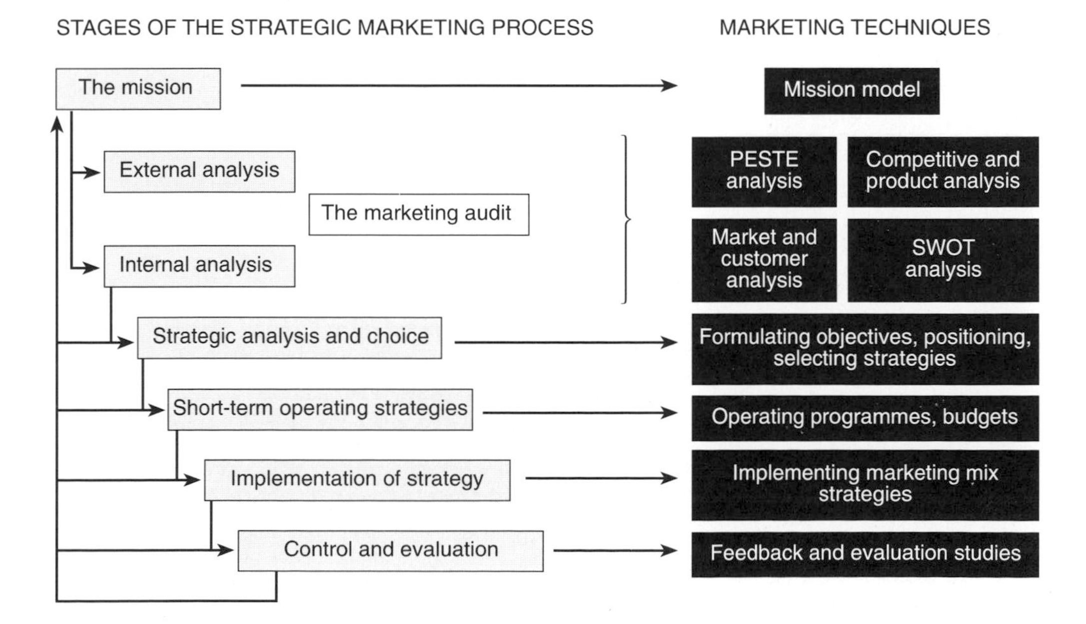

Fig. 9.3. Strategic marketing planning process. (Source: Bécherel, 1999.)

which enable the firm to: (i) understand specific markets; (ii) understand how the traveller makes decisions; (iii) assess the competition; and (iv) develop a theme or long-range strategy.

The emergence of alternative tourism, sustainable development, and other 'softer' and more ethical ways of viewing human–environment relationships have provided the impetus for viewing the tourism industry in a different light. Perhaps as a result of such paradigmatic transformation in the nature of tourism, a number of changes have occurred in tourism marketing practices. Past marketing tactics involved a concentration on the product and sales, while more recent views focused on the consumer, and later on the consumer in the context of society and ecology. However, we should also not discount the influence of competition and shifting markets in such changes, which have forced marketing strategists to be innovative but also sensitive to the needs of people and the resource base. In general, four stages can be identified in the transformation of marketing philosophies:

1. *Product-led marketing.* The assumption here was that by having a good quality service, the tourist would naturally buy the product. The focus was on the product itself.
2. *Sales-led marketing.* As productivity increased, as well as competition, the emphasis shifted towards the need to purchase the product. This approach parallelled the mass development of tourism and the need for volume to survive.
3. *Consumer-led marketing.* Tourists were placed at the centre of the efforts of the marketer. Techniques evolved to identify and understand the various wants and needs of the consumer. As well, customer-service techniques evolved to maintain competitive advantage of rivals.
4. *Consumer-led marketing with socio-ecological concerns.* In this final stage, developed through the latter part of the 1980s and early 1990s, marketing shifted to coincide with many of the societal views on the importance of community and ecology. Consumers demanded accountability of firms, their business practices and procedures, and were sensitive to the effects that they had on the social and ecological aspects of the destination. Travel firms quickly discovered that many market opportunities could be gained by 'greening' the product.

It is difficult to predict how marketing will influence and be influenced in the future. Although the environment continues to be hot and cold in regards to public opinion (especially when the public is asked to rank environment against jobs, health care, crime and other priorities), there appears to be a trend that has the public more willing to hold the service industry accountable for their actions. This means, in the words of Peattie (1995), that green marketing will be: (i) open-ended, instead of locked for the long term; (ii) it will place more emphasis on the natural world; (iii) the environment will have intrinsic instead of instrumental value; and (iv) it will have more of a global focus.

Programming Tip 9.1.

Brochures.

While handouts such as maps and brochures can be a valuable source of information for ecotourists, the service provider must recognize that many handouts may ultimately end up in the garbage. Niagara Nature Tours hands out such information sparingly, and only when it is absolutely essential for the education of participants. Any handouts should be on 100% recycled, unbleached paper.

Quality

While some organizations work to improve quality in the workplace, others strive to improve service. Both are one and the same, according to Albrecht (1992), who suggests that the bottom line is the implementation of measures to ensure that customers have beneficial experiences. In addition, service is not isolated to particular events or specific times, but rather to the entire experience. This will mean a service approach that goes well beyond simply being 'nice' to customers in selling a good or service, towards a combination of 'tangibles, intangibles, experiences, and outcomes designed to win the customer's approval and secure the right to survive and thrive in the marketplace' (Albrecht, 1992, p. 13). The implications for ecotourism service provision are such that operators must work hard to understand and manage the whole tour experience better from beginning to end. By placing the customer in the middle of a triangle of systems, strategy and people, the agency is able to deliver the appropriate level of value (Fig. 9.4). This strategy, according to Albrecht, involves people making sense out of what they do by: (i) understanding customer value; (ii) individually and collectively having a spirit of service; and (iii) implementing systems which are designed to support employees in their efforts to deliver value. Further to this, McCarville (1993) writes that a number of guidelines may be adopted which will enable the agency to develop a culture of quality. This, he suggests, must filter through the entire organization from senior administration right down to the front line:

> Brochures, program entrances, and telephone contacts, often the first contacts between the agency and the client, are examples of key encounter points. First impressions that set the stage for subsequent interactions with staff may determine how an entire agency is perceived by a client. Programmers must identify these encounter points and ensure that the experience is positive for their clients
>
> (p. 36)

In the past, tourism research has tended to examine service quality as separate from goods quality. As suggested earlier in this book, goods are evaluated by tangible features such as workmanship, durability and packaging; whereas services are marked by intangibility, perishability, heterogeneity and inseparability (see Chapter 1). The service orientation which is inherent in tourism demands that the host be responsible for creating and delivering services at the same time they are consumed (Vogt and Fesenmaier, 1995). However, services are not all the same, as suggested above, differing on the basis of a number of variables (Iacobucci and Ostrom, 1996), including variations across sectors of the tourism industry (e.g. hotels as compared

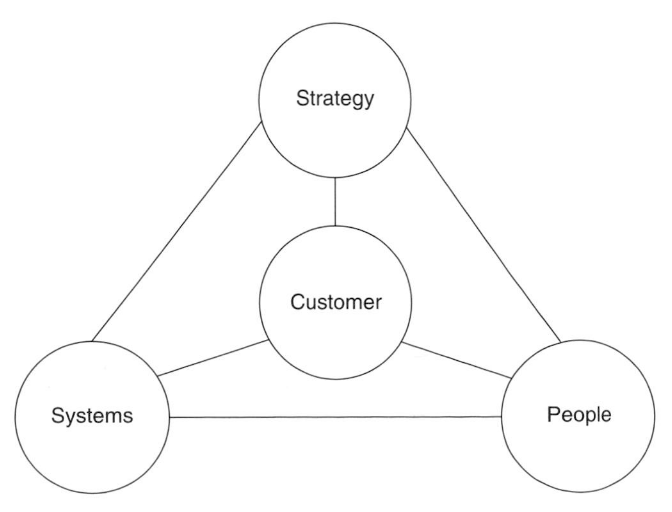

Fig. 9.4. The service triangle. (Source: Albrecht, 1992.)

to airlines) (Otto and Ritchie, 1996). Such differentiation would seem to suggest that service is not a unidimensional construct. Furthermore, there appear to be differences that exist in how customers and service providers perceive and rate service delivery performance, although the results seem to be mixed (Vogt and Fesenmaier, 1995). Also, various features of the tourism experience will be evaluated differently by different groups of people. This goes for technical quality aspects (e.g. size of room, provision of toiletries), as well as functional qualities of service (e.g. service provider's attitudes) (Reichel *et al.*, 2000). Attempting to exceed the expectations of tourists is a difficult one, especially given the differences that exist in tourist expectations within and between groups and across various sectors. For those ecotourists who have invested a great deal of time and money in an ecotour, and therefore have high expectations, operators will need to focus on delivering high-quality programmes which may be measured in costs and benefits, interpersonal relations, or other technical and functional terms. This will mean a conscious effort to be innovative, to get things right the first time, to improve continuously, to take pride, anticipate, eliminate mistakes, eliminate negatives, assume personal responsibility for quality and act as a team (Edginton and Edginton, 1993).

Staff Training

In attempts to maintain a competitive advantage in such a dynamic sector, ecotourism service providers, and the leaders they hire, should seek training which allows for performance at the highest levels possible (see Chapter 8). This is especially true for programmers and for employees who deal with clients on a day-to-day basis. This means knowing what they are showing (interpretation), as well as knowing some of the general attributes of the clients who they are serving. Research has shown that employees are only 50–60% effective (National Recreation and Parks Association, 1989). This has been attributed to factors such as a lack of a clear understanding of what is required; poor attitude and motivation of staff; and a lack of the appropriate knowledge and skill to do the job right. One of the absolutes in human resource management is that staff training should be ongoing (Jackson and Bissix, 1979). In addition, administrative and supervisory personnel should use all opportunities, including those which are positive and negative, as means by which to assist programme staff to become better leaders. Table 9.1 illustrates data from a 1984 study of US corporations in regards to the specific types of training and their frequency of use. (Given the date of the study, some items will have changed, e.g. word processing, in terms of their overall importance to employees and the organization as a whole.)

The recent publication of the book *Steps to Success: Global Good Practices in Tourism Human Resources*, by Williams and Watts (2002) provides a needed forum by which to address human resource issues that exist in the tourism industry. This book, as well as the periodical published by the World Travel and Tourism Human Resource Centre by the same name, highlights case studies and topics dealing with in-service training, front-line operations, management tools and practices, peer training, and specialty training, and many other topics. The challenge taken up by the World Travel and Tourism Council (WTTC) is to ensure that a well-trained workforce will be able to meet the dynamic expectations of visitors in a globally competitive marketplace. Recent issues of the periodical have focused on several ecotourism case studies including Borneo, Scotland and Ecuador, and the innovative mechanisms that have been employed to generate successful programmes in these places. Ecotourism operators may benefit directly from this publication, in addition to other sources that deal directly with the day-to-day human resource issues, such as training, which confront the tourism industry.

Job satisfaction

Although studies have found that a happy worker does not necessarily make a good worker, measures of worker satisfaction have been shown to be important in management's attempts to come to grips with absenteeism,

Table 9.1. Specific types of training.

Type	Percentage providing
New employee orientation	87.0
Performance appraisals	76.7
Time management	73.9
Leadership	73.2
New equipment operation	68.9
Hiring/selection processes	66.9
Product knowledge	64.3
Train-the-trainer	63.6
Interpersonal skills	63.5
Word processing	62.4
Problem solving	61.4
Goal setting	60.0
Listening skills	59.9
Stress management	58.5
Decision making	57.2
Safety	56.9
Team building	56.8
Delegation	56.8
Planning	55.2
Public speaking/presenting	55.0
Customer relations	53.1
Computer programming	53.0
Management information systems	52.5
Conducting meetings	50.9
Writing skills	50.5

National Recreation and Parks Association (1989).

turnover, and physical and mental health problems (Iaffaldano and Muchinsky, 1985), as well as counter-productive behaviours such as doing poor work, spreading rumours, damaging employer's property, stealing, and using alcohol or drugs (Jandt, 1995). Landy (1989) identifies a number of work factors and their effects on job satisfaction (Table 9.2). Interesting is the finding that mentally challenging work, personally interesting work and goal attainment, are all satisfying.

These results relate to the findings of Jandt (1995) who reported that although *service* workers (e.g. workers in the recreation and tourism sectors) felt overworked, they remained satisfied with their jobs. Jandt found that these employees were: (i) more proud than not of their service job; (ii) the job more matched their skills than not; (iii) the job was perceived to be important relative to other job types; (iv) the job was found to be as satisfying as one originally expected it to be; and (v) overall they were quite satisfied with the current job. Furthermore, he asked respondents to rank the following five items according to their level of importance in determining satisfaction: (i) expressing concern and warm feelings for others; (ii) having freedom and independence at work; (iii) knowing you are doing a service for others; (iv) the respect you receive from others; and (v) salary and benefits. He found that service workers reported freedom first, followed by respect, salary and benefits, expressing concern, and knowing you are doing a service job. This element of freedom is a critical leisure-defining construct, as suggested in Chapter 1. It would seem that the imposition of certain constraints (e.g. loss of decision-making ability) may remove one's sense of control (see Neulinger, 1974), and thus lead to less satisfying conditions within the work place. The service workers who were found to be more satisfied in Jandt's work were older college graduates in higher paying jobs. These individuals had more training, felt more management encouragement, and were afforded the opportunity to solve problems.

Table 9.2. Effects of various work factors on job satisfaction.

Work factors	Effects
Challenge	Mentally challenging work is satisfying
Physical demands	Tiring work is dissatisfying
Personal interest	Personally interesting work is satisfying
Reward structure	Equitable, accurate rewards are satisfying
Physical conditions	Satisfaction depends on the match between working condition and physical needs
Goal attainment	Conditions promoting goal attainment are satisfying
Self	High self-esteem conducive to job satisfaction
Other workers	Individuals will be satisfied with those others who help them attain rewards
Management	Dissatisfaction stems from conflicting roles imposed by management
Fringe benefits	Benefits do not have a strong influence on job satisfaction for most workers

Adapted from Landy (1989).

At present, there appear to be few studies that have examined the various aspects of job satisfaction among employees in the ecotourism industry. Anecdotal evidence points to the belief that many who choose to work in ecotourism do so because of the rewards that come from a love of the outdoors, interaction with people and the opportunity to travel. However, we can never assume that such intrinsic motivators will continue to maintain high levels of satisfaction within the organization. Consequently, service providers would be wise to consider many of the results uncovered in the business and management literature in order to appreciate more fully what it is that they may do to ensure continued levels of enjoyment.

Public Relations

One of the best ways for the ecotourism service provider to develop a positive profile in a community is to implement good, ethical programmes which are considerate of tourists, local people and the environment. Unfortunately, the good things that ecotourism operators do in and for host communities are not always communicated. In helping to spread the word about the positive initiatives of the organization, managers should consider the development of public relations strategies which allow the following: (i) dissemination of information about the organization which is designed to engender favourable attitudes towards the agency; and (ii) a means of providing the public with an ongoing understanding of the things that the agency is doing. The initiation of multiple lines of communication between the agency and the community (broadly defined) allows for the development of management policies and practices which may be conducive to public interest and good-will (Edginton *et al.*, 1998). Such policies may include the creation of codes of ethics, involvement of local people in the planning of the ecotourism business, and the development of environmental education programmes for the community. The following is a list of different public relations approaches that may be used to satisfy the goals of a recreation or tourism agency (adapted from Kraus and Curtis, 1990, p. 325):

1. To provide accurate information regarding the overall programme and offerings of the agency to the general public to overcome misunderstandings, false impressions, or lack of information about ecotourism.
2. To inform the public specifically about the services, facilities and programmes offered by the organization.
3. To impress the public with the values and benefits achieved by and through the organization.
4. To keep the public informed about any major development plans of the organization.
5. To tap into certain members of the community who have special knowledge about the natural and cultural history of the region, who may benefit from a relationship with the ecotour operator.
6. To help promote other forms of community involvement through the development of networks of individuals who have a vested interest in the environment.

7. To develop channels for two-way communication with the public at large in the form of meaningful dialogues on community social and environmental issues.

Kraus (1997) writes that there are a number of channels which may be used to accomplish the aforementioned goals, including: (i) information media, such as newspapers, magazines, television and radio; and (ii) interpersonal links, such as advisory groups, councils, appearances before civic groups, public meetings or other such events. Russell (1982) identifies two other tools which may be effective in casting a positive light on the organization. First, annual reports may be designed to inform the public about the constructive work of the agency over the course of the year. The report should include a summary of financial, facility and programme developments and accomplishments, with the aim of building both community and organizational confidence. Information kits, as the second tool, provide background information on the organization, its programmes, employees and facilities. This is a useful vehicle which allows the service provider to communicate with the public as well as to interface with other similar types of organization over common themes. A variety of items should be included in the kit, such as: a history of the organization, an organizational chart, biographical sketches of staff members, agency services, the annual report, reprints of newspaper articles or other media materials, and photographs of activities or services (Edginton *et al.*, 1980).

Budgeting

The budget is a must both for those wishing to start a business and those who are involved in a business. It is a plan that represents the operator's intentions and expectations through the allocation of funds to achieve desired outcomes (Finkler, 1992). The budget balances the forecasted expenditures (money going out) with revenues (money coming in). Expenditures include the various agency costs and resources needed to plan, develop, implement and monitor programmes. Broadly, this includes costs related to accommodations, travel, guide services, equipment, food and wages. However, the operator will incur a whole array of other expenses – using Canada as an example – including vacation pay for employees (0.057 multiplied by the total earnings); employment insurance (gross earnings multiplied by 0.037 multiplied by 1.4); Canada pension plan (subtract 130.76 from the gross earnings and multiply by 0.026); workers' compensation board (calculated by adding all of the gross earnings of the employees for the year and paying 0.95 on each C$100 paid out), insurance, rent, telephone, photocopying, accounting, staff development, advertising, programme supplies, office supplies, and perhaps inspections (for fire and health). Revenues, on the other hand, include tax exemptions, earned income (entrance fees, tour fees), and perhaps financial assistance in the form of grants or donations, for those agencies eligible for such.

The advantage of the budget is that it allows for: (i) the systematic review of the programme, including who or what is involved in running the programme; (ii) long-range planning (e.g. the purchase of vehicles to shuttle ecotourists); and (iii) change (e.g. the elimination of poor programmes, defined as those which lose money, have little participation, or which have decidedly few benefits to participants). Table 9.3 is an example of a pro forma developed by the Australian Department of Tourism, which is illustrative of the income and expenses that are typical in running an ecotourism business.

Although perhaps subject to debate, it is the author's belief that the programme budget should emerge as a result of programme needs, and not as a reflection of the amount of money that may be available in a common pool. This means that it is the programme itself, and all the various needs there within, which should drive the budget. Doing otherwise would severely bias decisions on the availability of resources which are needed and appropriate for numerous different activities and programmes which take place within the organization. Also, a good rule of thumb is to ensure that there is enough money in the budget to cover the salary of all employees for at least 3 months. This is a resource that the operator cannot afford to compromise!

Table 9.3. Components of a profit and loss statement.

Income
Sales
Rebates, discount, commission received
All other trading income
Total trading income
Expenses
Costs of goods sold
Opening stock
Plus purchases
Less closing stock
Costs of goods sold
Gross profit = income minus cost of goods sold
Overheads/operating expenses
Advertising and promotion
Bad debts
Bank and credit charges
Depreciation of equipment and vehicles
Electricity and gas
Insurances
Interest
Lease payments
Motor vehicle operating costs
Personal salary/drawings (if incorporated)
Postage
Rent on premises
Rates and taxes
Repairs and maintenance
Salaries, wages and taxes on same
Superannuation, long service leave
Telephone and fax
Travel and entertainment
Other expenses
Total overheads/operating expenses
Trading profit
= Gross profit less overheads/operating expenses
Net profit
= Remainder after tax considerations

Source: McKercher (1998).

Budgets can be prepared in any number of different ways. Common to many approaches, however, is consideration of the following steps for the purpose of making informed budget decisions: (i) gather data (e.g. numbers of tourists, facility and equipment inspections); (ii) compare data with past budgets (e.g. 3–5 years ago, include inflation); (iii) budget construction, including personnel, operations, capital expenditures; (iv) presentation of the budget; and (v) interpret and evaluate the budget.

Implementation Strategies

Unfortunately, there is often the belief among programmers that implementation occurs naturally as a result of effective planning. Conversely, implementation just doesn't happen, but rather should involve a series of processes, practices and techniques which allow the event to reach its full potential. In their book on health promotion programmes, McKenzie and Smeltzer (1997) state that programme implementation will occur successfully if consideration is given to: (i) the resources that are available for the programme (as suggested in the section on budgeting, above); and (ii) the setting for which the programme is intended – which is especially relevant to ecotourism. Programme implementation, therefore, may involve a significant outlay of resources, including time, personnel and capital. These resources are best viewed in the context of a specific strategy.

The various approaches used in programme implementation have been referred to as systems which are designed to allow programmers to understand the extent to which desired outcomes are produced by a particular preconceived approach to organizing a programme (Edginton *et al.*, 1998). The systems design allows for planning and implementation, but also for the review and control of the programmes. Implementation strategies are derived from the Program Evaluation and Review Technique (PERT) process developed by the US Navy in the late 1950s (Russell, 1982). PERT is essentially a flow chart which outlines the various steps that need to take place in moving from the conception of the project through to its evaluation, where the tasks to be completed occur in an ordered sequence over time.

The critical path method (CPM), a well used implementation strategy in the recreation field, is an adaptation of the PERT method that includes a time line and flow chart of events, but also an outline of how each component relates to another, and how and when various tasks will be accomplished. (The CPM is depicted in Fig. 9.5.) Like other implementation strategies, the CPM must be derived from an implementation checklist, which divides the programme into a number of major components (e.g. personnel, programme, marketing), and which are in turn broken down further into a number of specific tasks (e.g. personnel, tasks may include interviews, training, preparation of manuals, skills development, and so on). For these, a time line is constructed for the completion of each task, on the basis of priority or other relevant factors. (The checklist for the CPM presented in Fig. 9.5 is located in Table 9.4.)

Another implementation strategy, which has been used quite successfully in the recreation and leisure services field, is the Gantt chart, which is a visual management tool designed to illustrate tasks and time frames in a bar graph format (see Fig. 9.6). Each task is listed on the vertical axis, with the time frame listed on the horizontal axis. The result is a graphic illustration of the programme which quite easily allows the programmer to gain an understanding of all tasks in relation to others. The use of the Gantt chart or CPM does require some orientation, as well as a good deal of brainstorming on behalf of the programmer who must effectively identify all of the various tasks, and their associated time frames, which are required for the successful implementation of the programme. Furthermore, because of the nature of adaptability of the Gantt chart and CPM, they are equally valuable as programme evaluative mechanisms (see DeGraaf *et al.*, 1999).

McKenzie and Smeltzer (1997) identify two other programme implementation strategies which they find to be flexible enough to apply to many different programme situations. The first of these, by Parkinson and Associates (1982), suggests that there are three main ways in which to implement a programme: piloting, phasing-in and total programme, as outlined below.

1. *Piloting.* The advantages of such an approach include the opportunity to test the programme at initial stages, as well as to maintain close control of the programme. The approach is disadvantaged by having very few people involved, by not being able to test and meet various administrative and programme-related needs, and by being hard to generalize about results.
2. *Phased-in.* The advantages of this approach are: (i) that it is easier to cope with workload, and (ii) that it helps protect planners from getting in over their heads. Phasing-in can be completed by different programme offerings, by a limited number of participants, by choice of location, and by participant ability. Its disadvantage includes the fact that there are fewer people involved in the process.

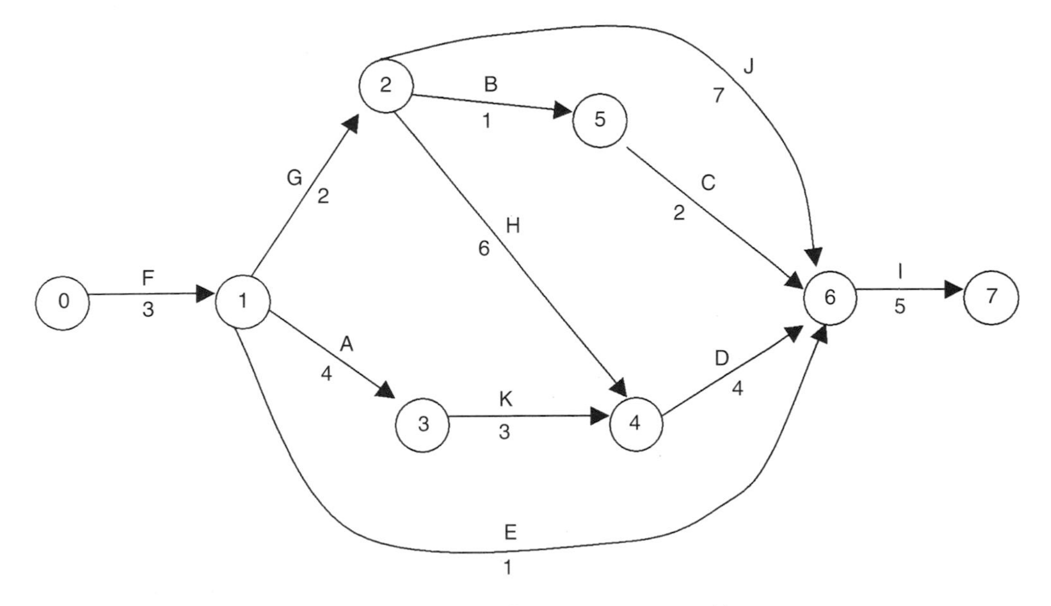

Fig. 9.5. Critical path for interpretive programme. (Source: Russell, 1982.)

Table 9.4. Checklist for implementation of an interpretive presentation.

Component	Task	Time required to complete (weeks)
Personnel	A. Ongoing training of staff in interpretation	4
Materials	B. Prepare scenery and props	1
	C. Install scenery and props	2
	D. Prepare costuming and other effects	4
Facilities	E. Modify site for presentation	1
Programme	F. Develop theme and objectives	3
	G. Identify or cast interpreters	2
	H. Rehearse	6
	I. Dress rehearsal	5
Publicity	J. Advertise programme in park/community	7
	K. Print audience song booklets and programmes	3

Source: Adapted from Russell (1982).

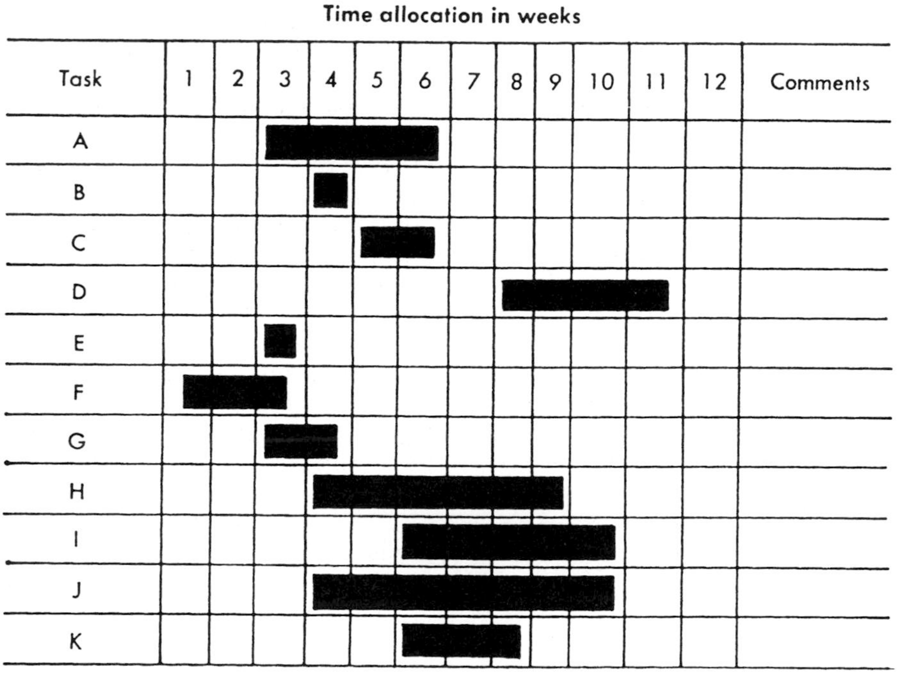

Time allocation in weeks

Task	1	2	3	4	5	6	7	8	9	10	11	12	Comments
A													
B													
C													
D													
E													
F													
G													
H													
I													
J													
K													

Fig. 9.6. Gantt chart for interpretive programme. (Source: Russell, 1982.)

3. *Total programme.* Advantages include the fact that more people are involved and evaluation is more meaningful based on the larger group. Disadvantages include the commitment that goes hand-in-hand with this approach, and no chance to test the programme beforehand.

McKenzie and Smeltzer note that programme planners should lean toward total programme implementation, but only after first taking steps to pilot and phase-in. Such an approach, although costly in time, would provide a significant degree of control over the programme and provide much needed information on the success of the programme through resource efficiency and client satisfaction. In this model, programme implementation becomes a process of programme building through the three approaches mentioned.

The second model discussed by McKenzie

and Smeltzer, is one developed by Borg and Gall (1989), which was originally presented as a research and development cycle. It is a ten-stage model (adapted in Table 9.4 to show only the last seven stages), which is particularly relevant to programme planning and the purposes of this book. The model is based on a series of field testing and revision stages which allow the programmer to secure a heightened level of acceptance and momentum of the programme. Its strength lies in the ability to change the programme at various stages according to the requirements of participants. It also allows the programmer to drop the programme altogether without a significant outlay of capital if, for example, participants are greatly dissatisfied with it after Step 3 (main field testing).

This comprehensive implementation strategy includes an element of programme building, which is strongly recommended by McKenzie and Smeltzer. While it appears to be about as thorough as one needs to be in the construction of a programme, the authors suggest that the process may be modified or scaled down to suit the needs of the service provider. Successive rounds of testing and data analysis will require a great deal of time and effort. The importance of the programme relative to other programmes offered, in light of time and other resources, will be an important consideration in assessing the level of detail required in using this approach. Having said this, it should be realized that those who follow through with the model as it is outlined above, in all likelihood stand a greater chance of developing more successful programmes later on.

Schedules and Itineraries

Tourism schedules usually refer to charts or screens which show departure and arrival times, as well as dates, fares and classes of various forms of transportation. In recreation, schedules refer to the length, duration, days of the week (or day), and time of day in which programme activities are offered (Russell, 1982). This latter interpretation of scheduling is quite similar to the tourism itinerary, which is basically the plan of a journey or a travel route (Metelka, 1990). However, itineraries can be much more detailed and include a great deal of information on the day-to-day events of the trip. This is often the case with longer trips (e.g. 10 days), where the itinerary needs to be very specific about the events of each day. Such is the case in Appendix 10, which is the itinerary for the 'Birds of Trinidad and Tobago' programme offered by Quest Nature Tours of Toronto.

Programmes may be scheduled on a number of different time frames, including seasonally, 21-days, 14-days, 10-days, weekly, daily and half-day bases. On a seasonal basis, for example, scheduling may be based on weather patterns and temperature (e.g. wet and dry seasons), market seasonality (high and low seasons), or migration patterns of animals (e.g. August may be the best time to view polar bears in Churchill, Manitoba, Canada). One of the definite constraints to the ecotourism industry in more northern climates such as Canada, is the seasonal nature of the industry. Visitation peaks in the warmest months of the year, leaving

Table 9.4. Major steps of Borg and Gall's research and development cycle adapted to programme planning.

1. *Preliminary field testing.* Programme given to a few individuals from the target population, with interview, observation and survey data collected and analysed.
2. *Main programme revision.* Based on results of preliminary field test.
3. *Main field testing.* Programme given to about twice as many individuals as in Step 1. Pre- and post-programme quantitative data collected; results examined with respect to programme objectives.
4. *Operational programme revision.* Based on results of the main field test.
5. *Operational field testing.* Programme given to about twice as many clients as Step 3. Interview, observation and survey data collected and analysed.
6. *Final programme revision.* Based on results of the operational field test.
7. *Dissemination and implementation.* Programme shared with others and implemented.

Original source: Borg and Gall (1989). Adapted from: McKenzie and Smeltzer (1997).

fewer opportunities for operators to attract the same volume of visitors in the colder months. While these operators may shift into other activities (e.g. snowshoeing), levels of visitation may still be constrained by the climate and by the dramatic drop in domestic and international tourism. Day, half-day or hourly programmes are usually scheduled tightly, with very little opportunity to participate in activities outside of the main programme offering, as demonstrated in Table 9.5.

Scheduling should also complement the existing programme format (e.g. class, drop-in, etc.) as well as make the best use of available resources (e.g. people, transportation, equipment, time possible). Not surprisingly, the schedule is one of the fundamental keys to the implementation of a successful programme. Ford and Blanchard (1993) write that outdoor pursuits are, in general, very time consuming, so the effort expended in scheduling is decidedly worthwhile. They illustrate that programme planners must be sensitive to the life-style patterns of potential participants and that there are a myriad of variables which must be juggled in seeking the best possible time for an event. These include the work and leisure patterns of participants, environmental conditions (e.g. river levels, natural area permits), qualified staff, facilities, transportation and lodging.

Service providers should also be sensitive to the fact that wildlife viewing during the warmest part of the day will not provide the ecotourist with as much of an opportunity for viewing success. The transition between night and day (dawn and dusk) are times when there is the most animal activity. Those who are interested in maximizing their opportunities to view in these places, will want the programme to be scheduled with this in mind. It should also be noted that physical exertion during the warmest times of the day, in countries with extreme heat, is not recommended. This brings to mind the importance of scheduling events to the capabilities and needs of the population served, as suggested above. Early morning recreation programmes will not be as popular for adolescents, in the same way that evening classes will not work for children. Once while conducting seminars in Lesotho, our group was invited to go horse-riding in the mountains. After about a 2-hour trip in rather cramped vehicles, through winding roads, we arrived at the ranch for our supposed 45-minute experience. The trip turned out to be 3-hour trek over very rough terrain. There were a number of elderly people in the class, who were absolutely exhausted at the end of the trip. Apparently there was some mis-communication between organizers which resulted in the extended trip. This was a case where: (i) the activity was mismatched with at least some of the group; and (ii) the scheduling of the main activity was grossly underestimated.

Table 9.5. The secret world of the platypus, Wildscapes Safaris, Australia.

Includes: Morning tea and return transportation.
Be part of a **spectacular** platypus **phenomena**! It is a ritual being cherished by visitors from around the world visiting tropical North Queensland. With a rainforest backdrop, the **platypus**, our most intriguing mammal, will be **revealed to you**. Observe and photograph this world famous 'living fossil' in its own undisturbed and **secluded natural habitat**. Described by researchers as the northern most distant playground in platypus habitat and the **best area to observe** their rare **daylight behavioural movements**.
Hear the billabong come to life with the mystic sounds of laughing kookaburras ... and reflect on these moments with a fresh cup of billy tea and tasty baked muffins.
Onwards from this exclusive quiet waterway retreat we **visit** other remarkable areas of the tropics, **quaint historical little towns** encircling **majestic rainforests** and the beauty of our open dry country revealing more **unique wildlife**.
Bring your camera and be part of this photographers and filmmaker's paradise.
Departs Cairns 6:30 a.m.
Returns 12:30 p.m.
Cost: $95.00

Source: http://www.wildscapes-safaris.com.au/wildscapes_safaris_tour_1_.htm

Programming Tip 9.2.

Providing maps for ecotourists.

Beyond the description of daily events in the itinerary, the service provider might think about providing ecotourists with maps of the regions in which they are travelling. Such visual daily planners may enable the ecotourist to get a relative sense of both space and place (distances, scale, and geographical features), and get away from the feeling of being 'herded' from one site to the next.

The itinerary might also include a description of the number of stops planned for the day, restaurants, parks and protected areas, amount of time on buses or other vehicles, safety considerations, contingency plans, amount of time at each site, the trip characteristics vs. client characteristics (e.g. an elderly group), and the amount of down or leisure time that will be offered each day. These elements should further be examined through the concept of pacing, which can be defined as 'the rate, amount, and sequence of activities to assure a realistic schedule and provide for a balance of free time, travel, rest, sightseeing and social events' (Metelka, 1990). First- and second-day activities may need to be less rigorous on trips of a week or longer to accommodate participants who may have travelled extensive distances to attend the ecotour (providing these individuals with the time to recover and adjust from their travels). Obviously, for those shorter 1- and 2-day programmes, service providers do not have the luxury to do the same.

Conclusion

This chapter examined the concept of programme implementation through a discussion of the programme life cycle, marketing, quality, staff training, public relations, budgeting, as well as a number of strategies that may be used to get the programme up and running. It was suggested that the implementation stage is often given little consideration because of the feeling that after a great deal of planning, the programme simply falls into place. However, there are a number of implementation strategies and tools that may be used to launch ecotourism programmes effectively. These include approaches developed by Parkinson and Associates (1982) and Borg and Gall (1989), in addition to approaches that have commonly been used by recreation practitioners, including the critical path method and the Gantt chart. The chapter concluded with a discussion of the link between schedules and itineraries, and how these enable the service provider to plan the logistical aspects of the programme.

10
Evaluation

Mirrors should reflect a little before throwing back images.

(Jean Cocteau)

Introduction

This chapter begins with an overview of evaluation, along with who should evaluate and what should be evaluated, in addition to a discussion of formative and summative evaluation approaches. Although evaluation is often thought of as a process that is employed at the end of a programme only, it is actually an ongoing process which enables service providers to control the events that go on within the programme setting. The bulk of the chapter focuses on a number of different evaluation models which, although not exhaustive, provide an overview of many evaluative options available to the service provider. A discussion of certification and accreditation is included in this chapter because of the evaluative basis of these concepts. This is followed by reference to environmental auditing and green taxes in order to examine some of the ecologically based evaluative mechanisms which have recently been put into place. The intent is to demonstrate that programmes have implications for people, as well as for the environment and, in the case of the latter, these should also be subject to scrutiny. The chapter concludes with a brief discussion on why programmes do not work, along with an analysis of programme revision and death.

The Scope of Evaluation

Worthen *et al.* (1997) suggest that among professional evaluators there is no commonly accepted definition of the term. Some, these authors say, cloud the meaning of evaluation by equating it solely with research or measurement, the satisfaction of objectives only, or professional judgement. The essence of their view of evaluation is couched in earlier definitions which view evaluation as judging the worth or merit of something. More specifically, the term can be defined as 'the identification, clarification, and application of defensible criteria to determine an evaluation object's value (worth or merit), quality, utility, effectiveness, or significance in relation to those criteria' (Worthen *et al.*, 1997, p. 5). In a leisure context, evaluations are undertaken by programmers to answer some of the questions about what they had hoped to be able to accomplish. These questions often surface in the form of pre-determined goals, whose effectiveness are ascertained through a process involving any number of qualitative and quantitative techniques. This information arms the service provider with the ability to demonstrate accountability in regards to finances, demand, benefits or the wise use of resources (Busser, 1990).

A number of authors have examined the various aspects (e.g. people and processes) of a programme that can be evaluated. Russell (1982) and Farrell and Lundegren (1993), for example, identify four main areas for evaluation, including: (i) administration (e.g. planning, organizing, staffing, training, directing work, exerting control and inspiring creativity); (ii) leadership or personnel (the qualifications, competence and performance of staff); (iii) programme content (meeting needs of participant, scheduling, transitions, goals and objectives, supervision, records, and so on); and (iv) physical properties (including design, construction and maintenance of facilities and equipment, meeting of facility standards, as well as the appropriateness of equipment to participant skills). A more recent treatment of evaluation (Edginton *et al.*, 1998), illustrates that evaluation may be examined from three general perspectives: the customer, the programme and the organization. These are more fully explained below.

1. Customer evaluation:
   - Examines customer satisfaction, and thus attains a reading on the benefits and enjoyment of programmes.
   - It measures change in customer behaviour, which may manifest itself into increased or enhanced levels of participation.
   - Allows for customer input, and is responsive to the needs of customers.
   - Encourages customer support, for the organization and as a means by which to contribute to their own leisure interests.
2. Programme evaluation:
   - Promotes leader/customer transactions, and the opportunity for bonding between the group and the programmer.
   - Develops sensitivity to the customer, people's feelings, differences in gender, age and ethnicity.
   - Determines programme design effectiveness, to people, settings, objects, rules and relationships.
   - Indicates needs for programme enhancement, what improvements are needed with regard to service delivery or facilities.
3. Organizational evaluation:
   - Links programme performance to budgetary allocations, what is desired by customers and what is available.
   - Focuses on more specific and concrete objectives, in better defining goals.
   - Helps in determining programme priorities, the costs and benefits of the programme, and other political decisions.
   - Aids in control, such as the establishment of performance standards.

This more complete breakdown allows for an appreciation of how evaluation may aid the organization in many capacities, especially in the face of shrinking budgets or other diminishing resources. The time and effort placed into evaluation yields a number of rewards, including the ability to: (i) be more accountable to the business itself (including employees) and the participants; (ii) effectively determine pros and cons of the programme; (iii) address quality control; (iv) plan for the future; and (v) revisit the programme goals and objectives which have been so important in guiding the shape of the programme. The hesitancy that operators have towards evaluations may be traced back to earlier stages of the programming cycle, particularly cases where goals and objectives have not been well articulated. In such cases evaluation becomes difficult because operators do not know what needs to be evaluated, the methods by which to accomplish their end, and why.

Service providers must also be prepared to conduct frequent personnel or performance appraisals, for those employed in the organization. These evaluations, according to Culkin and Kirsch (1986), measure and evaluate 'an employee's job-related behaviour and outcomes to discover how and why the employee is performing on the job at present and how the employee can perform more effectively in the future so that the employee, the organization, and society all benefit' (p. 166). Such an appraisal is usually performed at a pre-determined time, perhaps every 6 months, and may be required by law, depending on the jurisdiction. More specifically, however, such an evaluation may be done weekly, seasonally, formally and informally depending on the type of programme, setting and length. Evaluations are carried out by supervisors who are most familiar with the work of the employee, and the criteria for evaluation must be made available to the employee well in advance. Usually the evaluation is done on a five-point comparative scale which ranges from very superior to very inferior, or from outstanding to poor (unsatisfactory). Employees may be evaluated on the basis of personality, fairness, compassion, organization, leadership, report writing, various skills, facility upkeep, alertness, approachableness, team orientation, dependability, promotion of programme, how informed they are, problem solving ability, initiative, enthusiasm, absenteeism, lateness, emotional stability, and so on.

Who should evaluate?

Managers of an ecotourism organization will have to choose who among the staff is most qualified or suitable to conduct an evaluation. In small agencies it may be the owner; whereas in larger firms with more administrative levels it may be someone in middle management, such as a programme coordinator. There are certain advantages to selecting someone like a programme coordinator – or someone who is closest to the programme – because this saves the firm time and money. However, there are also potential disadvantages to adopting this strategy, including the introduction of bias, the perception that the actions of the evaluator may antagonize superiors, along with the consequences of negative evaluations to the evaluator him- or herself. On the other hand, managers may wish to have evaluations conducted by people from outside the organization in an attempt to gain objective insight into the positive and negative aspects of the programme and/or organization. This approach, however, is also subject to pros and cons. In the case of the former, consultants or experts may be less likely to offend the boss, less likely to accept administrative excuses, and add a fresh point of view to the programme and agency. However, they may also be threatening to employees and charge excessive amounts of money. Table 10.1 summarizes some of the advantages and disadvantages of internal and external evaluations (Henderson and Bialeschki, 1995, p. 73).

Formative and summative evaluation

One of the misconceptions of evaluation is that it is just a tool to be used at the end of

Table 10.1. Advantages and disadvantages of internal and external evaluations.

Internal evaluations	External evaluations
Advantages:	Advantages:
Knows the organization	More objectivity
Accessible to colleagues	Competence
Realistic recommendations	Experience
Can make changes	More resources
	Less pressure to compromise
Disadvantages:	Disadvantages:
Pressure to have positive results	Threat to employees
Difficult to criticize	Must get to know organization
May lack training	May disrupt organization
	May impose values
	Expensive

Source: Henderson and Bialeschki (1995).

the programme cycle. In fact, evaluation should used at stages where it is deemed to be most beneficial. This may be at the end of the programme, but also at selected points during programme development and implementation. Evaluation that occurs during programme planning and implementation is referred to as formative evaluation, and includes efforts to help the programmer to monitor progress towards the successful launch and completion of the programme (i.e. the identification of problems which may be resolved in an effort to improve the programme). For example, it may be used to see that programme funds are being spent wisely, to avoid the negative impact of poor programmes on a community or resource, or to fine-tune an interpretive programme at selected periods during an event. These formative evaluation results provide immediate feedback on the value of the interpretive tools, which may thus be built into the programme at later dates.

The intent of summative evaluation is to provide programme decision makers and consumers with judgements about the programme's worth and merit in relation to criteria that have been deemed important. Using the example on environmental education above, a summative evaluation would seek to determine the degree to which the curriculum improved the quality of the programme and the level of satisfaction of ecotourists, and thus the future value of the programme. The element of continuation is important, because it is a key concept in differentiating between formative and summative evaluations. While formative evaluation leads to decisions about programme development and implementation (modification and revision), summative evaluation leads to decisions about programme continuation, termination, expansion or adoption (Worthen *et al.*, 1997). These authors illustrate the relationship between formative and summative evaluation (Fig. 10.1) and the differences between both forms of evaluation (Table 10.2).

Models of Evaluation

In maximizing the chances for continued growth and success, businesses and other organizations will want to implement some form of evaluation. This is particularly true for the service sector which is based on quality and the satisfaction of consumers. The continued acceptance and formalization of evaluation has prompted the development of a number of different evaluation models to aid programme delivery. These range from non-systematic methods which rely on the judgement of one or more individuals in the organization, to those which are systematic and comprehensive and force the programmer to consider a number of steps in completing the evaluation process. Furthermore, some models are more appropriate for

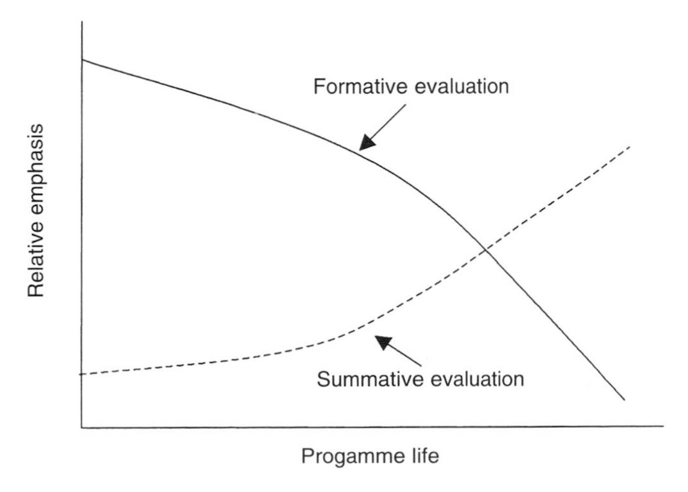

Fig. 10.1. Relationship between formative and summative evaluation. (Source: Worthen *et al.*, 1997.)

Table 10.2. Differences between formative and summative evaluation.

Variable	Formative evaluation	Summative evaluation
Purpose	To determine value or quality	To determine value or quality
Use	To improve the programme	To make decisions about the programme's future/adoption
Audience	Programme administrators and staff	Programme administrators and/or potential consumer or funding agency
By whom	Primarily internal, supported by external	External, supported by internal in unique cases
Major characteristics	Feedback so programme personnel can improve it	Feedback to enable personnel to decide to continue it, or consumers to adopt it
Design constraints	What information is needed? When?	What evidence is needed for major decisions
Purpose of data collection	Diagnostic	Judgemental
Measures	Sometimes informal	Valid and reliable
Data collection frequency	Frequent	Infrequent
Sample size	Often small	Usually large
Questions asked	What is working? What needs to be improved? How can it be measured?	What results occur? With whom? With what training? Under what conditions? At what cost?

Source: Worthen *et al.* (1997).

certain types of evaluation, and perhaps for different times within the programme cycle. The operator may thus consider a range of different approaches to accomplish his or her ends (see Carpenter and Howe, 1985, for an excellent overview of evaluation models).

Intuitive evaluation

This non-systematic approach to evaluation is used often and under many circumstances to make decisions on a variety of programme components. It is a 'gut-instinct' form of evaluation that is excellent for the day-to-day observations of the programme. Consequently, new information is easily integrated into the context of the programme during implementation. It is therefore an approach that is used much more extensively than other methods, because it is quick and inexpensive and does not put demands on employees. DeGraaf *et al.* (1999) write that at times these benefits come at a cost of reliability (consistency) and the concern that the evaluator might not address the real issues at hand. These authors also state that intuition is based on the assimilation of various conscious and unconscious stimuli that form an impression on our subconscious. The way we view the current, day-to-day state of affairs of our programme (e.g. numbers of participants, setting, equipment, staff) is compared with our expectations of what the programme should be like. Our judgements are thus intuitive comparisons between what is and what ought to be. While intuition has its place, its drawback is when it is used too often, especially when other approaches are more appropriate.

Evaluation by standards

Edginton *et al.* (1992) define evaluation by standards as an approach 'in which a set of local, state, national, program-specific, or federal standards are applied to the existing program or operations of an agency' (p. 382). These professionally developed standards, often developed by respected professionals in the field, are established to act as levels of achievement or goals (e.g. for safety purposes) which must be satisfied (Kraus and Allen, 1997). For example, canoeing operators would have to ensure that participants wear life jackets at all times while in the canoe, and that participants must be able to swim 50 m before being admitted to the programme. This method focuses more on administrative details than customer satisfaction, and the prescribed standards which are developed often reflect the values of the agency or professionals responsible for their development (e.g. Association for Experiential Education, which has established a whole series of standards for a number of outdoor recreation activities). A political jurisdiction that is in the process of finalizing its accreditation standards for ecotourism is Saskatchewan. This province has developed a whole range of professional standards which prescribe exactly what operators must do to ensure an appropriate standard of care (e.g. considerations of design, ecolodge considerations, enhancing the nature experience) in order to gain accreditation (see Fig. 10.2). Farrell and Lundegren (1993) write that this is one of the most common evaluation approaches used by those in the public domain. Other groups who develop standards include, for example, the American Camping Association, the Australian Ecotourism Accreditation Programme, and other agencies who maintain a vested interest in the proper and safe conduct of those working in the field.

Evaluation by goals and objectives

Otherwise known as the goal attainment method, this approach 'involves identification of goals and objectives and then an assessment to determine the discrepancy between the expected performance/outcome and the actual or realized performance/outcome' (Edginton *et al.*, 1992, p. 377). It is a method that is used after programme implementation, and one that yields information to be used during the next period of implementation. Its usefulness lies in its ability to be objective, especially when goals and objectives have been well written (it is much less useful if goals and objectives have not been well articulated). Furthermore, its usefulness is judged by the meaningfulness of evaluative criteria that have been selected by the professional (Carpenter and Howe, 1985). The following is a hypothetical example of how a goals attainment scale (GAS) may be used by a park agency.

Parks managers must attempt to accomplish a series of goals which relate to protection and use. Unfortunately, these uses are often incompatible – especially when the achievement of one goal may cause the failure of another. The result is that a number of well established goals must be articulated to ensure that people receive meaningful experiences, and wildlife and natural resources are protected for the future. For this hypo-

❑ yes	❑ no	❑ n/a	Does the operation include accommodation that directly involves and enhances the activities?
❑ yes	❑ no	❑ n/a	If portable accommodation facilities (tents/tipi) are used, are they designed for minimum impact?
❑ yes	❑ no	❑ n/a	In the case of tipi or other ethnic structures used as part of a cultural experience, are the materials and design authentic?
❑ yes	❑ no	❑ n/a	Does the accommodation facility fit the definition of an ecolodge?
❑ yes	❑ no	❑ n/a	Does the accommodation enhance the 'nature' experience?

Fig. 10.2. Accreditation standards for sustainable accommodation. From Ecotourism Society of Saskatchewan (2000).

thetical example, our park has the mandate of ensuring: (i) protection, (ii) education, (iii) recreational enjoyment, and (iv) forestry. The first step (A) in the GAS process is to establish goals and assign a weight to each goal relative to other organizational goals. Step B involves the establishment of levels of attainment, which are possible outcomes that are established for each of the goals (only one set of attainment criteria will be established for one of the aforementioned goals). In step C, at the beginning of the evaluation period, the programmer determines the current level of functioning by multiplying the weighted value (A) by the level of attainment (B), and adding the scores. This is done for each of the goals. Finally, in step D, the same process is followed at the end of the programme, with the purpose of determining differences in the achievement of goals over the course of the entire programme. This procedure is outlined below in Box 10.1.

Collectively, it is shown that there is an increase in goal attainment in three of four goals. This is particularly true for education, where there was a noticeable increase in goal attainment; little change in use and community; and no change in protection at the end of the programme.

Evaluation by professional judgement

External evaluators are often used to judge the worth of programmes in the field of recreation, as stated earlier. These experts typically use a set of external standards to measure the effectiveness of the programme under review, which come in the form of checklists, as in the case with the standards method of evaluation, but also on the basis of observations, interviews and a review of programme documents. This process yields peer evaluation and acceptance, as well as information on

Box 10.1.

Use of a goals attainment scale (GAS).

A. Establish goals and assign a weight to each goal:

	Weighted value
1. Protection	15
2. Education	5
3. Recreational enjoyment	10
4. Forestry	5

B. Establish level of attainment criteria: 5 levels (using only goal number 2 as an example)

1 = most unfavourable condition vs. 5 = most favourable condition

1. No forms of environmental education within the programme
2. Leaflets provided for participants
3. Leaflets and interpretive signs distributed throughout the park
4. Leaflets, interpretive signs, and videos shown every 2 hours
5. All of the above with an interpreter

C. At beginning of evaluation period, determine current level of functioning by multiplying ...
Weighted value (A) × Level of attainment (B) for each goal and add scores

D. Repeat step C at the end of the programme.

Level of attainment	Protection (15)	Education (5)	Use (10)	Community service (5)
1		5 [a]		
2			20 [a]	
3	45 [a] 45 [b]		30 [b]	
4		20 [b]		20 [a]
5				25 [b]

[a] Beginning of evaluation period: Total equals 90

[b] End of evaluation period: Total equals 120

participant outcomes and the basis for programme revision and improvement. The differences between the evaluation by standards, evaluation by goals and objectives, and evaluation by professional judgement approaches are summarized in Table 10.3 (developed by Carpenter and Howe, 1985).

Transaction observation method

This method of evaluation is completed by the programme participants themselves, often in conjunction with the service provider who is responsible for the distribution of the evaluative tool. This approach is based on the measurement of participant satisfaction/dissatisfaction, attitudes, behaviours, values, benefits, and other such variables. The effectiveness of these measures is magnified by the inclusion of sociodemographic data, which provides the programmer with the ability to link levels of satisfaction, for example, with age, gender, geographic location, income, levels of education, and occupation. Questions typically relate to whether scheduling of the programme is convenient to one's lifestyle, if a facility is adequate in terms of room and amenities, ratings on leadership, marketing of the programme, fees, interpretation and safety. An example of such a form is found in Fig. 10.3.

Communicative and interactive approach

The communicative and interactive approach is 'based on the involvement of individuals, community groups, special interest groups, and focus groups to gather information before the planning of a program' (Edginton *et al.*, 1992, p. 387). As ecotourism often involves a number of different stakeholder groups such as aboriginal groups, park staff, other land users, local people, this approach may hold merit in soliciting a level of support from these groups that would allow for the continued and successful development of ecotours in the region. This becomes an important step particularly if the ecotour operator is competing for the resources of the destination with these other groups. This approach allows for direct feedback, and lends itself well to both quantitative and qualitative analysis.

Evaluation of leisure experience

The following model developed by Little (1993) examines evaluation in the context of leisure experience (Fig. 10.4). The model is hinged on the notion, as stated earlier, that justification of programmes will depend not only on attendance and revenue, but also, and perhaps more importantly, on the value

Table 10.3. Taxonomy of models for programme evaluation.

Category	Purpose	Primary methodology	Outcomes
Evaluation by standards	Programme efficiency	Checklist	Comparison to standards
	Programme effectiveness and accountability	Checklist	Comparison to standards
Evaluation by goals and objectives	Congruence between goals and outcomes	Comprehensive planning process	Citizen input in goal determination
	Congruence between behaviour/objectives	Systems approach to programming	Indicators of degree of achievements
Evaluation by professional judgement	Professional experts judge and approve programmes	Observations, interviews and review of documents	Professional/peer acceptance and validation
	Thorough description, achievement of outcomes, and judgement of programme worth	Case-study technique	Detailed information on client outcomes as basis for programme revision and improvement

Source: adapted from Carpenter and Howe (1985).

ECOTOUR EVALUATION FORM

Newton County Resource Council
P.O. Box 513
Jasper, Arkansas 72641
Phone: (501) 446-5898

Please participate... *Your comments and suggestions are important in our constant effort to improve our ecotours. It will take only a few minutes to complete this short form. We appreciate your cooperation.*

Date of Ecotour: ____________ Name of Ecotour Guide: ____________

1. What did you enjoy about this ecotour? ____________
2. What, if anything, did you not enjoy? ____________
3. How could this ecotour be improved? ____________
4. Please write in below your opinion of the printed materials provided for this ecotour, using the following scale:
   5 = *Outstanding* 2 = *Fair*
   4 = *Good* 1 = *Poor*
   3 = *Average*

	Pre-tour information	Tour information
A. Amount of information	____	____
B. Usefulness and readability	____	____
C. Relevant Information	____	____
D. Design and graphics	____	____
E. Overall evaluation	____	____

5. Was the price of this ecotour an important factor when you chose it? ______ Yes ______ No
6. What factors do you consider when deciding how much to pay for an ecotour?

____

Figure 10.3 Ecotour Evaluation Form
Source: Adapted from ARCDC (1997)

This ecotour evaluation form, completed by Ozark Ecotours' participants, yields important marketing and program evaluation information to the Newton County Resource Council that supports in-house promotion and redesign efforts. The type of questions used in evaluation forms such as this should be driven by the planned end use of the gathered information and by the uniqueness of the ecotourism program developed in your community.

Fig. 10.3. Ecotour evaluation form. (Source: ARCDC, 1997.)

of services designed for the long-term improvement of society. Clawson and Knestch's five-stage total-experience model (see Chapter 1) is matched with Driver and Tocher's recreation experience continuum (pre-recreation engagement, choice to seek recreation, intervening conditions before and during participation, and post-recreation engagement conditions) in forming Model 1 of Fig. 10.4. Model II is based on the work of Iso Ahola (1980) (see Chapter 5), and involves the social–psychological aspects of leisure experience including the individual, the social situation, participation decisions and engagement assessments. The figure also includes a programme diagnosis component which allows for an assessment of the programme on the basis of formative evaluation (at stages II and III) and summative evaluation (at stages IV and V). This information is examined and later used as a diagnostic tool. In the words of Little (1993):

> The two models of leisure experience … show that conditions prior to engagement lead to a participation decision. In stages I to III of the experience, an individual's participation decision can result in participation, postponement of participation, or withdrawal from participation, depending upon the circumstances. During and after the leisure engagement, the individual assesses the consequences of the engagement. Positive outcomes (e.g., joy, enjoyment, fun, pleasure, satisfaction) are among the expected conditions after the leisure engagement. Whether positive or negative, the outcome of the leisure experience becomes part of an individual's past experience and influences future leisure definitions.
>
> (p. 28)

The model provides a theoretical basis from which to better understand the success of a programme on the basis of the engagement of participants and their willingness to continue with a programme or not. Although decidedly more conceptual than other models of evaluation, it nevertheless enables the service provider to structure some very detailed analyses on the basis of time, programme diagnosis and experience.

Importance–performance evaluation

In the previous model, Little talked about positive outcomes, including fun and satisfaction, as being derived from leisure participation. This perspective lays the groundwork for other intuitive and empirical measures of satisfaction. The importance–performance procedure (based on the work of Auld and Cuskelly, 1989) is one such measure which allows recreation and tourism practitioners to understand the importance that participants place on certain aspects of the programme, and how effective the programmer has been in accomplishing their ends. The first part of the process, the importance scale, requires participants to rate the importance of various aspects of the programme on a seven-point

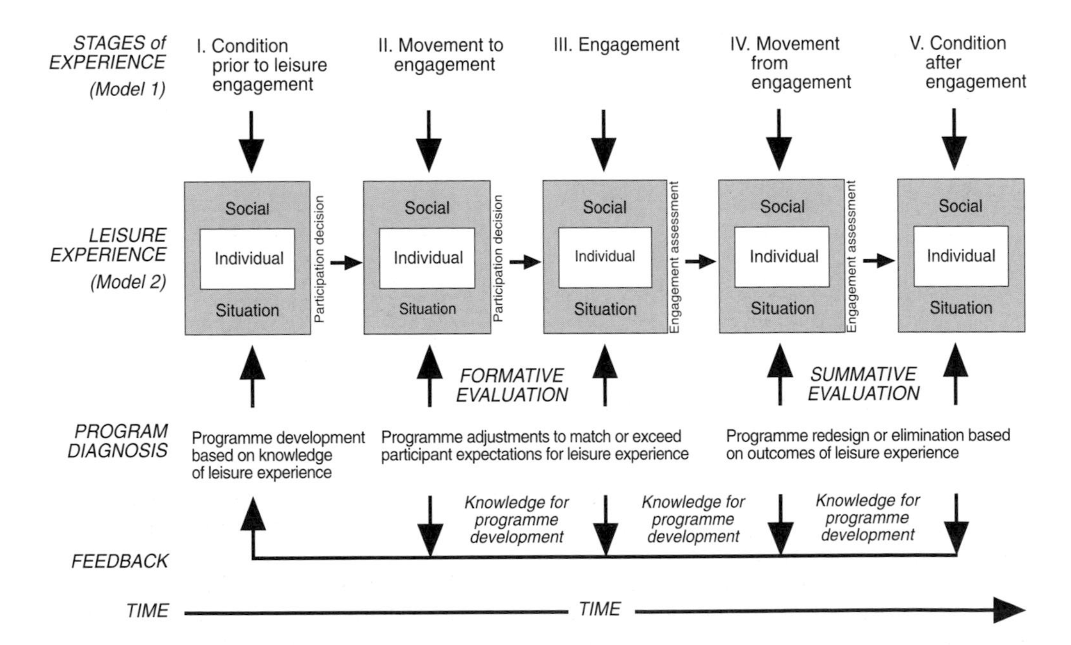

Fig. 10.4. Models of leisure experience used as a diagnostic tool for evaluation. (Source: Little, 1993.)

scale, ranging from 'Not important' to 'Very important'. Similarly, the second part of the process, the performance scale, also uses a seven-point scale, but asks participants to rate programme variables on the basis of 'Terrible' to 'Delighted'. An example of both scales adapted to ecotourism is given in Box 10.2, with calculations of mean scores for each of the two scales included.

Scores are then plotted on an action grid (Fig. 10.5), which contains the following four quadrants: (I) 'Keep up the good work', indicating programme features of the highest importance which receive the highest performance ratings; (II) 'Possible overkill', indicating those features which are of lower importance to tourists, but rating high on performance; (III) 'Low priority', including those features rated low on importance and performance by tourists; and (IV) 'Concentrate here', which includes features rated as high on importance and low on performance (said to be the most critical for programme evaluation). Consistent with past applications of the model, importance and performance axes are set at 5, representing *of importance* on the importance scale and *mostly satisfied* on the performance scale. Auld and Cuskelly (1989) write that the importance–performance evaluation is easy to use, analyse and interpret, with the added benefit of readily demonstrating which aspects of the programme are good and which are not.

Evaluation of operator ethics

Although ecotourism is often given the label of a more responsible form of tourism, justifiably or unjustifiably, there are surprisingly few mechanisms, approaches or theoretical perspectives in place which may be used to substantiate such a claim. Fennell (2000a) writes that the tourism industry in general could benefit from more formalized centres of tourism ethics in the same way business, the environment, medicine and law have

Box 10.2.

The importance–performance procedure.

Importance scale:	Not important		Somewhat important			Very important	
1. Quality of lodging	1	2	3	4	5	6	7
2. Knowledge base of guide	1	2	3	4	5	6	7
3. Pre-trip information	1	2	3	4	5	6	7
4. On-site transportation	1	2	3	4	5	6	7

Performance scale:	Terrible		Mixed			Delighted	
1. Quality of lodging	1	2	3	4	5	6	7
2. Knowledge base of guide	1	2	3	4	5	6	7
3. Pre-trip information	1	2	3	4	5	6	7
4. On-site transportation	1	2	3	4	5	6	7

Feature	Quadrant	MIR	MPR
Quality of lodging	Good work	6.58	6.36
Knowledge base of guide	Low priority	3.97	4.24
Pre-trip information	Concentrate	6.32	4.08
On-site transportation	Possible overkill	3.84	5.38

MIR, mean importance rating; MPR, mean performance rating.

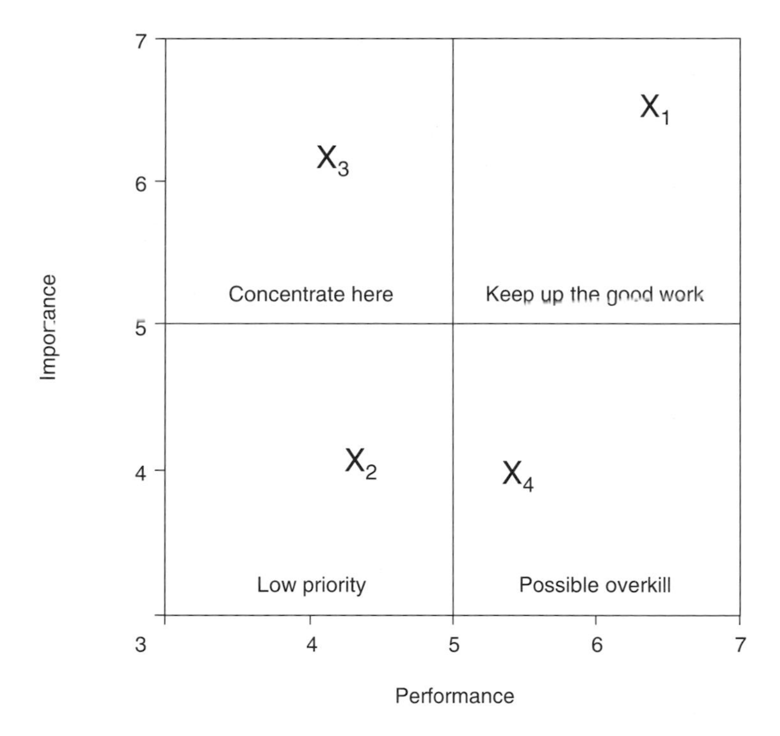

Fig. 10.5. Mean importance–performance ratings. X_1, quality of lodging; X_2, knowledge of guide; X_3, pre-trip information; X_4, on-site transportation. (Source: adapted from Auld and Cuskelly, 1989.)

benefited from these institutes of applied ethics. One of the initiatives of such an enterprise for ecotourism would be to evaluate the ethical behaviour of ecotourism operators, with the overall goal of developing a database of service providers regionally, nationally and perhaps internationally. A standard survey mechanism may be used to examine the ethical orientations of different service providers, such as the one in Fig. 10.6 (from Fennell, 2001). This would provide the added benefit of allowing consumers to gain an appreciation of the ethical nature of service providers along a number of themes, including general ethics, local people, environmental education, operator professionalism, contribution to conservation, and accommodation and transportation.

The reader is urged to consult the many recreation programming books available to peruse other evaluation methods which may not be listed here. At the same time, however, the approaches highlighted in this book are felt to be among those which are most frequently used by practitioners. The vast number of evaluative models that are available provide the programmer with the opportunity to adopt an evaluation approach which is most practical given available resources and time. The important thing, as outlined earlier, is to evaluate. In doing so it completes the programming cycle, allowing for critical information to be fed back into the next programme incarnation for the purpose of implementing better services – which is the topic of the next section.

Evaluation Details and Issues

Details

The results generated from the use of one or many of the aforementioned evaluation models will need to be carefully considered by management. As suggested in the section on needs assessment in Chapter 5, assessments and evaluations should not be under-

Programming Tip 10.1.

Participant evaluations.

Quest Nature Tours sends a 'Welcome Home' postcard to clients who are still away on a Quest trip. On arriving home, the client will discover the postcard, which asks the client to outline highlights of the trip, whether expectations were met, or other aspects which may improve the programme. The responses from participants usually come in the form of personal, hand-written letters which often include very personal and thought-provoking responses. The postcard thus helps Quest to further develop the personal relationship with the client, which has enabled Quest to attain a 65% rate of repeat visitation. Quest avoids the use of on-site questionnaires which Barry Griffiths, owner of Quest, feels compel participants to focus on negative aspects of the trip, even if such aspects have been relatively well received by participants.

This evaluation form includes many of the most frequently cited aspects and concerns related to an ecotour experience. Please rate your ecotour operator's ethical behaviour on how well they satisfied each of the criteria below, from a 'Superior', 'Adequate' or 'Inferior' perspective.

Operator: ______________________ Tour dates: ______________________

Tour description: ____

Evaluation criteria	Operator's ethical behaviour			
	Superior	Adequate	Inferior	N/A
General ethics:				
Respect shown to animals				
Respect shown to plants				
Operator's treatment of financial matters related to the tour				
Respect shown to local people				
Operator following a code of ethics				
Respect shown to ecotourists				
Local people:				
Operator employing local/indigenous people as guides or front-line personnel				
Operator efforts to employ local/ indigenous people in middle or upper management positions				
Local people were in control of the decision making in relation to ecotourism in their community				
Local people owned and ran the ecotour operation				
Environmental education:				
The interpretation of the natural world				

Fig. 10.6. Evaluation of ecotour operator ethics. (Source: Fennell, 2001b.)

Evaluation criteria	Operator's ethical behaviour			
	Superior	Adequate	Inferior	N/A
Knowledge of plant and animal life				
Level of pre-trip information provided to the client				
Level of on-site information provided to the client				
Operator/guide had formal education in ecotourism or ecotourism-related fields				
Operator professionalism:				
Operator/guide was accredited or certified				
Knowledge of land use and park policies				
Knowledge of the site or region itself (e.g. location, history, weather)				
A low guide-to-participant ratio				
Language proficiency of guides				
Degree of safety of the experience				
Understanding of the importance of minimizing tourism impacts				
Contribution to conservation:				
Monetary (e.g. park fees, support of local nature groups)				
Physical (e.g. removal of garbage, tree planting)				
Accommodation and transportation:				
Reliance on 'green' hotels				
Use of hotels that are locally owned				
Use of hotels that are built with local materials				
Use of hotels that have a low carrying capacity				
Use of vans/buses that use cleaner petroleum				
Over reliance of vehicles at ecotour sites				

Fig. 10.6. *Continued*

taken if there is a lack of motivation to do so, if resources are lacking, or if the right questions have not been asked. Failing to place a needed level of energy behind such initiatives will result in poorly conceived methods and answers to questions, which have little use. (See Chapter 5 for an overview of techniques which may be employed in conducting needs assessments and evaluations, as well the process involved in constructing reports.)

Foremost, the evaluation equips the operator with the required information to gauge the significance of the programme. With this information in hand, the service provider has what is needed to decide the fate of the programme on the basis of a number of different options, including: (i) continue the programme with no modification; (ii) continue the programme with modifications; (iii) end the programme and repeat it at some other time; (iv) end the programme and modify it at a later date; or (v) end the programme altogether. Obviously in cases where the programme has been quite favourably evaluated, little or no modification is necessary (although like any true professional the service provider will naturally want to tinker with the programme to make it better). At the other end of the spectrum negative evaluations would mean that the programme, especially if it has been negatively viewed in the past, should be shelved or removed completely. However, as Carpenter and Howe (1985) quite rightly suggest, there needs to be a very tight relationship between programme revision and needs assessment (the beginning of the next programme cycle). Without assessing the clients' needs from time to time, programmers will be unable to know if the demand for a terminated programme can be rekindled. The link between the evaluation and the needs-assessment cycle is effectively articulated by these authors who write that 'Program revision allows you to be immediately responsive to your clientele as a result of your evaluation findings, while needs assessment keeps you informed of the changing demands for leisure programs and services, enabling you to remain responsive over the long run' (p. 172). Programme demise or death for those which are unpopular may simply mean that there may be a time in the future when the programme is able to generate enthusiasm.

Whoever is responsible for making programme change – if change is required – it is important to include not only the people who are closest to the programme (e.g. guides), but also those in upper management. This latter group is especially important as they are the ones who will financially and administratively support such changes. Proposed changes will come in the form of recommendations which may have a whole range of implications to the organization, clients and other programmes. Although primarily programme-based, these recommendations may also attempt to address a number of other important financial, political or community-based issues that have a direct bearing on the programme, or on the organization in general.

Henderson and Bialeschki (1995) remind us that no matter who conducts the evaluation, how, and with what means, issues and dilemmas will almost certainly arise. In an effort to control these inconveniences as much as possible, these authors offer the following rather lengthy list of points which will be useful in helping to avoid many of the downfalls of evaluation projects. These points, some of which have been discussed in this chapter and in Chapter 5, are as follows (adapted from Henderson and Bialeschki, 1995, pp. 310–312):

1. Make sure you have the skills and knowledge necessary to do the type of evaluation needed. If not, get help or get someone else to do the evaluation.
2. Be clear about the purpose of the evaluation. Make sure stakeholders are also clear about these purposes as well.
3. From an administrative point of view, make sure the expense of an evaluation in terms of time and effort is comparable to the value received from doing the evaluation.
4. Consider how both internal and external evaluations can be used from time to time. Using only one or the other has some drawbacks.
5. Remember that evaluation does not only occur at the end of the programme (summative), but may occur as an assessment or as a formative evaluation.
6. Try not to allow bias, prejudice, preconceived perceptions, or friendship to influence evaluation outcomes.
7. Make sure external evaluators enter into an evaluation thoroughly understanding the organization and its limitations.
8. Select and use evaluation research methods and measurement instruments that are specifically related to the criteria – this entails a great deal of planning at the beginning of the task.
9. Think about the logical timing of the evaluation. Keep in mind when it can best be

done within an agency and when people involved are most likely to be receptive to data collection as well as the reporting of conclusions and recommendations.

10. Continually think about how the criteria, evidence and judgement phases of the project fit together. Each succeeding component should be a natural outgrowth of what has gone before.

11. Careless collection of data results when the wrong instrument and poorly written questions that are inappropriate to respondents are used.

12. A well written survey is necessary to avoid pitfalls related to reliability and validity. Avoid the use of technical jargon so that the sample will not be biased toward more highly educated groups.

13. When doing any evaluation using quantitative data, keep in mind issues related to randomization.

14. Consider sample selection carefully to avoid bias, including how you may motivate people to participate. Before collecting data think about the size of the sample, its representativeness, and the desired response rate.

15. Data analysis should be considered before the project is begun. The type of analysis(es) will be a function of the types of questions asked.

16. Make conclusions and recommendations based only on the data collected from the project. Do not claim more than you have evidence to support.

17. Ethically, it is very important to address both positive and negative findings, the latter of which may need to be handled more carefully than the former.

18. Be open to all of the results of the project. This forces the evaluator to consider information that falls outside the realm of the expected.

19. Write concise, complete, well planned reports, which enable the evaluator to effectively communicate results through a variety of means – written and graphic.

20. Disseminate results as soon as possible. Nothing kills enthusiasm for an evaluation like having it drag on for months or years.

Issues

It is an unfortunate fact that many programmes fail because of the things that service providers did not do to make it better; the things they did do but failed; or simply because the programme has run its course. The failure of a programme is especially frustrating if similar programmes are thriving in other agencies. It then becomes essential to evaluate more fully the practices of the agency, including the management of resources, experience of employees, training regimes, logistical expertise, and understanding of the market. Perhaps the service provider has simply ignored the signs that indicate poor programme performance, and not made the appropriate adjustments for the future. Problems may stem from what is or is not being done at the upper management level, such as asking employees to take on too much responsibility (e.g. bus and hotel bookings as well as interpretation). It may also be the guide who is problematic, for example, restricting most of his/her time to three or four ecotourists on the trip instead of the whole group. In other cases there will be failure to communicate between levels of the organizational hierarchy, with the ultimate consequence of not making connections, missing buses, failing to book rooms or restaurants, or other matters. Still other problems may come about from the inability of the service provider to adequately handle participant problems that may or may not result from the ecotourism operation itself.

It is a widely believed adage that: 'the customer is always right'. Those who lead trips, however, know first-hand that this is not always the case. Trip leaders also know that they must walk a fine line between knowing what is right and 'negotiating' with participants who may think they are right. There is little to be gained in attempting to win an argument with a customer; a notion which relates well to the position of Mitchell (1992), who suggests that the client is not someone to argue with or match wits with. In cases that are isolated and infrequent, the leader may wish to simply brush off comments which are negative or simply wrong and proceed without confrontation. However, in situations where an individual is continually disruptive, the leader may wish to pull him or her aside and be very clear about the roles and responsibilities of leaders and followers (see Chapter 8). This may well extend to a confrontation before the group, which may be an important step in maintaining the respect of the group and representing the concerns of the group members, who may have also grown tired of the frequent outburst by the offending party. In order to be as up-front as possible in identifying and dealing with these predicaments, Van der Wagen and Davies (1998) identify a general problem-solving flow chart for the myriad and inevitable issues that continually bombard those who work in the service sector, including the need to respond immediately to the various needs of customers. The stages in this typology include:

- *Define the problem.* Defining the problem, which in many cases will solve the problem immediately.
- *Analyse the problem.* This includes an analysis of the extent and severity of the problem, helping to clarify thinking and acting as a precursor to the next stage.
- *Develop a range of solutions.* Here there is an attempt to arrive at a solution that is easier to implement than other options.
- *Test the solutions.* Solutions can be tested for their short-term and long-term implications. The best solution is the one which will most effectively satisfy the customers' needs, usually as quickly as possible.
- *Take action.* As the most important stage, the intent is to fix the problem, but often the situation is so complex and threatening that no action is taken at all (e.g. cancelling a trip).
- *Follow through.* In cases where customers have been dissatisfied or complained, this is the most important process, and one which can turn a customer into a loyal supporter. Examples may include offering free merchandise, like a shirt or discounts on other trips.

This is quite similar to the customer complaints action plan established by Edginton and Edginton (1994) in Fig. 10.7, which is a process that ranges from acknowledgement to the review of the organization's commitment to service quality. This latter framework differs from the preceding one in its initial stages, which focus on affirmation or acknowledgment of the importance of the customer's complaint. This includes the involvement of the customer in gathering information, in identifying correctable options, and the importance of holding the organization accountable for its actions. Applied to any of the hypothetical situations stated above, the service provider would acknowledge the complaint; gather information from customers and the leader; be sincere in following through with the plan to resolve the issue (which may be either to set the guide straight or change the guide); follow up with the clients to ensure their satisfaction; and finally to review the organization's policies in regards to client–company relations.

Debriefing

Part of the discussion in Chapter 7 included a section on briefing the client through the dissemination of pre-trip information which was designed to allow the participant to prepare adequately for the ecotour. At a more specific or applied level, the service provider will need to brief participants at appropriate times during the course of the trip, usually in the morning before embarking on their adventures and at various times of the day. Briefing will include information on what is to be expected at the attraction, key information on species, communities, or other phenomena, which allows the ecotourist to maximize their on-site experience. While it is understandably important to brief participants, it is equally important to debrief them as well. Debriefing or processing, as it is also known, occurs both at the end of a programme experience, but also during the event itself (Bisson, 1999).

Priest and Gass (1999) have identified six discrete generations of leader facilitation skills which accentuate the evolution of outdoor and adventure learning. These authors suggest that in the 1940s and 1950s techniques typically involved 'letting the experience

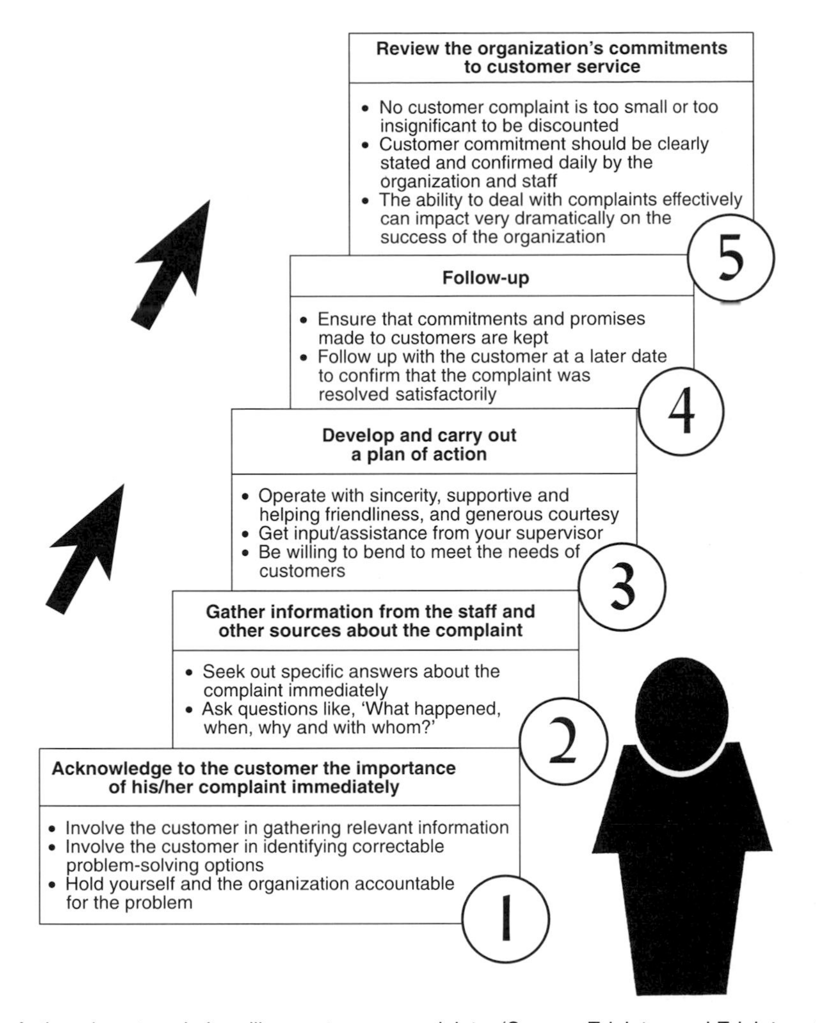

Fig. 10.7. Action plan steps in handling customer complaints. (Source: Edginton and Edginton, 1994.)

speak for itself' (i.e. clients left to sort out their own personal insights) and 'speaking on behalf of the programme' (i.e. learning by telling, where the leader interprets the experience for the clients). More recent approaches allow participants to learn through reflection which may be induced through a variety of questions posed by the leader. In the words of Priest and Gass (1999), 'In this model, clients learn under the guidance of a questioning leader, who helps the group members to discover their own learning. Although numerous methods of debriefing exist, the most common approach in North America is the group discussion' (p. 216).

The ecotourism service provider thus has the opportunity of enriching the experiences of his/her clients by stopping and taking the time to ask some provocative questions related to the attraction(s) of interest. Questions might include: 'What ecological relationships exist in the park?', 'What is the impact of overuse of the region?', 'What was learned from the experience?', 'How should visitors act in minimizing impacts?', and 'Who knows of similar situations in other regions of the world?' Ecotourism guides may wish to continually refine their group discussion techniques to allow ecotourists to learn more fully from their experiences (see Chapter 6 for a discussion of environmental education and interpretation).

Certification and Accreditation

In general, certification has been referred to as the 'process by which an individual is tested and evaluated in order to determine his or her mastery of a specific body of knowledge, or some portion of a body of knowledge' (American Society of Association Executives, in Morrison *et al.*, 1992, p. 33). Closely tied to certification is accreditation, which Issaverdis (2001) defines as 'programmes that provide a means of establishing the extent to which a business offering tourism experiences meets industry nominated standards. The programme encourages the delivery of consistently high quality products and promotes continuous improvements' (p. 583). As a point of differentiation, accreditation focuses on programmes and institutions, while certification applies to individuals (see Morrison *et al.*, 1992). A more recent view of the two concepts suggests that certification, in the context of tourism, 'refers to a procedure that assesses, monitors, gives written assurance, and awards a marketable logo to a business, attraction, destination, tour, service, service provider (such as a naturalist guide), process, or management system that meets specific standards'; and that accreditation is 'the process of qualifying, endorsing and "licensing" entities that perform certification of businesses, products, processes, or services. In other words, an accreditation programme certifies the certifiers' (Honey and Rome, 2001).

Certification, as regards outdoor recreation programme leadership and implementation, has been very much about experience and expertise. Those who have certifications are often the ones who are most capable of leading successful trips. However, at the same time, there are individuals who have been participating in outdoor-based activities for years, have excellent skills, but have not attained specific levels as set forth by a certifying body. In such cases, what is needed is a mechanism to determine competency, without punishing the individual who has a wonderful skill set but who lacks the formal certification that other less experienced leaders may have. Ewert (1989) has developed a certification model (Fig. 10.8) for outdoor leaders which gauges a person's leadership attributes in three ways: through an objective examination, experience and a selected jury. This information would be used by the certifying body to determine eligibility for certification, at which point the leader would be subject to a fee structure and later to periodic reviews by participants.

This three-pronged approach acknowledges that certification is only one of a number of avenues available for guaranteeing competence. Ewert illustrates that conferences, research, training programmes, journals and informal controls, such as reputation and information exchange, are also important in demonstrating competence in any profession. He further cites at least 20 institutions (e.g. Association of Experiential Education) and journals (e.g. *Journal of Experiential Education*) which distribute information relative to competencies in the field of outdoor adventure pursuits. This, it is argued, is important as a means by which to effectively develop a core body of knowledge, which may ultimately set adventure pursuits aside as a profession. The same could ultimately be established for ecotourism, based on the emerging literature (e.g. the newly created *Journal of Ecotourism*), development of regional certifying bodies, and an assessment of competence based on experience in the field, among other types of controls.

The concept of accreditation in ecotourism has generated a great deal of interest of late, both by practitioners and academics. Unfortunately, such programmes lag far behind the outdoor recreation field, where organizations such as American Mountain Guides Association and the American Camping Association have been developing programmes for years. The enthusiasm for accreditation seems to have been catalysed as a result of the development of Australia's National Ecotourism Accreditation Programme (NEAP). McArthur (1997) writes that in the mid 1990s there was a growing number of stakeholders lobbying for ecotourism accreditation in Australia. The programme became more of a reality after a survey of some 500 ecotour operators in 1995, where there was resounding support for the idea. Sustainability appears to be a central tenet of the

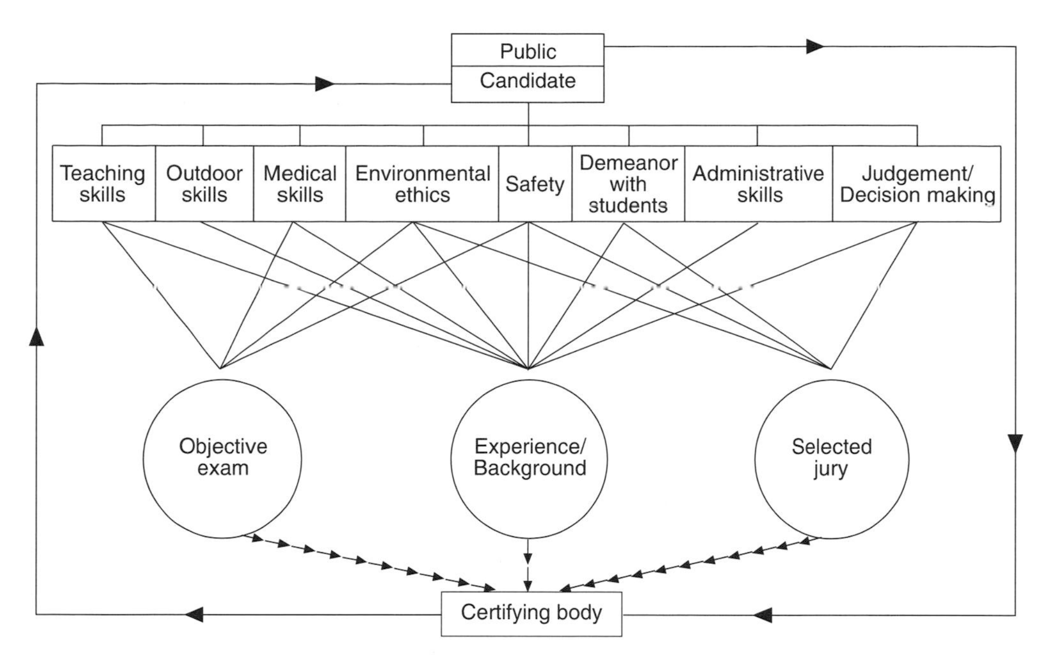

Fig. 10.8. Certification model for outdoor leaders. (Source: Ewert, 1989.)

accreditation process, with the hope that in all manner of business practice, operators develop exceptional programmes which are beneficial economically, socially and ecologically. (See Ecotourism Society of Saskatchewan, 2000 for a similarly devised accreditation programme for that province.) This self-regulating programme is based on four monitoring procedures: (i) applicant honesty in completing application documents; (ii) feedback from clients; (iii) feedback from referees (random selection of operators); and (iv) through random audits. Operators are given a core level of accreditation if they satisfy the programme's basic criteria. However, the system encourages operators to implement measures beyond the standards of the core criteria in earning advanced standing. Accreditation in Australia is based on the following eight criteria (from McArthur, 1997):

1. Focus on personally experiencing natural areas.
2. Integrates into each experience opportunities to learn.
3. Represents best practice for sustainable tourism.
4. Contributes to the conservation of natural areas.
5. Provides ongoing contributions to local communities.
6. Involves and is sensitive to different cultures.
7. Consistently meets client expectations.
8. Marketing is accurate and leads to realistic expectations.

A more recent incarnation of Australia's NEAP programme is NEAP2 (renamed Nature and Ecotourism Accreditation Program) which is designed to certify programmes, not companies, as a result of the belief that even within companies there can be massive deviations between different programme offerings. Another change is the incorporation of Nature Tourism programmes under the umbrella of the concept. Whereas in the past, operators were given one of two levels of certification, Ecotourism and Advanced Ecotourism, the third Nature Tourism category allows services that are natural areas based and sustainably oriented (but do not fit the stricter definition of ecotourism) to be recognized.

The jury is still out on the success of Australia's accreditation process. As suggested by Issaverdis (2001) there appears to be a significant need for research to demonstrate the value of accreditation products over non-accredited products. This gets back to the questions raised earlier on the pros and cons of certification for guides. Although certification provides a recognized seal of approval, we should not forget that there are countless other experienced and professional guides operating in the field who see no need to accumulate certifications. The real measure of success for certification and accreditation schemes is in the many tangible benefits they have on the industry and its various stakeholder groups, as well as the success of the reporting and audit processes. In this regard, studies comparing the programme offerings of accredited and non-accredited operations would greatly help in achieving a better sense of programme excellence. Furthermore, documented evidence on the merits of self-regulation as the principal vehicle driving the accreditation process may instill more confidence that the process does in fact maintain the natural and cultural heritage values of the regions in question (see Wearing, 1995, for a discussion of the pros and cons of accreditation). Furthermore, accreditation bodies should continue to look at other models that exist in other regions, and fields, throughout the world. For example, the Ontario Camping Association's (1990) Guidelines for Accreditation for Resident Camps, provides an excellent model of how *mentorship* can aid in allowing camps to reach their full potential as accredited bodies. The Ontario Camping Association (OCA) incorporates the following measures into their process (pp. i–5):

1. The owner/operator or designate applies for Provisional Membership.
2. The Membership Committee determines that the application is complete.
3. The Board approves the camp as a Provisional Member in the first (mentor) year of a 2-year accreditation process.
4. A mentor is assigned to the camp by the Standards Committee.
5. The mentor contacts the camp and assists the camp by answering questions, making suggestions, and encouraging involvement in the OCA.
6. The mentor (or another OCA member) visits the camp during the following summer and goes through the established Standards Visit process (a complete tour of the camp).
7. All visitation forms are returned to the office.
8. The Standards Committee reviews the Answer Sheet and informs the camp of any areas of concern (e.g. health, facilities, equipment, programmes, transportation, administration).
9. The mentor continues to maintain contact with the camp during the following winter.
10. The accreditation visit takes place in the second summer. A different, experienced OCA visitor is assigned by the Standards Committee. All forms are returned to the office.
11. The Standards Committee reviews the Answer Sheet. A camp must attain a minimum of 90% in each section of the 'Guidelines For Accreditation' and subscribe to all the applicable major standards. The camp is then eligible for approval by the Board of the OCA as an Accredited Camp and the owner/operator as a Camp Member.
12. If the camp does not maintain a minimum of 90% or does not subscribe to all the applicable major standards, a review will be made by the Standards Committee. This review will involve personal contact with the owner/operator of the camp and the visitor. The contents of the review, the evaluation of the Standards Committee, and its recommendation will be brought before the OCA Board. The Board must act in one of the following ways:

- The camp is accredited and the owner/operator becomes a Camp Member.
- The camp is not accredited and the owner/operator remains a Provisional Member or reverts back to Provisional status.
- The camp maintains accreditation but must be revisited the following year.
- The Board requires the owner/operator to show evidence of the camp's compliance with the 'Guidelines For Accreditation' before accreditation is given.

13. An accredited camp is visited every 4 years or when there is a significant change in the camp ownership, site or programme or if the Board determines that a visit is desirable.

The element of mentorship is important because it enables both the OCA and the camp itself to be consistent on what is required for proper conduct. Visits by mentors, which form the core of the OCA's accreditation programme, ensure that high standards (over 400 of them) of practice are in place at all OCA member camps. Because these camps deal with children, there is virtually no reliance on a client feedback, which is so important to the Australian system. Where these systems deviate from the OCA procedures is perhaps in the lower level of interaction between the granting body and the operator. For those granting bodies faced with literally hundreds of applications, the challenge to provide mentorship, in the way of the OCA, will be significant. One certification programme faced with such a problem is the Costa Rican Certification for Sustainable Tourism (CST) system (at present it just certifies the accommodation sector, with plans to branch out into restaurants, transportation and operators), which appears to be one of the most highly regarded of its kind in the world (Honey and Rome, 2001). After an initial site visit by Costa Rica Tourism Institute staff, a formal assessment of 153 criteria, and a written evaluation, candidates are supposed to be subject to a surveillance audit every 6 months. Unfortunately, of the 171 hotels in Costa Rica who had signed up by mid-2001, only 51 had been certified. The remainder had not been audited, or if site visits had been completed, written evaluations had not been submitted by auditors. Perhaps the inclusion of a mentorship programme in the way it is used in the OCA, above, may help with such a serious constraint to the Costa Rican system. In addition, this may be aided though the regionalization of the programme to make it more manageable. These considerations, along with a focus on other models, should be considered by jurisdictions in arriving at a process that is thought to be most appropriate for the region.

Environmental Programmes

In general, the certification programmes outlined above are developed on the basis of two principal methodologies: those which are process-based, using environmental management systems (EMS) such as environmental auditing and environmental impact assessments in programme implementation; and those which are performance-based, and which use economic, social, and environmental standards derived from external sources (see Honey and Rome, 2001). Many jurisdictions, however, will use a combination of the two.

Environmental auditing, for example, is a process-based management tool (EMS) that allows for the systematic documentation and evaluation of an organization's equipment and management philosophies and practices, which are geared towards the protection of the natural environment. Auditing may come in the form of water management or transportation systems, for example, which have been put in place to meet regulatory controls and/or to comply with the policies of the organization for the purpose of improving the environmental practices of a business. Diamantis and Ladkin (1999) write that environmental auditing can be used by tourism firms to provide benchmarks, i.e. to allow organizations to compare their practices with those of their competitors, in striving to achieve excellence (see also Issaverdis, 2001). In the context of the EMS developed for recreation and tourism by Tribe *et al.* (2000), the audit is designed to:

- determine whether the EMS is effective (management);
- check environmental performance against stated targets outlined in the policy and programme (environmental performance); and
- check that significant environmental impacts are being addressed.

These authors state that the creation of a checklist of items to audit enables the organization to keep track of progress, while at the same time acting as a simple guide to follow.

Table 10.4 is an example of audit items which have been identified by Tribe *et al.* (2000), under the categories of policy, site review, programme, and operations. Outcomes are described as 'Yes', 'No', or 'Not applicable', and information which pertains to 'Evidence and location' is recorded for each of the items (not included in the table).

Environmental systems, such as the EMS, have evolved with the purpose of adhering to the standards of a number of international and national organizations, which are geared towards more environmentally sound manufacturing. The International Organization for Standardization (ISO), for example, a world federation that is designed to facilitate trade, has developed a number of standards designed to support sustainable development (among other initiatives through the family of ISO 14000 standards). Because of its international derivation, it can be used across and within different industries. However, its strength is also its weakness. The system appears to be too general in addressing the myriad social, ecological and economic issues that persist within a region, and it only demands that improvements in environmental management be reported within the organization itself. Conversely, the European Union's Eco Management and Auditing Scheme (EMAS) is a voluntary form of regulation which has been initiated for businesses interested in EMS certification (Honey and Rome, 2001). Citing Krut and Gleckman (1996), Honey and Rome suggest that EMAS is designed to encourage firms to improve their environmental performance continually, through the production of an initial environmental impact assessment and by disclosing their environmental improvements in annual reports. Also, while the ISO system usually includes government representation from participating members, the EMAS was developed by public and private stakeholders, it includes public access to information, environmental proceedings, assurance of regulatory compliance, it requires third party verification and inde-

Table 10.4. Audit checklist.

Audit item	Outcome Yes/No/Not applicable	Evidence and location
Policy		
Is there a written policy document?		
Commitment to comply with environmental legislation?		
Has the policy been disseminated to staff?		
Site review		
Is responsibility for carrying out review assigned?		
Is there a background of the site report?		
Have start/finish dates been set?		
Programme (management)		
Has the firm got a strong environmental culture?		
Have significant impacts been prioritized?		
Have environmental targets been set?		
Operations		
Are there procedures to operationalize the programme?		
Are results of environmental projects recorded?		
Are all results from operations monitored?		
Audit and review		
Has audit and review procedure been established?		
Does the audit assess the EMS itself?		
Has an audit report been completed?		

EMS, environmental management system.

pendent audits every third year, and is performance-based against the base-line review, which is also publicly reported (Honey and Rome, 2001).

The Green Management Assessment Tool developed by Eagan and Joeres (1997) is a tool which focuses on the facility-based level of operations, and measures progress towards the achievement of sustainable development. The authors state that new ways to measure environmental performance are required as companies have begun to realize the advantages of thinking environmentally, and the consequences of ignoring the environmental impacts of their activities. The Environmental Audits for Sustainable Tourism, or EAST, developed by Hagler Bailly (1997) is another example of an environmental auditing programme which is geared towards generating savings in water, energy, solid wastes, and so on, in the hotels of Jamaica. According to Smith (G., 1998), the EAST project has successfully certified the first two hotels worldwide under Green Globe's internationally based environmental standards, based on the 1992 Earth Summit's Agenda 21. Hagler Bailly developed an audit protocol covering basic areas of concern in addition to an audit guide for the collection of information necessary to evaluate hotel operations.

Although seen by some as the most progressive move by members of the industry to become more environmentally minded, others are more cautious in their assessment of auditing. Honey and Rome (2001), for example, write that certification is often awarded on the basis of EMS setup and not necessarily on implementation. Mowforth and Munt (1998) feel that it is unrealistic to think that environmental auditing, with a focus on the environment and society, would change the operating motives of large tourism firms. These authors cite cases where the environmental policies of tourism firms have emerged to be relatively meaningless as a result of the power struggles which continuously exist in the industry (and which influence such firms). These authors feel that environmental auditing has more value as a tool by which to fine-tune the system of operations, instead of one that has a more overall significance. This point is taken up by Honey and Rome who suggest that process-based management systems *by themselves*, are insufficient to generate sustainable tourism practice – a point which will be revisited in the final chapter of the book.

Green taxes

Green taxes, which have been examined quite thoroughly by politicians and academics over the past two decades (see OECD, 1975) are, simply stated, charges for using the environment. One of the best examples of the green tax phenomenon is the carbon tax, designed to minimize the level of greenhouse gas emissions. In general, prices for licences and permits are set at a rate that would force polluters to internalize the full measure of damage that they inflict on the environment (Goodin, 1999). Such costs include the effects on air, water, ozone, people and other members of the biosphere. The value of a green tax is that it forces companies to acknowledge the impacts that they have on the environment as a consequence of their way of doing business. The chief criticisms, however, are that green tax prices may be set too low as a function of not appropriately assessing costs, as well as the feeling that once environmental costs have been paid, there is no other means by which to curtail the polluting behaviour of the offender. In many cases it has been suggested that the green taxes levied are only a fraction of what the offending party makes financially on a day-to-day basis. In such cases it pays for the offending party to continue to pollute. The green tax viewed from this context equates to basically selling rights to destroy nature (Goodin, 1999).

In tourism there is an increasing number of examples of the implementation of green taxes imposed on those, usually tourists, who use natural resources. Croall (1995) cites examples where the imposition of such taxes has worked, including a 1% tax of evening meals by a hotelier in Dartmoor, a £10 conservation levy on the bill of all customers by the owner of a mountaineer school on the Isle of Skye, and a levy on the use of car parks with money being earmarked for conservation initiatives. In other cases, the implementation of a sixfold increase in the tourist levy to the Great Barrier Reef Marine Park was met with hostility and opposition (Elliot, 1997). It was suggested that in order to implement the tax, better avenues of communication needed to be opened to ensure that all parties were privy to proposed changes. In one final case, an eco-tax was proposed in the Balearic Islands by a newly elected left-wing/green government. The proposal came about in response to the massive damage to the islands through rapid tourism development, which had been encouraged through the previous political regime. The tax was designed for restoration, conservation, and, hopefully, to reduce the number of tourists. While there appears to be general support for the idea, unfortunately, hoteliers and tour operators, as representatives of the international tourism industry (proposed agents to collect the tax), have been opposed to the idea due to the perception of lost earnings (see Morgan, 2000b). Here is a case where those who benefit most from the industry are unprepared to support the destination, despite the potential for some very positive environmental and social assistance – which would no doubt help the industry in the long run.

There appear to be a number of cases where such taxes have been used in an ecotourism context. Some of the most responsible ecotour operators have built a green tax into their fees for the purpose of contributing to conservation. Tactically this is probably better accomplished upfront in the overall fee of the ecotour, to avoid having ecotourists contribute continuously throughout the course of the tour. This may also provide the possibility of having ecotourists contribute at other times of the trip under their own accord, in cases where they are given such an opportunity to do so, and where there does not appear to be a tie to the service operator.

Conclusion

The esteem given to evaluation will most certainly differ significantly among ecotourism service providers. While some may view it as a worthwhile exercise that will help salvage poorly conceived programmes or make excellent programmes better, others may view it as a massive waste of time and other resources. The fact is that many tourism and recreation agencies conduct evaluations infrequently (e.g. every 5 years), while others are much more rigorous in their use of such tools. Evaluation should not be viewed as a special function to be employed in difficult times, but rather as part of the programme cycle. Just as programmes need to be planned and implemented, so too should they be evaluated as part of a continuing process. The purpose of this chapter was to introduce the reader to many of the different evaluation models and approaches used in the service sector. These range from very simple methods such as intuition, to more complex models involving more time and effort. It was suggested that evaluation is not simply a task that occurs at the end of a programme, but one that may take place during planning and implementation. These formative evaluations are diagnostic in nature, allowing the programmer to understand what is working, what is not, and how aspects of the programme may be improved.

The chapter also drew a link between evaluation and needs assessment, with the former enabling the service provider to be immediately responsive to client needs, and the latter allowing for an understanding of client needs over the long run. Finally, the chapter examined accreditation, environmental auditing and green taxes as evaluative mechanisms which have been used in many different industries, including tourism. The hope is that such environmental evaluations become more a fabric of the way in which tourism companies do business, which may entail a great deal of intervention from regional, national and international organizations.

11

Integrated Ecotourism Programme Planning: a Synthesis

He who exercises government by means of his virtue may be compared to the north polar star, which keeps its place and all the stars turn towards it.

(Confucius)

General Conclusion

In the first chapter of this book it was noted that tourism and recreation differ little from an experiential standpoint. This is especially true of outdoor recreation and ecotourism which share the out-of-doors as a setting for a great number of nature-based activities. The differences between tourism and recreation discussed were those related to space (e.g. the movement of people from a generating to a destination region in the case of tourism), as well as perhaps the sense that tourism is considered more a vehicle to generate foreign exchange. At the same time, however, the service ideals which were once so dominant in recreation have eroded too, as a result of administrative and programming approaches which are much more strongly marketing-based. This point was stressed by Shultz *et al.* (1988), who were indeed timely in identifying the attrition of the service ethic in recreation. This, they suggest, is as a result of the move away from a public service provision model towards one based on the heavy involvement of private sector entities who market their numerous products to consumers. The same was said about the journals and conferences in recreation (by these authors), which have moved away from a focus on better serving the leisure needs of people, to an approach that concentrates on marketing strategies catering to the wants of those who choose to pay. So, even though much has been said about the benefits of the recreational approach to programming in this book, here too we often see an erosion of service values. Fortunately, there appears to be a strong history and foundation of service in recreation that counterbalances approaches that are too heavily weighted on economics. At the risk of being overly simplistic and general, can we say the same about tourism?

Tourism is a business and will remain a business. Consequently, attempts to pry it away from its profit motive roots would be fruitless. At the same time, however, there needs to be the realization that financial success is not completely contingent upon the development of a solid business plan and the adoption of responsible business and financial practices. Accepting this, the ecotourism operator does him/herself a disservice by ignoring other non-economic models, frameworks and approaches that may aid in the development of excellent ecotourism programmes. This includes recreation programme planning. which has long acted as a template for the creation of countless leisure expressions; which allows service providers to be accountable to participants and the industry; and which provides a way of benefiting financially through the wise and controlled use of resources. This means that the chances of being successful – experientially and financially – are increased through the operationalization of the programming model. This is especially important in ecotourism given the level of competition that has emerged over the course of the last decade (Australia has well over 700 ecotourism operators). Some ecotourism service providers will be fortunate to occupy a defined niche in the marketplace, which may thus guarantee a good level of success. Sooner or later, however, others will follow with similar ideas. (This was the case in Costa Rica where a successful canopy tour company has had to deal with other companies who were able to run identical operations, with similar names, as a result of some very 'relaxed' laws in the country.) Over the long term it may well be the financial management, level of service quality, and programming skills that sets one operator apart from others. To the author, the issue is one of balance between the market culture end of the spectrum and the experiential end, which will enable ecotourism service providers to offer more enjoyable programmes for a very discerning clientele.

In putting this ecotourism market culture in perspective, the author recently had a discussion with several Canadian industry representatives in an attempt to plan for a major ecotourism conference for the year 2002. Taking time to peruse the proposed sessions and schedule of events, the author found that it was decidedly product and marketing based. This, it was argued, was to provide 'what the industry wants'. The suggestion for the inclusion of topics related to programming for more balance, including accreditation, guiding, ethics, resource management, among others was met with a good deal of hostility. It brought to mind a recent e-mail transmission from representatives of a number of NGOs based in Thailand and Malaysia, who raised some serious concerns over the United Nation's promotion of the International Year of Ecotourism, without first making an adequate assessment of the nature of the ecotourism industry and its multi-dimensional effects. Ecotourism continues to be a strong international force, and without adequate planning and leadership at all levels of decision making, it runs the risk of pulling away from its initial promise to be a force for education, local development and conservation, to one that is just another vehicle for big business to make money at the expense of local people and the environment. The point of this example is to illustrate that unfortunately there does not appear to be much understanding or acceptance of how newer or different approaches to the provision of ecotourism services might ultimately benefit the industry, beyond the conventional 'product psychology' that seems to exist universally.

Although some would argue that we have yet to define the concept of ecotourism effectively, there is at least some consensus of its general attributes, including sustainability, education, local benefits and conservation. The definition set forth in this book is thought to represent these general themes, but also a few other concepts which hopefully will become increasingly valued in ecotourism. These include the intrinsic nature of ecotourism, the notion that it should be participatory, a focus on ethics, and the need to avoid stressing the biotic and abiotic conditions of the environment (see Chapter 2). All are meant to reinforce the need for ecotourism to be a contributor to the land ethic. The hope is that such principles become articulated in government policies, and further operationalized in the field. The problem most pressing in this regard is that a whole host of operators exist around the world, practising ecotourism based on many very different definitions of the concept. Unfortunately this contributes little consistency in programme content, and probably a significant degree of confusion within an expanding market, the bulk of which remains very under-educated about what is and is not ecotourism. Better control and organization within the field (e.g. through the numerous eco-labelling programmes that seek to nationalize and internationalize ecotourism goods and services) will allow ecotourism to play a significant role in demonstrating how sustainability may be realized at various scales and within different settings. It must be able to accomplish this if it is to remain as it has been theorized. If it fails, it runs the risk of becoming mired down in a stew of inconsistency and a lack of clarity. Furthermore, the present inability to isolate the concept effectively has been problematic to operators and jurisdictions who are concerned with the management of areas in relation to ecotourist expectations. For this reason a great deal of space in the book was devoted to describing the environments in which ecotourism occurs, including public and private protected areas, as well as some of the inherent resource-management issues. These include impacts, carrying capacity, norms, community development, as well as some of the management responses needed to address such issues. The dynamic relationship that exists between ecotourists, operators and the natural setting, is one which will continue to challenge resource managers and policy makers for some time.

One of the most obvious outcomes of this book has been a recognition of the value of adopting systematic planning methods to accomplish a variety of goals. This requires the recreation and tourism service provider to embark on a cyclical planning tour involving: (i) a consideration of various programme philosophies, theories and approaches; (ii) an assessment of participant and agency needs; (iii) an extensive programme planning process, including an understanding of programme areas, formats, settings, mobility, lodging, as well as trip planning and leadership; (iv) programme implementation, which may include marketing, quality, training, budgeting and one or more strategies to implement the programme; and (v) an evaluation of the event, through the use of one or more widely accepted evaluation tools. This general systems approach is not only applicable to recreation and tourism, but has been applied in many related disciplines, including marketing and environmental management. The commonality means that there is an opportunity to develop links between these discrete categories (recreation, marketing and environmental management) in the development of more holistic programmes for ecotourism.

Integrated Ecotourism Programme Planning

Integrated ecotourism programme planning (see Fig. 11.1) is premised on the notion that ecotourism, like other forms of tourism and outdoor recreation, can be viewed as a component part of a broader overall social–ecological system, which encompasses the vast range of human–environment relationships (i.e. those social political, legal, marketing and economic ones). One paradigm that has emerged from this continuum or matrix of relationships (since these relationships are never linear) is sustainability, which has been widely regarded as the most appropriate model for positive global socio-economic change. Defined as a process or an ethic, sustainability appears to be the most realistic paradigm by which to clearly define the possibilities and limits of ecotourism. The sustainability paradigm has also strongly influenced a number of different governance regimes (described here in the context of institutions, policies and management regimes) which continue to provide the structural basis for decision making in society as a whole. Flexibility is one of the fundamental dimensions of the governance model, which allows for collaborative management structures, the support for regional diversity, and the encouragement of citizen engagement, at many scales.

On the basis of its embeddedness in social and ecological systems, its link to sustainability, and its potential to evolve and develop through different forms of governance (the 'Broad environmental realm' in Fig. 11.1), ecotourism stands as a phenomenon which may be subject to more holistic, integrated interpretations. This is particularly the case given the book's focus on a systematic programme-planning perspective in ecotourism. The systems approach therefore provides the means by which to examine the 'Programme realm' not only from a 'recreation' programming perspective, but also from environmental management and marketing perspectives, which are also subject to the same systematic processes (as outlined earlier in the book). Indeed, there appears to be a move to be as integrative and adaptive as possible in resource management, in recognizing that many social and natural phenomena need to be examined as a linked whole, using processes and techniques that cut across different sectors and scales. Figure

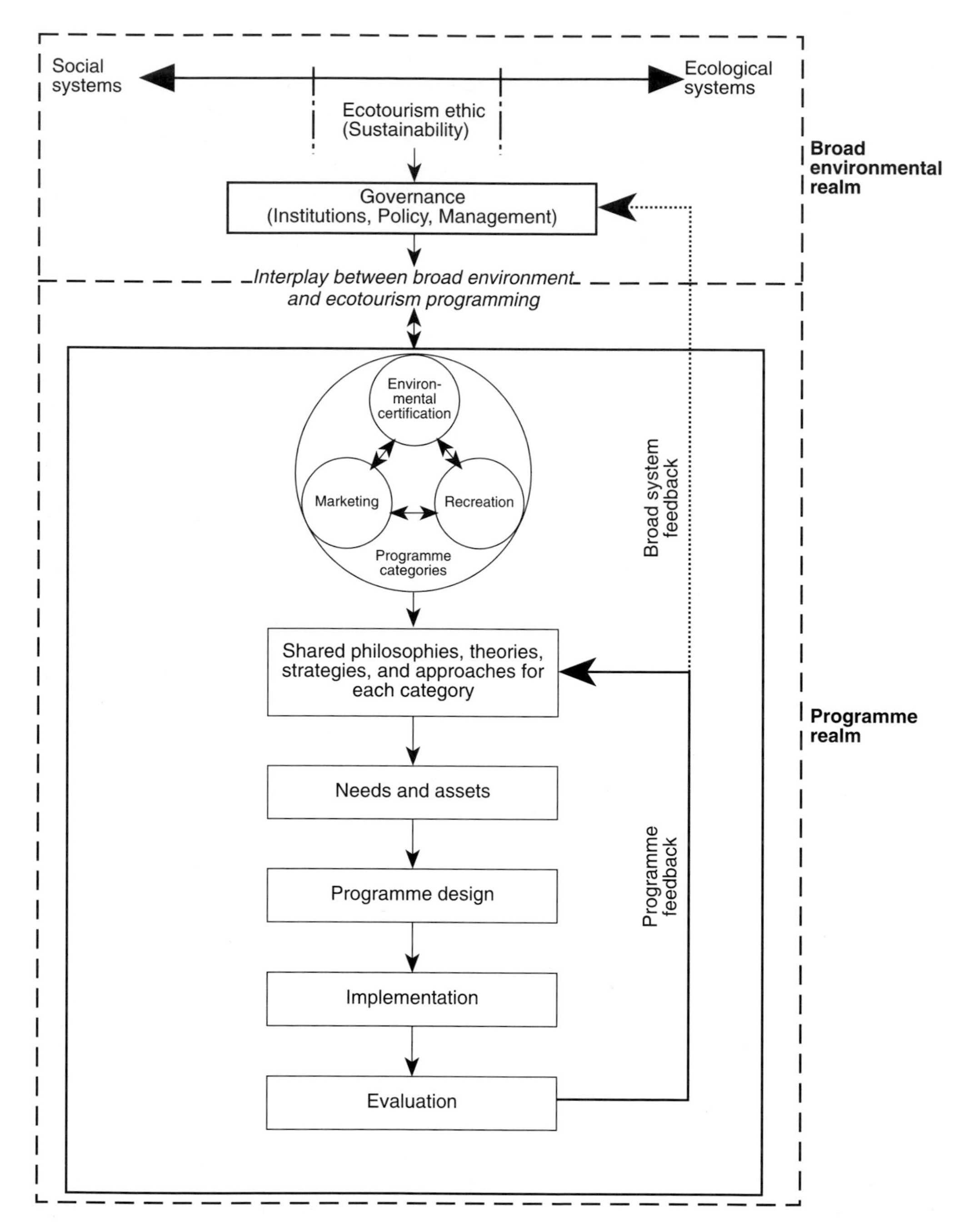

Fig. 11.1. Integrated ecotourism programme planning.

11.1 positions the ecotourism operator as a microcosm within this broader overall system of stakeholders, each interacting within a broad environment. In drawing this link there is the belief that governments, tourism associations, community groups, ecotourism operators and other interested parties might better understand and articulate the social, ecological and economic values that underlie ecotourism. Each of the component parts of Fig. 11.1 are elaborated upon below.

Social and ecological systems

As suggested in Chapter 2, the systems approach is a perspective that allows for the understanding of the various components and interrelationships within an entity. Scientists and policy makers are just beginning to understand that natural and social systems do not occur in isolation (e.g. that natural scientists are not necessarily the best or only scientists to advise on natural policy), as originally thought, but rather as integrated systems, with social and natural characteristics. These systems have built in feedback mechanisms which allow the system to adapt on the basis of positive and negative economic, social and ecological inputs. Holling *et al.* (1998) discussed the importance of resilience in the context of the capacity of a system to absorb disturbance. These authors suggest that the loss of resilience will move systems closer to thresholds, perhaps causing them to flip from one equilibrium state to another (see, for example, the literature on social and ecological impacts of tourism), with associated impacts on social and ecological realms.

Theoretical perspectives which allow researchers to examine ecotourism from the combined systems viewpoint may provide a solid foundation from which to build more diverse and more fully integrated structures. This book talked about human ecology as an example of the application of ecological concepts to the study of the relations between people and their physical and social environment (Johnston, 1987). Human ecologists attempt to describe systems and to explain the interconnections between people and environments which sustain the system and keep it equilibrated (Porter, 1978). Although human ecology appears to have a greater level of applicability to local or site-specific situations, it also provides the necessary impetus to tackle issues related to underdevelopment, colonialism and the tropics, e.g. areas that are either too hot, too cold, too dry or too high for most people (Porter, 1978). Given this local focus, human ecology could serve ecotourism research in areas where tourism is newly developed and where societies are culturally and/or economically marginalized, especially in light of the rapid growth of ecotourism and adventure tourism in such areas.

Complexity was also examined as a theoretical perspective that emphasizes the link between social and ecological systems. Complexity is viewed as the antithesis of reductionism and of the linear, equilibrium theory of nature, which suggests that there is a predictable pattern of events that determine cause and effect relationships (see Perrings *et al.*, 1995). Complexity theory is hinged on the notion that social systems cannot be analysed independently of natural systems, and that these need to be studied as a linked whole (Berkes and Folke, 1998). This may require the application of several conceptual frameworks, and the synthesis of these, in attempting to resolve problems that cut across different sectors at different scales. Ecotourism, and tourism in general, appears to be one such phenomenon that requires the application of many different approaches in attempting to examine complex systems.

Ecotourism ethic

There would appear to be value in interpreting ecotourism from an ontological position, which would thus enable the examination of some very basic assumptions about this concept. These might include the nature of our relationship with the environment, the nature of human nature, and the nature of human relationships. The exercise of peeling back the layers of reality would enable the ecotourism industry to formulate values and beliefs that may be common across the various sectors of the field. Humans by nature are self-interested and competitive (Edney, 1980). This self-interest often acts as a barrier to associations of mutual benefit within the community, as individuals seek to get more out of the relationship than their counterparts. Competition and self-interest therefore are perhaps what create such a formidable challenge within all industries, as well as the so-called 'green' industries of which ecotourism is a part.

Environmentalists have demonstrated that there are alternatives to the conventional technocentric ways of the Western world. One of the main alternatives, ecocentrism, is an extreme view which values harmony, responsibility and sensitivity to natural laws. This view appears to be sound in theory, but difficult to operationalize and feared by decision makers as a model that requires the complete deconstruction of society. Those who are more conservative, instead call for a rational approach that attempts to create balance between the practical and the passionate. This call for balance is not new. Leopold felt that science – the dispassionate and methodological realm – and humanism – the aesthetic and romantic – views of nature should never operate in isolation, but rather as a necessary coalition, and a symbol that all things have value beyond their use as resources for humanity. Leopold's call for a land ethic explored the notion that acts of responsibility (ethics) were in part resultant from the recognition and understanding of options, which must be weighed according to oneself, one's family, community and the planet.

As one of many different paradigms that attempt to explain the relationship between humans and nature, sustainable development provides a philosophical and practical means by which to address some of the most fundamental questions regarding the future, resource use, people, communities, and plants and animals. Robinson *et al.* (1990) suggest that its shortcomings are based largely on the feeling that it should be a solution to all of our problems, when realistically it should be best viewed as a process. This process is important in helping to guide our planning and management efforts at many scales (Nelson, 1991). Perhaps even more important, it is critical to view sustainable development as an ethic, with the understanding that all of our decisions have social, economic and ecological repercussions. This is especially true for ecotourism which, in principle, is supposed to be one of the greenest forms of travel. Although much has been written on the pros and cons of sustainable development for tourism and ecotourism, it still appears to be the most rational approach to the ethical development of the ecotourism industry. However, sustainable approaches to ecotourism may be aided by the considerable tensions that flourish between those more concerned with development versus those who would take a stronger ecocentric viewpoint.

Governance

Governance includes the institutions, policies and management practices which articulate any number of different approaches to achieving sustainability. Consequently governance, at least the type that is most appropriate for ecotourism, is derived from the sustainable development paradigm. There is formal governance, which includes government, institutions and policies as outlined above, but also informal governance, including the ethics, morals, values and norms that guide formal structures (in the words of Rousseau: 'only good men will obey good laws'). By nature, governance should be interactive, involving cooperation between public, private and non-governmental organizations. Such collaborative or cooperative management, when embedded in environmental policy, allows for the potential to further achieve environmental objectives (Glasbergen, 1998). Environmental governance, as a sub-category of governance, is shaped by a number of different models. Two which have particular relevance to ecotourism, include (Glasbergen, 1998, p. 3):

- *Regulatory control.* This model assigns a key role to governments. They are seen as regulators of the process of change. In this model, setting standards and imposing stringent rules constitute the mechanisms of change.
- *Cooperative management.* This model assigns a key role to collaborative relations between governments, mediating non-governmental organizations and private interests. The mechanism for change lies in communication and dialogue, the

results of which are laid down in voluntary agreements among participants. Here, stakeholders are urged to contribute their knowledge and various perceptions regarding the issue. Resolution is achieved through negotiation, which reflects a common analytical and normative framework through: (i) shared definitions of the issue(s); (ii) common desires as regards the issue; and (iii) full and equal contribution to the problem.

Good governance may require the implementation of integrated management strategies, which incorporate the use of multiple methods and sectors in matching the scale of decision making with the scale of the environment. This may be achieved through the application of philosophy and principles related to sustainability (at a broad level), ecosystem management, as well as a precautionary approach. The precautionary principle, for example, prompts us to confront the many shortcomings of science, as well as the role of ethics, economics, and politics in decision making. Decisions that directly affect the environment, including the available research on various discharges and procedures, need to be made in consultation with many interested stakeholders. The precautionary principle has direct relevance to ecotourism because it addresses many of the same ethical and impact-oriented concerns about the sector, and tourism in general, raised in the literature. Precaution is a perspective that places stock in the ecological consequences of human action, but it is more than that. It is also a social construct which places responsibility on policy makers for the decisions that they make. Preventative action may be realized if the various stakeholders in tourism planning and development adhere to common goals. From a programming perspective, these goals may include the insulation of local people from many of the socio-economic dislocating characteristics of the tourism industry, putting the environment first, and spreading economic benefits as widely as possible. Ecotourism programming would also benefit from an adaptive management scheme, allowing governmental decision makers and/or service providers to learn continually from feedbacks from the environment. This data allows for the continual modification of programmes and policies in generating managerial approaches that respond to the continuing diversity of changing conditions within the environment (Norgaard, 1994).

The collectivity of these various institutional arrangements, policies and management approaches, is required to provide constructive tension between the rich array of human–human and human–environment relationships that take place within our communities. Ecotourism, as we have seen, does not occur unisectorally, but rather as a land use that is multidimensional and integrated within a broader environment. These relationships occur on a number of different spatial and temporal contexts that include interpersonal, community, regional, national and international levels. Important in approaches to governance is the move from principles to practice, such that there are tangible outcomes which may evolve from the implementation of methods. The example used earlier in the book illustrates that community and resource decisions within an aboriginal community might best be made by the community itself. On the other hand, issues related to the direct and indirect effects of air and water pollution might best be made through international institutions.

Most obvious is the need for a greater degree of interest and support on behalf of governmental and industry decision makers, who must place high priority on the adoption of an interdisciplinary approach to problem solving. Ecotourism, and the environment in general, must compete against a whole host of other issues (including jobs, education, health care and crime) which are given higher priority by government and society. While political debates are often won and lost on these other issues, governments cannot become too pro-environment, especially at the risk of ostracizing industry. Unfortunately this often relegates the environment and tourism as secondary issues, with the latter given priority only during times of economic downturns. Such was the case in the recent 'skyjacking' of planes and subsequent destruction of the World Trade

Center towers and parts of the Pentagon in the USA on 11 September 2001, when there was a plea by governments for citizens to continue to travel in order to kick-start the economy and restore confidence in air travel.

Interplay between broad environment and ecotourism programming

The junction between the broad environmental realm and the programme realm in Fig. 11.1 may manifest itself as the confluence of a number of policies and management approaches which are regionally or nationally derived (e.g. ecotourism policies related to carrying capacities and land use), and which are a direct reflection of the various social, economic and ecologically based regimes adopted in the jurisdiction. It may also include the policies and management styles which have been developed at the ecotourism organizational level, which naturally fit within the broader policy and institutional frameworks of the region. Accordingly, there is an element of scale that is important to consider at this level of the conceptual framework. It would appear logical to envision a two-way flow of information between governance and the various ecotourism programmes that exist within the jurisdiction, which provide the impetus for further refinement of policies and management for the field.

Main programme categories

This text has dealt almost exclusively with a systematic recreation programme-planning process for ecotourism. However, we have seen that systematic planning occurs in the areas of marketing and environmental management as well. In the case of the former, strategic marketing includes a mission, internal and external evaluation, analysis and choice, short-term operating strategies, implementation strategies, and control and evaluation (Chapter 9). Similarly, environmental management systems also take place through systematic planning processes involving many of the same characteristics (Chapter 10). The environmental implications of tourism and some forms of outdoor recreation may be compounded as compared to some recreation programmes, which are much shorter in duration and less reliant on an extensive network of support mechanisms. Tribe *et al.* (2000) write that the use of systematic environmental management systems (EMS) allow managers to minimize negative impacts, while maximizing positive impacts. The EMS, as described by Tribe *et al.*, involves environmental policy, site review, the programme (a formal agenda for action), operations, and audit and review. This system follows very closely other environmental certification programmes such as ISO 14001 and EMAS, which include environmental policy, planning, implementation, action and management review steps in leading in general to an improved situation. These various planning processes have been included in the context of this book in order to contribute to a larger, more inclusive type of programme planning which acknowledges that ecotourism is a complex phenomenon that takes place under very dynamic social, ecological and economic conditions. This challenges tourism planners to be as innovative and multidisciplinary as possible in addressing the many needs and issues of such a complex industry.

Gunderson *et al.* (1995) suggest that all living systems, social and ecological, are nested in time and space at a variety of different scales, and characterized by numerous non-linear feedbacks. Terming these 'panarchies', each level is said to go through a cycle of growth, maturation, destruction (or death) and renewal. Just as these living systems go through a systematic process, so too do goods and services, as we have seen. However, whereas natural cycles may take hundreds of years, economic cycles are based on much shorter periods of time (say 15 years). Although linked in process, Clark (1973) has illustrated that ecological cycles and market cycles are incongruent in a number of ways. For example, ecological preservation is endangered, from a market perspective because: (i) the future value of slow-growing ecological resources are discounted to nothing because of market pressures; and (ii) slow-producing ecological resources are viewed as non-renewable because the exploiter will make more money depleting the resource and re-investing in other areas.

Viewed panarchically, integrated ecotourism programming allows for the amalgamation of marketing, environmental management and recreation/tourism activity systems within a common and comprehensive, systematic programme planning approach (the same horizontal level). The panarchy concept also allows for ecotourism to be nested vertically within an environment of many other ecotourism programmes at numerous scales, as well as within the social, economic and ecological fabric of conditions of the region. The dynamic nature of this integration in time and space provides positive and negative feedback which dictates the behaviour of the ecotourism system in the future.

Shared philosophies, theories, strategies and approaches

Programming is founded on a great number of different philosophies, theories, strategies and approaches (as outlined in Chapter 4). These shape the development of ecotourism programmes and how we view them as satisfying various human needs and wants. This book has attempted to identify specific philosophies and values that will be most instrumental in ecotourism programming. Sustainable development, for example, is critical to the success of ecotourism. Yet if governments do not value sustainable development and, more specifically, ecotourism operators, then what chance does ecotourism have to make a difference? The servant leadership perspective to programming and leadership was selected specifically because it considers values as critical to the programming process. Where such a perspective may be elaborated upon for our purposes in this book, is in its view of the natural environment, and how people should view the natural world. Other values of importance include respecting nature and local people, principled negotiations and policies, equity, education, minimum impact, treasuring natural and cultural heritage, and reducing stress on human and ecological systems. These values need to be articulated in the mission and vision statements written by service providers, as well as within goals and objectives that naturally flow from these prime directives. Service providers must also develop strategies and approaches that are the practical manifestations (marketing, activity programming and environmental management) of the agency's core programme delivery values. Again, depending on the philosophical orientation of the service provider, such approaches will lean more heavily towards community development, social marketing or social planning. These will in turn influence decisions regarding programme approaches, but typically involve human services, environmental/aesthetic/preservationist, and the sustainable development (Inskeep, 1991) approaches.

Needs and assets

Although difficult to isolate, ecotourists, like other tourists, travel to satisfy many different needs. These may be fundamental needs such as novelty and escape, or more directed, such as the desire to see the great beasts of Africa before they are less accessible or even gone from the wild forever. There appears to be general consensus in the recreation programme literature that needs assessment forms the first stage of the programming cycle. A properly conducted assessment allows for the development of programmes that lead to satisfying experiences, and the ability to modify or create new programmes to be as timely as possible. Service providers are unlikely to have the luxury of testing their programmes over many years and many populations without some degree of financial success. This makes it essential to conduct accurate and timely needs assessments for the purpose of knowing what clients want from a programme. These may be specific ecotours which focus on certain environments, activities, or species, or general ecotours which are designed to be all encompassing.

At the same time, service providers should strive to learn as much as possible about the characteristics of the regions in which they operate, especially in cases where programmers are exploring new programmes in new places. A documented overview of the attractions and other resources will help with the decision to initiate programming in the region under consideration. Although not dealt with extensively, the business plan, which precedes these processes depending on the stage of business development, will enable the programmer to take account of a broad range of internal and external information, in addition to acting as a tool for decision making, integrating ideas, analysis, performance, raising finance and assessing ideas.

Programme design

Conventional recreation programming is based on the identification of programme areas (e.g. hiking or dog sledding) and programme formats (e.g. trips or instructional). To this, however, we must add considerations related to: (i) programme setting, such as backcountry, rural or urban settings; (ii) programme mobility, including motorized and non-motorized forms; and (iii) programme lodging, which includes a variety of fixed-roof and non-fixed-roof dwellings. All five of these combined will allow the ecotourism service provider to develop a programme which fits with the environmental resources of the destination and the expectations of the clientele. This information, along with the information developed in the needs and assets stage, provides the service provider with an opportunity to develop a number of programme alternatives for consideration. Criteria will be selected, through brainstorming, which allows the individual to construct these scenarios for the purpose of identifying the most successful programme. Such criteria may include available resources, the political situation in the community, risk management, local codes, environmental impact, and so on. The nature of the alternative selected will provide the necessary information from which to build an educational or interpretive component in keeping with the expectations of clientele. The effectiveness of the interpretive element may be contingent upon the ability of the programmer to plan effectively before the event. This may include an overview of the topic, theme, message, the design technique, the experience, the audience, objectives, site, time length, time of day, props, promotion, costs, and so on (see Chapter 6).

Programme design also demands the detailing of many logistical considerations to ensure the health and safety of participants. Indeed, risk management continues to generate much enthusiasm these days as a relatively underutilized tool for the tourism field. Information on the destination and travel procedures will greatly aid the traveller, especially those travelling great distances or those unfamiliar or inexperienced with backcountry areas. The service provider is thus obligated to know about the physical conditions of the environment in which the tour takes place, including weather and other hazards such as plants and animals. Guides who are well trained in 'people' skills and hard or technical skills will be important in helping to ensure that trips are safe and enjoyable. This has been stressed in books in related fields, including adventure programming (see Priest and Gass, 1999). These guides may be formally trained through certification programmes, or informally trained with years of valuable experience. The professionalism of the field (i.e. accreditation and certification) will continue to be a topic of debate among academics and practitioners, as it is in Australia, which appears to be the furthest along in terms of accreditation and guide training.

Implementation

Once planned, programmes do not just happen, but must be presented through a process of implementation, involving one or more techniques such as piloting, phasing-in, or total immersion (programme testing in one form or another, allowing for a successful launch). This becomes particularly important when significant cost is invested in the programme. Consideration must be given to a number of associated aspects of the programme, including the programme life cycle, marketing, quality, staff training, public relations and budgeting. The green marketing initiative, with a focus on social and ecological concerns, is a response to the sensitivities of consumers in regard to the products they purchase. Efforts have been made to implement programmes to minimize the negative effects of the tourism industry, and these efforts, whether proven or not, have been communicated to the market. This, according to Peattie (1995), has meant that tourism marketers have had to be more responsive, open-ended and global in their approaches; and according to Bécherel (1999), that marketing needs to employ a systems approach with many feedback mechanisms enabling changes to reflect changing circumstances with the market and marketplace.

Evaluation

During the course of the last 10–15 years, programme evaluation in recreation has generated more interest because of the need for service providers to be accountable at a number of scales. Evaluation may be conducted internally and/or externally, and take place at various times during programme planning and implementation, and/or at the end of the programme. Each has benefits, and must be selected according to what is to be evaluated and why. Indeed, the 'why' is extremely important, and service providers should not attempt to evaluate whether the right questions have been asked or whether there are not the resources – time, human, and financial – to undertake such a task. Furthermore, service providers need to be aware of a number of different models of evaluation that may be used to accomplish a variety of different ends.

Feedbacks

Systems, as stated earlier, are characterized by built-in feedback mechanisms which allow the system to 'learn'. By this it is meant that through successive iterations, decision makers can make subtle or extensive changes to the system based on the available economic, social and ecological inputs and the managerial or institutional processes to interpret the data. Figure 11.1 illustrates a two-tiered feedback mechanism built into the conceptual framework. The 'Programme feedback' allows for evaluative information to be fed back into the philosophies, theories, strategies and approaches that are common to marketing, recreation and environmental programme categories. This follows a rather traditional model of systematic programme planning in recreation. However, Fig. 11.1 also illustrates a second feedback level which branches on from programme evaluation into the 'Broad environment realm', and particularly the level of governance. This infers that perhaps the very best evaluative information from the service providers who are on the 'cutting edge' may provide the impetus for change at the policy and management levels. Such institutions, if they are to be truly integrative and adaptive, and therefore open to innovative models of governance, must be prepared to shift policy on the basis of the dynamic nature of public, private and/or non-governmental ecotourism operations in the field.

Final Thoughts

Ecotourism continues to be criticized for its short-sightedness in relation to sustainability, local control, community development, lack of concern for biodiversity conservation, metropolitan domination, and focus on the profit motive. These are all important aspects of ecotourism which must continue to be debated to ensure equitable balance throughout the industry. Due in part to its infancy as a field of study and as an industry (depending on one's perspective), theory has not caught up to practice, but similarly, practice has not been subject to a great deal of scrutiny. What one will find then is the provision of many types of ecotourism services, many of which have been planned without any established set of criteria beyond a business plan and probably a great deal of energy and enthusiasm to make a go of it. This book has sought to illustrate that the recreation field, particularly programming, has much to offer ecotourism in terms of theory and practice. However, it is not just recreation (a field which has drawn on a great deal of research from other fields) that offers guidance but also marketing, environment and resource management, and ecology. As such, researchers and practitioners alike must be motivated to consume the rich theoretical offerings from other disciplines, which can only make those working in the ecotourism field more knowledgeable and better prepared to address the many challenges that lie ahead.

References

Abbey-Livingston, D. and Abbey, D. (1982) *Enjoying Research?* Ministry of Tourism and Recreation, Toronto.

Ahmed, Z.U., Krohn, F.B. and Heller, V.L. (1994) International tourism ethics as a way to world understanding. *The Journal of Tourism Studies* 5(2), 36–44.

Albrecht, K. (1992) *The Only Thing that Matters: Bringing the Power of the Customer into the Centre of Your Business.* HarperBusiness, New York.

Alderman, C. (1992) The economics and the role of privately owned lands used for nature tourism, education and conservation. Paper presented at the *4th World Congress on Parks and Protected Areas*, Caracas, February.

Alreck, P.L. and Settle, R.B. (1985) *The Survey Research Handbook.* Irwin, Homewood, Illinois.

American Camping Association (1984) *Camp Standards with Interpretations for the Accreditation of Organized Camps.* American Camping Association, Bradford Woods, Indiana.

Ammon, R. (1997) Risk management process. In: Cotten, D.J. and Wilde, T.J. (eds) *Sport Law for Sport Managers.* Kendall/Hunt, Dubuque, Iowa.

Andereck, K.L. (1993) The impacts of tourism on natural resources. *Parks and Recreation* June, 26–32, 86.

Arai, S. (1999) Voluntary Associations as Spaces for Democracy: Toward a Critical Theory on Volunteers. PhD Dissertation, University of Waterloo, Waterloo, Ontario, Canada.

[ARCDC] Arkansas Resource Conservation and Development Council, Inc. (1997) *Ecotourism Development Manual.* Words Out, Harrison, Arkansas.

Archer, B. and Cooper, C. (1999) The positive and negative impacts of tourism. In: Theobald, W.F. (ed.) *Global Tourism*, 2nd edn. Butterworth Heinemann, Oxford, pp. 63–81.

Arlen, C. (1995) Ecotour, hold the eco. *U.S. News and World Report* May 29, 61–63.

Arrow, K., Bolin, B., Costanza, R., Dasgupta, P., Folke, C., Holling, C.S., Jansson, B., Levin, S., Maler, K.G., Perrings, C. and Pimental, D. (1995) Economic growth, carrying capacity, and the environment. *Science* 268, 873–879.

Association of Experiential Education (1993) *Manual of Accreditation Standards for Adventure Programs.* Association of Experiential Education, Boulder, Colorado.

Auld, C.J. and Cuskelly, G. (1989) Program evaluation made easy: a case study using importance performance evaluation. *The ACHPER National Journal* December, 13–17.

Bailey, R.G. (1998) *Ecoregions Map of North America.* USDA Forest Service, Fort Collins, Colorado.

Barnes, R.D. (1980) *Invertebrate Zoology.* Saunders College, Philadelphia.

Battisti, G. (1982) Central places and peripheral regions in the formation of a theory on tourist space. In: Singh, T.V. and Kaur, J. (eds) *Studies in Tourism Wildlife Parks Conservation.* Metropolitan, New Delhi.

Bécherel, L. (1999) Strategic analysis and strategy formulation. In: Vellas, F. and Bécheral, L. (eds) *The International Marketing of Travel and Tourism.* Macmillan Press, London, pp. 37–106.

Bennett, M.M. (1996) The marketing mix: tourism distribution. In: Seaton, V.A. and Bennett, M.M. (eds) *The Marketing of Tourism Products: Concepts, Issues and Cases.* International Thomson Business Press, London.

Bennis, W. and Nanus, B. (1985) *Leaders: the Strategies for Taking Charge.* Harper and Row, New York.

Benveniste, E. (1971) *Problems in General Linguistics*. Translation by Meek, M.E., Coral Gables, Florida.
Berkes, F. (1984) Competition between commercial and sport fisherman: an ecological analysis. *Human Ecology* 12(4), 413–429.
Berkes, F. and Folke, C. (1998) Linking social and ecological systems for resilience and sustainability. In: Berkes, F. and Folke, C. (eds) *Linking Social and Ecological Systems: Management Practices and Social Mechanisms for Building Resilience*. Cambridge University Press, Cambridge, pp. 1–25.
Bezruchka, S. (1988) There's more to malaria prevention than pills. *Real Travel* Winter, 11–12.
Bilodeau, M. and Trembley, P.A. (1996) Himalayas forever? *Ecodecision* 20 (spring), 46–48.
Bisson, C. (1999) Sequencing the adventure experience. In: Miles, C.J. and Priest, S. (eds) *Adventure Programming*. Venture Publishing, State College, Pennsylvania, pp. 205–214.
Black, R., Ham, S. and Weiler, B. (2000) *Ecotour Guide Training in Less Developed Countries: Some Research Directions for the 21st Century*. Working paper No. 07/00, Department of Management, Monash University, Australia.
Blake, B. and Becher, A. (2001) *The New Key to Costa Rica*. Ulysses Press, New York.
Blamey, R.K. (2001) Principles of ecotourism. In: Weaver, D.B. (ed.) *The Encyclopedia of Ecotourism*. CAB International, Wallingford, UK, pp. 5–22.
Blangy, S. and Nielson, T. (1993) Ecotourism and minimum impact policy. *Annals of Tourism Research* 20(2), 357–360.
Bloudin, M.S. and Connor, E.F. (1985) Is there a best shape for nature reserves? *Biological Conservation* 32, 277–288.
Boehmer-Christiansen, S. (1994) The precautionary principle in Germany – enabling government. In: O'Riordan, T.O. and Cameron, J. (eds) *Interpreting the Precautionary Principle*. Earthscan, London, pp. 31–61.
Bordieu, P. (1986) The forms of capital. In: Richardson, J. (ed.) *Handbook of Theory and Research for the Sociology of Education*. Greenwood Press, New York, pp. 241–258.
Borg, W.R. and Gall, M.D. (1989) *Educational Research: an Introduction*, 5th edn. Longman, New York.
Bowler, P.J. (1993) *The Norton History of the Environmental Sciences*. Norton, New York.
Bradshaw, J. (1972) The concept of social need. *New Society* 19, 640–643.
Brady, M. (2000) Eco-tourism's split personality: nature friendly travel? *Financial Post*, Tuesday, 28 March, E3.
Brandon, K. (1993) Basic steps toward encouraging local participation in nature tourism projects. In: *Ecotourism: a Guide for Planners and Managers*. The Ecotourism Society, North Bennington, Vermont, pp. 134–151.
Brewer, R. (1979) *Principles of Ecology*. Saunders College, Philadelphia.
Britton, S. (1982) The political economy of tourism in the Third World. *Annals of Tourism Research* 9(3), 331–358.
Brock, T.D. (1979) *Biology of Microorganisms*, 3rd edn. Prentice-Hall, Englewood Cliffs, New Jersey.
Bryan, H. (1977) Leisure value systems and recreational specialization: the case of trout fishermen. *Journal of Leisure Research* 9(3), 174–187.
Bryan H. (1979) Conflict in the great outdoors. *University of Alabama Sociological Studies* 4, 1–98.
Buckley, R. (1994) A framework for ecotourism. *Annals of Tourism Research* 21(3), 661–665.
Buckley, R. and Pannell, J. (1990) Environmental impacts of tourism and recreation in national parks and conservation reserves. *The Journal of Tourism Studies* 1(1), 24–32.
Bucy, D. (1991) Tourism vs. Environmental integrity and livability: using interpretive planning to solve tourism management problems. *Legacy* 2(4), 10–14.
Bull, A. (1991) *The Economics of Travel and Tourism*. Pitman, Melbourne.
Bunce, M. (1998) Thirty years of farmland preservation in North America: discourses and ideologies of a movement. *Journal of Rural Studies* 14(2), 233–247.
Burch, W. (1984) Much ado about nothing – some reflections on the wider and wilder implications of social carrying capacity. *Leisure Sciences* 6(4), 241–251.
Burkey, T.V. (1988) Extinction in nature reserves: the effect of fragmentation and the importance of migration between reserve fragments. *Oikos* 55(1), 75–81.
Burton, F. (1998) Can ecotourism objectives be achieved? *Annals of Tourism Research* 25(3), 755–758.
Busser, J.A. (1990) *Programming: For Employee Services and Recreation*. Sagamore, Champaign, Illinois.
Butler, J.R. (1993) Interpretation as a management tool. In: Dearden, P. and Rollins, R. (eds) *Parks and Protected Areas in Canada: Planning and Management*. Oxford University Press, Toronto, pp. 211–224.

Butler, R.W. (1980) The concept of tourist area cycle of evolution: implications for management of resources. *The Canadian Geographer* 24, 5–12.

Butler, R.W. (1991) Tourism, environment and sustainable development. *Environmental Conservation* (18)3, 201–209.

Butler, R.W. (1993) Tourism – an evolutionary perspective. In: Nelson, J.G., Butler, R. and Wall, G. (eds) *Tourism and Sustainable Development: Monitoring, Planning, Managing*. University of Waterloo, Waterloo, Ontario, pp. 27–44.

Butler, R.W., Fennell, D.A. and Boyd, S.W. (1992) *The POLAR Model: A System for Managing the Recreational Capacity of Canadian Heritage Rivers*. Environment Canada, Ottawa.

Campbell, A., Devine, M. and Young, D. (1990) *A Sense of Mission*. Pitman Press, London.

Canadian Parks/Recreation Association (1992) *The Benefits of Parks and Recreation*. CP/RA Gloucester, Ontario.

Canadian Red Cross Society (1994) *First Aid: the Vital Link*. Mosby Lifeline, Toronto.

Canadian Tourism Commission (1997a) *Adventure Travel and Ecotourism: the Challenge Ahead*. CTC, Ottawa, Canada.

Canadian Tourism Commission (1997b) *A Profile of Canada's Travel Agencies and Tour Operators*. CTC, Ottawa, Canada.

Canadian Tourism Human Resource Council (1996) *Canadian Tourism Human Resource Council*. CTHRC, Ottawa, Canada.

Cantril, J.G. (1996) Perceiving environmental discourse: the cognitive playground. In: Cantrill, J.G. and Oravec, C.L. (eds) *The Symbolic Earth: Discourse and Our Creation of the Environment*. The University Press of Kentucky, Lexington, Kentucky.

Carpenter, G.M. and Howe, C.Z. (1985) *Programming Leisure Experiences: a Cyclical Approach*. Prentice-Hall, Englewood Cliffs, New Jersey.

Carson, R. (1987) *Silent Spring*. Houghton Mifflin, Boston, Massachusetts.

Castro, Fabio de, and Begossi, A. (1996) Fishing at Rio Grande (Brazil): ecological niche and competition. *Human Ecology* 24(3), 401–411.

Catton, W.R. (1986) Carrying capacity and the limits to freedom. Paper presented for the Social Ecology session 1, *XI World Congress on Sociology*. New Delhi, India, 18 August.

Catton, W.R. (1987) Social and behavioral aspects of the carrying capacity of natural environments. *Behavior and the Natural Environment* 6, 269–306.

Cerruti, J. (1964) The two Acapulcos. *National Geographic Magazine* 126(6), 848–878.

Chadwick, D.H. (1995) Safari to Botswana. *National Geographic Traveler* 12(6), 82–93.

Chapin, M. (1990) The silent jungle: ecotourism among the Kuna Indians of Panama. *Cultural Survival Quarterly* 14(1), 42–45.

Chase, A. (1987) How to save our national parks. *The Atlantic Monthly* July, 35–44.

Chemers, M. and Ayman, R. (eds) (1993) *Leadership Theory and Research*. Academic Press, San Diego, California.

Christaller, W. (1963) Some considerations of tourism location in Europe: the peripheral regions-underdeveloped countries-recreation areas. *Regional Science Association Papers* No. 6, 95–105.

Christenson, J.A., Fendley, K. and Robinson, J.W. (1989) In: Christenson, J.A. and Robinson, J.R. (eds) *Community Development in Perspective*. Iowa State University Press, Ames, Iowa, pp. 3–25.

Chubb, M. and Chubb, H.R. (1981) *One Third of Our Time? An Introduction to Recreation Behavior and Resources*. John Wiley & Sons, New York.

Churchill Northern Studies Centre (1999) http://www.brandonu.ca/CNSC/polar.htm. Site accessed 1 June 1999.

Clark, C.W. (1973) The economics of overexploitation. *Science* 181, 630–634.

Clark, K.S. (1998) Safety in the outdoor environment. In: Dougherty, N.J. IV, (ed.) *Outdoor Recreation Safety*. Human Kinetics, Champaign, Illinois, pp. 3–10.

Clawson, M. and Knetsch, J.L. (1978) *Economics of Outdoor Recreation*. Johns Hopkins University Press, Baltimore, Maryland.

Clift, S. and Page, S.J. (1996) Introduction: tourism and health. In: Clift, S. and Page, S.J. (eds) *Health and the International Tourist*. Routledge, London.

Cohen, E. (1972) Toward a sociology of tourism. *Social Research* 39(1), 164–182.

Coleman, J.S. (1988) Social capital in the creation of human capital. *American Journal of Sociology* 84 (supplement), S95–S119.

Collins, V.R. (2000) *Becoming a Tour Guide: the Principles of Guiding and Site Interpretation*. Continuum, London.

Coole, L., Wiselka, M.J. and Nicholson, K.G. (1989) Malaria prophylaxis in travellers from Britain. *Journal of Infection* 18(3), 209–212.

Corbett, R. and Findlay, H. (1998) *Your Risk Management Program: a Handbook for Sport Organization.* Centre for Sport and Law, Ottawa.

Cordes, K.A. and Ibrahim, H.M. (1996) *Applications in Recreation and Leisure for Today and the Future.* Mosby, Toronto.

Cossar, J.H., Reid, D., Falcon, R.J., Bell, E.J., Riding, M.H., Follett, E.A.C., Dow, B.C., Mitchell, S. and Grist, N.R. (1990) A cumulative review of studies on travellers, their experience of illness and the implications of their findings. *Journal of Infection* 21, 27–42.

Costanza, R. and Folke, C. (1997) Valuing ecosystem services with efficiency, fairness, and sustainability as goals. In: Daily, G. (ed.) *Nature Services.* Island Press, Washington, DC, pp. 49–68.

Cotten, D.J. (1997) Exculpatory agreements or waivers. In: Cotten, D.J. and Wilde, T.J. (eds) *Sport Law for Sport Managers.* Kendall/Hunt, Dubuque, Iowa.

Crabtree, A.E. and Black, R.S. (n.d.) *Ecoguide Program: Guide Workbook.* Ecotourism Association of Australia, Brisbane.

Crick, M. (1989) Representations of international tourism in the social sciences: sun, sex, sights, savings, and servility. *Annual Review of Anthropology* 18, 307–344.

Croall, J. (1995) *Preserve or Destroy: Tourism and the Environment.* Calouste Gulbenkian Foundation, London.

Crompton, J.L. (1981) How to find the price that's right. *Parks and Recreation* March, 32–34.

Crompton, J.L. (1985) Marketing: neither snake oil nor panacea. In: Witt, P. and Goodale, T. (eds) *Recreation and Leisure: Issues in an Era of Change.* Venture Publishing, State College, Pennsylvania, pp. 175–194.

Crompton, J.L. and Richardson, S.L. (1986) The tourism connection where private and public leisure services merge. *Parks and Recreation* October, 38–44, 67.

Csikszentmihalyi, M. (1975) *Beyond Boredom and Anxiety.* Jossey-Bass, San Francisco, California.

Csikszentmihalyi, M. and Csikszentmihalyi, I.S. (1990) Adventure and the flow experience. In: Miles, J.C. and Priest, S. (eds) *Adventure Education.* Venture Publishing, State College, Pennsylvania, pp. 149–156.

Culkin, D. and Kirsch, S.L. (1986) *Managing Human Resources in Recreation, Parks and Leisure Services.* Macmillan, New York.

Curtis, R. (1998) *The Backpacker's Field Manual: a Comprehensive Guide to Mastering Backcountry Skills.* Three Rivers Press, New York.

Danford, H. and Shirley, M. (1970) *Creative Leadership in Recreation*, 2nd edn. Goodyear Publishing, Santa Monica, California.

Davidson, T.L. (1985) Strategic planning: a competitive necessity. Paper presented at the 16th Annual TTRA Conference, *The Battle for Market Share: Strategies in Research and Marketing*, Palm Springs, California, 9–12 June.

Davis, C.D., Hill, G.E. and LaForge, R.W. (1985) The marketing/small enterprise paradox: a research agenda. *International Small Business Journal* 3(3), 31–42.

Dawood, R. (1992) Staying healthy abroad. In: Dawood, R. (ed.) *Travellers' Health: Staying Healthy Abroad*, 3rd edn. Oxford University Press, New York, pp. 1–7.

Dearden, P. and Rollins, R. (1993) The times they are a-changin. In: Dearden and Rollings, R. (eds) *Parks and Protected Areas in Canada: Planning and Management.* Oxford University Press, Toronto, pp. 1–16.

Dearden, P. and Sadler, B. (1989) Themes and approaches in landscale evaluation research. In: Dearden, P. and Sadler, B. (eds) *Landscape Evaluation: Approaches and Applications.* University of Victoria, Victoria, British Columbia, pp. 3–18.

DeGraaf, D.G., Jordan, D.J. and DeGraaf, K.H. (1999) *Programming for Parks, Recreation, and Leisure Services: a Servant Leadership Approach.* Venture Publishing, State College, Pennsylvania.

Department of Conservation (2000) *Developing Ecotours and Other Interpretive Programs: A Guidebook for Planning, Designing, Promoting and Conducting Ecotourism Activity Programs.* Department of Conservation, Perth, Western Australia.

Dernoi, L.A. (1981) Alternative tourism: towards a new style in North–South relations. *Tourism Management* 2, 253–264.

Devall, B. and Sessions, G. (1985) *Deep Ecology: Living as if Nature Mattered.* Peregrine Smith, Salt Lake City, Utah.

Diamantis, D. and Ladkin, A. (1999) 'Green' strategies in the tourism and hospitality industries. In: Vellas, F. and Bécheral, L. (eds) *The International Marketing of Travel and Tourism.* Macmillan Press, London.

Diamond, J.M. (1975) The island dilemma: lessons of modern biogeographic studies for the design of nature reserves. *Biological Conservation* 7, 3–15.

Diamond, J.M. and May, R.M. (1976) Island biogeography and the design of natural reserves. In: May, R.M. (ed.) *Theoretical Ecology: Principles and Applications.* W.B. Saunders, Philadelphia, pp. 163–186.

Dillman, D. (1978) *Mail and Telephone Surveys: the Total Design Method.* John Wiley & Sons, New York.

Dolli, N. and Pinfold, J.F. (1997) Managing rural tourism businesses: financing, development and marketing issues. In: Page, S.J. and Getz, D. (eds) *The Business of Rural Tourism: International Perspectives.* International Thomson Business Press, London, pp. 38–60.

Doxey, G.V. (1975) *A causation theory of visitor–resident irritants; methodology and research inference.* In: *The Impact of Tourism,* Sixth Annual Conference Proceedings of the Travel and Tourism Research Association, San Diego, California, pp. 195–198.

Drake, S.P. (1991) Local participation in ecotourism projects. In: Whelan, T. (ed.) *Nature Tourism: Managing for the Environment.* Island Press, Washington, DC, pp. 132–163.

Driver, B.L. and Peterson, G.L. (1986) The values and benefits of outdoor recreation. An integrated overview. In: *A Literature Review: President's Commission on America's Outdoors.* Government Printing Office, Washington, DC, pp. 1–9.

Driver, B.L., Brown, P.J. and Peterson, G.L. (1991a) Research on leisure benefits: an introduction to this volume. In: Driver, B.L., Brown, P.J. and Peterson, G.L. (eds) *Benefits of Leisure.* Venture Publishing, State College, Pennsylvania, pp. 3–12.

Driver, B.L., Tinsley, H.E.A. and Manfredo, M.J. (1991b) The paragraphs about leisure and recreation experience preference scales: results from two inventories designed to assess the breadth of the perceived psychological benefits of leisure. In: Driver, B.L., Brown, P.J. and Peterson, G.L. (eds) *Benefits of Leisure.* Venture Publishing, State College, Pennsylvania, pp. 263–287.

Dryzek, J. (1997) *The Politics of the Earth: Environmental Discourses.* Oxford University Press, Oxford.

Dubos, R. (1976) Symbiosis between the earth and humankind. *Science* 19, 459–462.

Duchesne, M., Côté, S.D. and Barrette, C. (2000) Responses of woodland caribou to winter ecotourism in the Charlevoix Biosphere Reserve, Canada. *Biological Conservation* 96, 311–317.

Dyess, R. (1997) Adventure travel or ecotourism? *Adventure Travel Business* April, 2.

Eagan, P.D. and Joeres, E. (1997) Development of a facility-based environmental performance indicator related to sustainable development. *Journal of Cleaner Production* 5(4), 269–277.

Eagles, P.F.J. (1992) The travel motivations of Canadian ecotourists. *Journal of Travel Research* 31(autumn), 3–7.

Eber, S. (1992) *Beyond the Green Horizon: a Discussion Paper on Principles for Sustainable Tourism.* Tourism Concern, UK.

Ecotourism Society of Saskatchewan (2000) *Saskatchewan Ecotourism Accreditation System.* Regina, Saskatchewan.

Ecotourism Society (1993) *Ecotourism Guidelines for Nature Tour Operators.* The Ecotourism Society, North Bennington, Vermont.

Edginton, C.R. and Edginton, S.R. (1993) Total quality program planning. *Journal of Physical Education Recreation and Dance* October, 40–42, 47.

Edginton, C.R. and Hanson, C.J. (1976) Appraising leisure service delivery. *Parks and Recreation* March, 44–45.

Edginton, C.R., Compton, D.M. and Hanson, C.J. (1980) *Recreation and Leisure Programming: A Guide for the Professional.* Saunders, Philadelphia.

Edginton, C.R., Hanson, C.J. and Edginton, S.R. (1992) *Leisure Programming: Concepts, Trends, and Professional Practice.* Brown & Benchmark, Dubuque, Iowa.

Edginton, C.R., Hanson, C.J., Edginton, S.R. and Hudson, S.D. (1998) *Leisure Programming: a Service Centred and Benefits Approach.* WCB/McGraw-Hill, Boston.

Edginton, S.R. and Edginton, C.R. (1994) *Youth Programs: Promoting Quality Services.* Sagamore Publishing, Champaign, Illinois.

Edington, J.M. and Edington, M.A. (1986) *Ecology, Recreation and Tourism.* Cambridge University Press, Cambridge.

Edney, J. (1980) The commons problem. *American Psychologist* 35, 131–150.

Edwards, S., McLaughlin, W.J. and Ham, S.H. (2000) *Comparative Study of Ecotourism Policy in the Americas – 1998: US and Canada.* Department of Resource Recreation and Tourism, University of Idaho, Moscow, Idaho.

Edwards, Y. (1980) The interpretation experience. *Parks* 6, 12–13.

Ehrlich, P.R. and Ehrlich, A.H. (1992) The value of biodiversity. *Ambio* 21(3), 219–226.

Elliot, J. (1997) *Tourism: Politics and Public Sector Management.* Routledge, London.

Ellis, G. and Rademacher, C. (1986) Barriers to recreation participation. In: *A Literature Review of the President's Commission on Americans Outdoors*, Motivations. US Government Printing Office, Washington, DC, pp. 33–50.

Entrikin, N.J. (1980) Robert Park's human ecology and human geography. *Annals of the Association of American Geographers* 70(1), 43–58.

Environmental Management Services (n.d.) *The Green Book of Tourism: a Study of the Environmental Protection Activities of Germany's Tourist Industry.* Federal Ministry for the Environment, Nature Conservation and Nuclear Safety, Berlin.

Erikson, E. (1963) *Childhood and Society*, 2nd edn. Norton, New York.

Erwin, T.L. (1991) How many species are there?: Revisited. *Conservation Biology* 5(3), 330–333.

Evernden, N. (1985) *The Natural Alien: Humankind and Environment.* University of Toronto Press, Toronto.

Ewert, A. (1985) Why people climb: the relationship of participant motives and experience level to mountaineering. *Journal of Leisure Research* 17(3), 241–250.

Ewert, A. (1989) *Outdoor Adventure Pursuits: Foundations, Models, and Theories.* Publishing Horizons, Columbus, Ohio.

Ewert, A. and Shultis, J. (1997) Resource-based tourism: an emerging trend in tourism experiences. *Parks & Recreation* September, 94–103.

Falk, I. and Kilpatrick, S. (2000) What *is* social capital? a study of interaction in a rural community. *European Society for Rural Sociology* 40(1), 87–110.

Farrell, B.H. (1999) Conventional or sustainable tourism? No room for choice. *Tourism Management* 20, 189–191.

Farrell, P. and Lundegren, H.M. (1993) *The Process of Recreation Programming.* Venture, State College, Pennsylvania.

Faulkner, B. (1999) Developing strategic approaches to tourism destination marketing: the Australian experience. In: Theobald, W.F. (ed.) *Global Tourism.* Butterworth-Heinemann, Oxford, pp. 297–316.

Fennell, D.A. (1990) A profile of ecotourists and the benefits derived from their experience: a Costa Rican case study Masters Thesis, University of Waterloo, Waterloo, Ontario, Canada.

Fennell, D.A. (1996) A tourist space–time budget in the Shetland Islands. *Annals of Tourism Research* 23(4), 811–829.

Fennell, D.A. (1997) *Managing Recreational Use of the Clearwater River. Prepared for Saskatchewan Environment and Resource Management.* University of Regina, Regina, Saskatchewan.

Fennell, D.A. (1998) Ecotourism in Canada. *Annals of Tourism Research* 25(1), 231–234.

Fennell, D.A. (1999) *Ecotourism: an Introduction.* Routledge, London.

Fennell, D.A. (2000a) Tourism and applied ethics. *Tourism Recreation Research* 25(1), 59–70.

Fennell, D.A. (2000b) What's in a name? Conceptualizing natural resource-based tourism. *Tourism Recreation Research* 25(1), 97–100.

Fennell, D.A. (2000c) Ecotourism on trial: the case of billfish angling as ecotourism. *Journal of Sustainable Tourism* 8(4), 341–345.

Fennell, D.A. (2001a) Ecotourism: a content analysis of definitions. *Current Issues in Tourism* 4(5), 403–421.

Fennell, D.A. (2001b) Research areas and needs. In: Weaver, D.B. (ed.) *The Encyclopaedia of Ecotourism.* CAB International, Wallingford, UK.

Fennell, D.A. (2002) The Canadian ecotourist in Costa Rica: ten years down the road. *International Journal of Sustainable Development* (in press).

Fennell, D.A. and Butler, R.W. (2002) A human ecological approach to tourism interaction. *International Journal of Tourism Research* (in press).

Fennell, D.A. and Malloy, D.C. (1999) Measuring the ethical nature of tourism operators. *Annals of Tourism Research* 26(4), 928–943.

Fennell, D.A. and Prezeclawski (2002) Ethics and host communities. In: Singh, S., Timothy, D. and Dowling, R. (eds) *Tourism and Host Communities.* CAB International, Wallingford, UK.

Fennell, D.A. and Smale, B.J.A. (1992) Ecotourism and natural resource protection: implications of an alternative form of tourism for host nations. *Tourism Recreation Research* 17(1), 21–32.

Fennell, D.A. and Weaver, D.B. (1997) Vacation farms and ecotourism in Saskatchewan, Canada. *Journal of Rural Studies* 13(4), 467–475.

Finkler, S.A. (1992) *Budgeting Concepts for Nurse Managers*, 2nd edn. W.B. Saunders, Philadelphia.

Flora, J. (1998) Social capital and communities of place. *Rural Sociology* 63(4), 481–506.

Flora, C.B. and Flora, J. (1993) Entrepreneurial social infrastructure: a necessary ingredient. *The Annals of the American Academy of Political and Social Science* 529, 48–58.

Fodness, D. (1990) Consumer perceptions of tourist attractions. *Journal of Travel Research* 28(4), 3–9.

Folke, C., Berkes, F. and Colding, J. (1998) Ecological practices and social mechanisms for building resilience and sustainability. In: Berkes, F. and Folke, C. (eds) *Linking Social and Ecological Systems: Management Practices and Social Mechanisms for Building Resilience.* Cambridge University Press, Cambridge.

Foot, D. and Stoffman, D. (1996) *Boom Bust & Echo.* Macfarlane Walter & Ross, Toronto.

Ford, P. (1981) *Principles and Practices of Outdoor/Environmental Education.* John Wiley & Sons, New York.

Ford, P. and Blanchard, J. (1993) *Leadership and Administration of Outdoor Pursuits.* Venture Publishing, State College, Pennsylvania.

Forgey, W.W. (1990) *Travelers' Medical Resource: a Guide to Health and Safety Worldwide.* ICS Books, Merrillville, Indiana.

Friel, M. (1998) Marketing. In: Thomas, R. (ed.) *The Management of Small Tourism and Hospitality Firms.* Cassell, London.

Game, M. (1980) Best shape for nature reserves. *Nature* 287, 630–631.

Garrod, B. and Fyall, A. (1998) Beyond the rhetoric of sustainable tourism? *Tourism Management* 19(3), 199–212.

Gifford-Jones, W. (1988) Water purifier cup is a must. *Kitchener-Waterloo Record (Canada)* 18 February.

Gilchrist, N.L. (1998) Winter hiking and camping. In: Dougherty, N.J. IV (ed.) *Outdoor Recreation Safety.* Human Kinetics, Champaign, Illinois, pp. 67–84.

Gilmore, G.D. and Campbell, M.D. (1996) *Needs Assessment Strategies for Health Education and Health Promotion.* WCB Brown and Benchmark, Madison, Wisconsin.

Glasbergen, P. (1998) The question of environmental governance. In: Glasbergen, P. (ed.) *Co-operative Environmental Governance: Public-Private Agreements as a Policy Strategy.* Kluwer Academic Publishers, London, pp. 1–18.

Goeldner, C.R., Ritchie, J.R.B. and McIntosh, R.W. (2000) *Tourism: Principles, Practices, Philosophies.* John Wiley & Sons, New York.

Gold, S.M. (1994) Behavioral approach to risk management. *Parks and Recreation* November, 34–36.

Goodin, R.E. (1999) Selling environmental indulgences. In: Dryzek, J.S. and Schlosberg, D. (eds) *Debating the Earth: the Environmental Politics Reader.* Oxford University Press, Oxford.

Gössling, S. (1999) Ecotourism: a means to safeguard biodiversity and ecosystem functions? *Ecological Economics* 29, 303–320.

Goudie, A. (1990a) *The Nature of the Environment.* Basil Blackwell, Oxford.

Goudie, A. (1990b) *The Human Impact on the Natural Environment.* Basil Blackwell, Oxford.

Graburn, N. (1989) Tourism: the sacred journey. In: Smith, V.L. (ed.) *Hosts and Guests: the Anthropology of Tourism*, 2nd edn. University of Pennsylvania Press, Philadelphia, pp. 21–36.

Graefe, A.R., Vaske, J.J. and Kuss, F.R. (1984) Resolved issues and remaining questions about social carrying capacity. *Leisure Sciences* 6(4), 497–507.

Gray, G. and Guppy, N. (1999) *Successful Surveys: Research Methods and Practice.* Harcourt Brace, Toronto.

Green, M.J. and Paine, J. (1997) State of the world's protected areas at the end of the twentieth century. Paper presented at the *IUCN World Commission on Protected Areas Symposium.* Albany, Australia, 24–29 November.

Greenall Gough, A. (1990) Environmental education. In: McRae, K. (ed.) *Outdoor and Environmental Education.* The MacMillan Co., Melbourne, pp. 60–72.

Greenleaf, R. (1977) *Servant Leadership.* Paulist Press, Toronto.

Gunderson, L., Holling, C.S. and Light, S. (1995) Breaking barriers and building bridges: a synthesis. In: Gunderson, L., Holling, C.S. and Light, S. (eds) *Barriers and Bridges to the Renewal of Ecosystems and Institutions.* Columbia University Press, New York.

Gurung, G., Simmons, D. and Devlin, P. (1996) The evolving role of the tourist guide: the Nepali experience. In: Butler, R.W. and Hinch, T. (eds) *Tourism and Indigenous Peoples.* International Thompson Business Press, London.

Hagler Bailly (1997) *Environmental Audits for Sustainable Tourism.* Hagler Bailly, Arlington, Virginia.
Hall, C.M. (1992) Adventure, sport and health tourism. In: Weiler, B. and Hall, C.M. (eds) *Special Interest Tourism.* Belhaven Press, London, pp. 141–158.
Hall, C.M. (2000) *Tourism Planning: Policies, Processes and Relationships.* Prentice Hall, London.
Hall, C.M. and Jenkins, J. (1995) *Tourism and Public Policy.* Routledge, London.
Hall, C.M. and McArthur, S. (1996) Strategic planning: integrating people and places through participation. In: Hall, C.M. and McArthur, S. (eds) *Heritage Management in Australia and New Zealand: the Human Dimension.* Oxford University Press, Oxford, pp. 22–36.
Hall, C.M. and McArthur, S. (1998) *Integrated Heritage Management: Principles and Practice.* The Stationery Office, London.
Hall, C.M. and Page, S.J. (1999) *The Geography of Tourism and Recreation: Environment, Place and Space.* Routledge, London.
Hall, P. (1970) A horizon of hotels. *New Society* 15, 389, 445.
Ham, S. and Weiler, B. (2000) *Six principles for tour guide training and sustainable development in developing countries.* Working paper No. 102/00, Department of Management, Monash University, Australia.
Hamilton-Smith, E. (1981) The concept of recreational specialisation. In: Mercer, D. (ed.) *Outdoor Recreation: Australian Perspectives.* Sorrett Publishing, Malveru.
Hammitt, W.E. and Cole, D.N. (1987) *Wildland Recreation: Ecology and Management.* John Wiley & Sons, New York.
Hanna, G. (1986) *Safety Oriented Guidelines for Outdoor Education Leadership and Programming.* The Canadian Association for Health, Physical Education and Recreation, Ottawa.
Hanna, G. (1991) *Outdoor Pursuits Programming: Legal Liability and Risk Management.* The University of Alberta Press, Edmonton.
Hardin, G. (1968) The tragedy of the commons. *Science* 162, 1243–1248.
Harroy, J.P. (1974) A century in the growth in the 'national park' concept throughout the world. In: Elliot, H. (ed.) *Second World Conference on National Parks.* IUCN, Switzerland.
Hays, S. (1959) *Conservation and the Gospel of Efficiency.* Harvard University Press, Cambridge, Massachusetts.
Hendee, J.C., Stankey, G.H. and Lucas, R.C. (1990) *Wilderness Management.* International Wilderness Leadership Foundation, Golden, Colorado.
Henderson, K.A. and Bialeschki, M.D. (1995) *Evaluating Leisure Services: Making Enlightened Decisions.* Venture Publishing, State College, Pennsylvania.
Hetzer, N.D. (1965) Environment, tourism, culture, *LINKS* (July) (Reprinted in *Ecosphere* (1970) 1(2), 1–3).
Higgins, B.R. (1996) The global structure of the nature tourism industry: ecotourists, tour operators, and local businesses. *Journal of Travel Research* 35(2), 11–18.
Hills, T. and Lundgren, J. (1977) The impact of tourism in the Caribbean: a methodological study. *Annals of Tourism Research* 4(5), 248–267.
Holling, C.S., Berkes, F. and Folke, C. (1998) Science, sustainability and resource management. In: Berkes, F. and Folke, C. (eds) *Linking Social and Ecological Systems: Management Practices and Social Mechanisms for Building Resilience.* Cambridge University Press, Cambridge, pp. 342–362.
Holland, S.M., Ditton, R.B. and Graefe, A.R. (1998) An ecotourism perspective on billfish fisheries. *Journal of Sustainable Tourism* 6(2), 97–116.
Holloway, J.C. (1994) *The Business of Tourism*, 4th edn. Pitman, London.
Honey, M. (1999) *Ecotourism and Sustainable Development: Who Owns Paradise?* Island Press, Washington, DC.
Honey, M. and Rome, A. (2001) Certification and Ecolabelling. Institute for Policy Studies, Washington, DC. Available online at www.ips-dc.org
Horwich, R.H., Murray, D., Saqui, E., Lyon, J. and Godfrey, D. (1993) Ecotourism and community development: a view from Belize. In: *Ecotourism: a Guide for Planners and Managers.* The Ecotourism Society, North Bennington, Vermont.
Hudson, S.D. (1988) *How to Conduct Community Needs Assessment Surveys in Public Parks and Recreation.* Horizons, Columbus, Ohio.
Hultsman, J. (1995) Just tourism: an ethical framework. *Annals of Tourism Research* 22(3), 553–567.
Hultsman, J., Cottrell, R.L. and Hultsman, W.Z. (1998) *Planning Parks for People.* Venture Publishing, State College, Pennsylvania.
Hunt, S., Wood, V. and Chonko, L. (1989) Corporate ethical values and organizational commitment in marketing. *Journal of Marketing* 53, 79–90.

Hunter, C. (1995) On the need to re-conceptualise sustainable tourism development. *Journal of Sustainable Tourism* 3(3), 155–165.

Hunter, C. (1997) Sustainable tourism as an adaptive paradigm. *Annals of Tourism Research* 24(4), 850–867.

Hunter, C. and Green, H. (1995) *Tourism and the Environment: a Sustainable Relationship?* Routledge, London, pp. 14, 28, 35.

Husbands, W. (1983) Tourist space and tourist attraction: an analysis of the destination choices of European travellers. *Leisure Sciences* 5(4), 289–307.

Hvenegaard, G.T. and Dearden, P. (1998) Ecotourism versus tourism in a Thai national park. *Annals of Tourism Research* 25, 700–720.

Iacobucci, D. and Ostrom, A. (1996) Perceptions of services. *Journal of Retailing and Consumer Services* 3(4), 195–212.

Iaffaldano, M.T. and Muchinsky, P.M. (1985) Job satisfaction and job performance: a meta-analysis. *Psychological Bulletin* 97, 251–273.

[IAMAT] International Association for Medical Assistance to Travellers (1995) *World Malaria Risk Chart.* IAMAT, Guelph, Canada.

Industry Canada (n.d.) *Your Business Plan.* Industry Canada, Ottawa.

Inskeep, E. (1991) *Tourism Planning: An Integrated and Sustainable Development Approach.* John Wiley & Sons, New York.

Institute of Medicine (1991) *Disabilities in America: Toward a National Agenda for Prevention: Summary and Recommendations.* In: Pope, A. and Tarlov, A. (eds) National Academy Press, Washington, DC.

Iso Ahola, S.E. (1980) *The Social Psychology of Leisure and Recreation.* W.C. Brown, Dubuque, Iowa.

Iso Ahola, S.E. (1982) Toward a social psychological theory of tourism motivation: a rejoinder. *Annals of Tourism Research* 9(2), 256–262.

Issaverdis, J.-P. (2001) The pursuit of excellence: benchmarking, accreditation, best practice and auditing. In: Weaver, D.B. (ed.) *The Encyclopedia of Ecotourism.* CABI, Wallingford, UK, pp. 579–594.

Iwand, W.M. (n.d.) *Better Environment – Better Business: Tourism and Environmental Compatibility as Practised by a Tour Operator.* Touristik Union International, Hannover.

Jackson, J.M. (1965) Structural characteristics of norms. In: Steiner, I. and Fishbein, M. (eds) *Current Studies in Social Psychology.* Holt, Reinhart & Winston, New York, pp. 301–309.

Jackson, M. and Bissix, G. (1979) *A Manual for Outdoor Recreation Programming.* Department of Culture, Recreation and Fitness, Halifax, Nova Scotia.

Jandt, F.E. (1995) *The Customer is Usually Wrong!* Park Avenue, Indianapolis, Indiana.

Jansen-Verbeke, M. and Dietvorst, A. (1987) Leisure, recreation, tourism: a geographic view on integration. *Annals of Tourism Research* 14(3), 361–375.

Jensen, C.R. (1995) *Outdoor Recreation in America.* Human Kinetics, Champaign, Illinois.

Johnson, K. and Bruya, L. (1995) Kayak programming: a risk management approach. *Parks and Recreation* February, 49–59.

Johnston, M.E. (1992) Facing the challenges: adventure in the mountains of New Zealand. In: Weiler, B. and Hall, C.M. (eds) *Special Interest Tourism.* Belhaven Press, London, pp. 159–170.

Johnston, R.J. (ed.) (1986) *The Dictionary of Human Geography*, 2nd edn. Blackwell, Oxford.

Johnston, R.J. (1987) *Geography and Geographers.* Edward Arnold, New York.

Jones, H. (1972) Gozo – the living showpiece. *Geographical Magazine* 45(1), 53–57.

Jordan, D.J. (1996) *Leadership in Leisure Services: Making a Difference.* Venture Publishing, State College, Pennsylvania.

Jorgenson, D.L. (1989) *Participant Observation: A Methodology for Human Studies.* Sage, Beverly Hills, California.

Kahle, L. (1983) Dialectical tensions in the theory of social values. In: Kahle, L. (ed.) *Social Values and Social Change.* Praeger, New York, pp. 275–283.

Kao, R.W.Y. (1992) *Small Business Management: a Strategic Emphasis.* Holt, Rinehart and Winston of Canada, Toronto.

Katz, R.L. (1974) Skills of an effective administrator. *Harvard Business Review* 52, 5.

Kelly, J.R. and Godbey, G. (1992) *The Sociology of Leisure.* Venture Publishing, State College, Pennsylvania.

Kennedy, D.W., Smith, R.W. and Austin, D.R. (1991) *Special Recreation: Opportunities for Persons with Developmental Disabilities.* W.C. Brown, Dubuque, Iowa.

Killan, G. (1993) *Protected Places: a History of Ontario's Provincial Parks System.* Dundurn Press, Toronto.

Kimmel, J.R. (1999) Ecotourism as environmental learning. *The Journal of Environmental Education* 30(2), 40–44.
Klausner, S.Z. (1971) *On Man in his Environment*. Jossey-Bass, San Francisco, California.
Knowles, M. (1980) *The Modern Practice of Adult Education*. Cambridge University Press, New York.
Knowles, T. and Grabowski, P. (1999) Strategic marketing in the tour operator sector. In: Vellas, F. and Bécheral, L. (eds) *The International Marketing of Travel and Tourism*. Macmillan Press, London, pp. 249–262.
Knudson, D.M. (1980) *Outdoor Recreation*. Macmillan Publishing, New York.
Knudson, D.M., Cable, T.T. and Beck, L. (1995) *Interpretation of Cultural and Natural Resources*. Venture, State College, Pennsylvania.
Kodis, M. (1996) The last word on DEET. *Backpacker Magazine* 24(7), 24–29.
Kohlberg, L. (1984) *The Psychology of Moral Development: the Nature and Validity of Moral Stages*. Harper & Row, San Francisco, California.
Kotler, P. (1994) *Marketing Management: Analysis, Planning, Implementation, and Control*, 8th edn. Prentice Hall, Englewood Cliffs, New Jersey.
Kotler, P., Armstrong, G., Cunningham, P.H. and Warren, R. (1996) *Principles of Marketing*, 3rd edn. Prentice Hall, Scarborough, Ontario.
Kouzes, J. and Posner, B. (1987) *The Leadership Challenge*. Jossey Bass, San Francisco, California.
Kraus, R. (1977) *Recreation Today: Program Planning and Leadership*, 2nd edn. Goodyear Publishing, Santa Monica, California.
Kraus, R. (1997) *Recreation Programming: a Benefits-driven Approach*. Allyn and Bacon, Toronto.
Kraus, R. and Allen, L. (1997) *Research and Evaluation in Recreation, Parks and Leisure Studies*, 2nd edn. Gorsuch Scarisbrick, Publishers, Scottsdale, Arizona.
Kraus, R. and Curtis, J. (1990) *Creative Management in Recreation, Parks, and Leisure Services*, 5th edn. Times Mirror/Mosby, St Louis, Missouri.
Krejcie, R.V. and Morgan, D.W. (1970) Determining sample size for research activities. *Educational and Psychological Measurement* 30, 607–610.
Kretchman, J.A. and Eagles, P.F.J. (1990) An analysis of the motives of ecotourists in comparison to the general Canadian population. *Society and Leisure* 13(2), 499–507.
Kretzmann, J.P. and McKnight, J.L. (1993) *Building Communities from the Inside Out*. ACTA Publications, Chicago, Illinois.
Kreutzwiser, R. (1989) Supply. In: Wall, G. (ed.) *Outdoor Recreation in Canada*. Wiley, Toronto, pp. 19–42.
Krueger, R.A. (1988) *Focus Groups*. Sage, Beverly Hills, California.
Krut, R. and Gleckman, H. (1996) *ISO 14001: a Missed Opportunity for Sustainable Global Industrial Development*. Earthscan Publications, London.
Kuss, F.R., Graefe, A.R. and Vaske, J.J. (1984) *Recreation Impacts and Carrying Capacity: A Review and Synthesis of Ecological and Social Research*. Technical report submitted to the National Parks and Conservation Association, Washington, DC.
Laarman, J.G. and Durst, P.B. (1987) Nature travel and tropical forest. FPEI Working Paper Series, Southeastern Center for Forest Economics Research, North Carolina State University, North Carolina.
Landy, F.J. (1989) *Psychology of Work Behavior*, 4th edn. Brooks/Cole, Pacific Grove, California.
Langholz, J. and Brandon, K. (2001) Privately owned protected areas. In: Weaver, D.B. (ed.) *The Encyclopedia of Ecotourism*. CAB International, Wallingford, UK, pp. 303–314.
Lawton, L.J. (2001) Public protected areas. In: Weaver, D.B. (ed.) *The Encyclopedia of Ecotourism*. CAB International, Wallingford, UK, pp. 287–302.
Lee, M.A. and Solomon, N. (1991) And that's the way it is: media coverage of the environment has surely increased, but is it doing justice to the issues or doing the bidding of media owners? *E Magazine* Jan–Feb, 38–43, 95–67.
Leighly, J. (1987) Ecology as metaphor: Carl Sauer and human ecology. *The Professional Geographer* 39 (4), 405–412.
Lemke, J. (1995) *Textual Politics: Discourse and Social Dynamics*. Taylor & Francis, London.
Leopold, A. (1991) *A Sand County Almanac*. Ballantine Books, New York.
Lew, A. (1987) A framework of tourist attraction research. *Annals of Tourism Research* 14(4), 553–575.
Lewis, M.W. (1992) *Green Delusions: an Environmentalist Critique of Radical Environmentalism*. Duke University Press, Durham, North Carolina.
Liddle, M.J. (1997) *Recreation Ecology: the Ecological Impact of Outdoor Recreation and Ecotourism*. Chapman & Hall, London.

Lincoln, R.J. and Boxshall, G.A. (1990) *The Cambridge Illustrated Dictionary of Natural History.* Cambridge University Press, Cambridge.
Lindberg, K. (1991) *Policies for Maximising Nature Tourism's Ecological and Economic Benefits.* World Resources Institute, Washington, DC.
Little, S.L. (1993) Leisure program design and evaluation: using leisure experience models as diagnostic tools. *Journal of Physical Education, Recreation and Dance* October, 26–29, 33.
Lothian, W.F. (1987) *A Brief History of Canada's National Parks.* Supply and Services Canada, Ottawa.
Lovejoy, T.E. (1992) Looking to the next millennium: climate change, pollution, and growing visitation will increasingly challenge world parks. *National Parks* January/February, 41–44.
Lucas, R.C. (1964) Wilderness perception and use: the example of the Boundary Waters Canoe Area. *Natural Resources Journal* 3(3), 394–411.
Luhrman, D. (1997) WTO Manila Meeting, Internet communication, 23 May, WTO Press and Communications.
Lundberg, D.E. (1976) *The Tourist Business.* CBI, Boston.
MacArthur, R.H. (1957) On the relative abundance of bird species. *Proceedings of the National Academy of Sciences USA* 43, 293–294.
MacArthur, R.H. and Wilson, E.O (1963) An equilibrium theory of insular zoogeography. *Evolution* 17, 373–387.
MacCannell, D. (1989) *The Tourist: a New Theory of the Leisure Class.* Schocken Books, New York.
MacKinnon, B. (1995) Beauty and the beasts of ecotourism. *Business Mexico* 5(4), 44–47.
Malloy, D.C. and Fennell, D.A. (1998) Codes of ethics and tourism: an exploratory content analysis. *Tourism Management* 19(5), 453–461.
Mand, C. (1967) *Outdoor Education.* J Lowell Pratt, New York.
Mannell, R.C. and Kleiber, D.A. (1997) *A Social Psychology of Leisure.* Venture, State College, Pennsylvania.
Manning, T. (1996) Tourism: where are the limits. *Ecodecision* 20(spring), 35–39.
Manning, T. (1999) Indicators of tourism sustainability. *Tourism Management* 20, 179–181.
Margerison, J. (1998) Business planning. In: Thomas, R. (ed.) *The Management of Small Tourism and Hospitality Firms.* Cassell, London, pp. 101–116.
Mason, J.G. (1960) *How to be a More Effective Executive.* McGraw-Hill, New York.
Mason, P. and Mowforth, M. (1995) *Codes of Conduct in Tourism.* University of Plymouth, Department of Geographical Sciences Occasional Paper No. 1.
Mathieson, A. and Wall, G. (1982) *Tourism: Economic, Physical and Social Impacts.* Longman, New York.
Matthews, S. (1997) When green is mean. *In Focus* 23, (no page number).
May, R.M. (1992) How many species inhabit the Earth? *Scientific American* October, 42–48.
Mayo Bay Camps and Harmony Studios (n.d.) *Ecotourism at Mayo Bay.* US Virgin Islands.
McArthur, S. (1997) Introducing the National Ecotourism Accreditation Program. *Australian Parks and Recreation* 33(2), 30–34.
McAvoy, L. and Lais, G. (1999) Programs that include persons with disabilities. In: Miles, J.C. and Priest, S. (eds) *Adventure Programming.* Venture, State College, Pennsylvania, pp. 403–414.
McCarville, R. (1993) Keys to quality leisure programming. *Journal of Physical Education, Recreation and Dance* October, 32–39.
McDonald, C.D. (1996) Normative perspectives on outdoor recreation behavior: introductory comments. *Leisure Sciences* 18, 1–6.
McIntosh, R., Goeldner, C. and Ritchie, J. (1995) *Tourism: Principles, Practices, Philosophies,* 7th edn. John Wiley & Sons, New York.
McKenzie, J.F. and Smeltzer, J.L. (1997) *Planning, Implementing, and Evaluating Health Promotion Programs: a Primer,* 2nd edn. Allyn and Bacon, Toronto.
McKercher, B. (1993) The unrecognized threat to tourism: can tourism survive 'sustainability'? *Tourism Management* 14(2), 131–136.
McKercher, B. (1996) Differences between tourism and recreation in parks. *Annals of Tourism Research* 23(3), 563–575.
McKercher, B. (1998) *The Business of Nature-Based Tourism.* Hospitality Press, Melbourne.
McLaren, D. (1998) *Rethinking Tourism and Ecotravel.* Kumarian Press, West Hartford, Connecticut.
Merry, W. (1994) *St. John Ambulance Official Wilderness First-aid Guide.* McClelland & Stewart, Toronto.
Metelka, C.J. (1981) *The Dictionary of Tourism.* Merton House, Illinois.
Metelka, C.J. (1990) *The Dictionary of Hospitality, Travel and Tourism.* Delmar, Albany, New York.

Middleton, V. and Hawkins, D. (1998) *Sustainable Tourism: a Marketing Perspective.* Butterworth-Heinemann, Oxford.

Mieczkowski, Z.T. (1981) Some notes on the geography of tourism: a comment. *Canadian Geographer* 25, 186–191.

Mill, R.C. and Morrison, A.M. (1985) *The Tourism System: an Introductory Text.* Prentice-Hall, Englewood Cliffs, New Jersey.

Miller, G.T. and Armstrong, P. (1982) *Living in the Environment.* Wadsworth International Group, Belmont, California.

Miller, K.R. (1976) Global dimensions of wildlife management in relation to development and environmental conservation in Latin America. Proceedings of *Regional Expert Consultation on Environment and Development.* Bogota, Colombia, 5–10 July. FAO, Santiago, Chile.

Miller, K.R. (1989) *Planning National Parks for Ecodevelopment: Methods and Cases from Latin America.* Peace Corps, Washington, DC.

Mintzberg, H. (1973) *The Nature of Managerial Work.* Prentice-Hall, Englewood Cliffs, New Jersey.

Mitchell, B. (1989) *Geography and Resource Analysis.* Longman, New York.

Mitchell, G.E. (1992) *Ecotourism Guiding: How to Start your Career as an Ecotourism Guide.* G.E. Mitchell Institute of Travel Career Development, Florida.

Mitchell, L.S. (1969) Recreational geography: evolution and research needs. *The Professional Geographer* 11(2), 117–118.

Mitchell, L.S. and Smith, R.V. (1985) Recreational geography: inventory and prospect. *The Professional Geographer* 37(1), 6–14.

Morgan, D. (2000a) Adventure tourism activities in New Zealand: perceptions and management of client risk. *Tourism Recreation Research* 25(3), 79–89.

Morgan, H. (2000b) A taxing time. *In Focus* 37, 6–7.

Morrison, A., Hsieh, S. and Wang, C.-Y. (1992) Certification in the travel and tourism industry: the North American experience. *The Journal of Tourism Studies* 3(2), 32–40.

Morrison, A., Rimmington, M. and Williams, C. (1999) *Entrepreneurship in the Hospitality, Tourism and Leisure Industries.* Butterworth Heinemann, Oxford.

Morton, K. (1997) *Planning a Wilderness Trip in Canada and Alaska.* Rocky Mountain Books, Calgary, Alberta.

Mowby, R.I. (2000) Tourist perceptions of security: the risk–fear paradox. *Tourism Economics* 6(2), 109–121.

Mowforth, M. and Munt, I. (1998) *Tourism and Sustainability: New Tourism in the Third World.* Routledge, London.

Munt, I. and Higinio, E. (1993) Belize – eco-tourism gone awry. *In Focus* 9, 18–19.

Murdoch, C.P. and Trygstad, C.W. (1993) Wilderness first aid. In: Schad, J. and Moser, D.S. (eds) *Wilderness Basics: the Complete Handbook for Hikers and Backpackers.* The Mountaineers, Seattle, pp. 188–204.

Murphy, J.F. (1975) *Recreation and Leisure Service: a Humanistic Perspective.* W.C. Brown, Dubuque, Iowa.

Murphy, P.E. (1983) Tourism as a community industry: an ecological model of tourism development. *Tourism Management* 4(3), 180–193.

Murphy, P.E. (1985) *Tourism: a Community Approach.* Routledge, New York.

Murray, E.F. (1964) *Motivation and Emotion.* Prentice Hall, Englewood Cliffs, New Jersey.

Myers, N. (1990) Mass extinctions: what can the past tell us about the present and the future? *Global and Planetary Change* 82, 175–185.

Myers, N. (1993) Biodiversity and the precautionary principle. *Ambio* 22(2–3), 74–79.

Nash, R. (1982) *Wilderness and the American Mind*, 3rd edn. Yale University Press, New Haven, Connecticut.

National Park and Recreation Association (1989) *Staff Training and Development for Park, Recreation, and Leisure Service Organizations*, 2nd edn. NRPA, Alexandra, Virginia.

Naylon, J. (1967) Tourism – Spain's most important industry. *Geography* 52, 23–40.

Nelson, J.G. (1991) Sustainable development, conservation strategies and heritage. In: Mitchell, B. (ed.) *Resource Management and Development.* Oxford University Press, Toronto, pp. 246–267.

Nelson, J.G. (1994) The spread of ecotourism: some planning implications. *Environmental Conservation* 21(1), 248–255.

Neulinger, J. (1974) *Psychology of Leisure: Research Approaches to the Study of Leisure.* Charles C. Thomes, Springfield, Illinois.

Neuman, W.L. (1997) *Social Research Methods: Qualitative and Quantitative Approaches*, 3rd edn. Allyn and Bacon, Toronto.
Newsome, D., Moore, S.A. and Dowling, R.K. (2002) *Natural Area Tourism: Ecology, Impacts and Management*. Channel View Publications, Clevedon.
Newton, K. (1997) Social capital and democracy. *American Behavioural Scientist* 40, 575–586.
Norgaard, R.B. (1994) *Development Betrayed: the End of Progress and a Coevolutionary Revisioning of the Future*. Routledge, New York.
Northwest Arkansas Resource Conservation and Development Council Inc. (1997) Ecotourism Development Council. Northwest Arkansas Resource Conservation and Development Council, Inc, Harrison, Arkansas.
Obua, J. and Harding, D.M. (1996) Visitor characteristics and attitudes towards Kibale National Park, Uganda. *Tourism Management* 17(7), 495–505.
[OECD] Organisation for Economic Co-operation and Development (1975) *The Polluter Pays Principle*. OECD, Paris.
Olokesusi, F. (1990) Assessment of the Yankari Game Reserve, Nigeria. *Tourism Management* 11, 153–162.
Ontario Camping Association (1990) *Guidelines for Accreditation for Resident Camps*. Ontario Camping Association, Toronto.
Ontario Ministry of Consumer and Commercial Relations (1995) *Travel Tips*. Ministry, Toronto.
Ontario Ministry of Tourism and Recreation (1985a) *Planning Recreation: A Manual of Principles and Practices*. Government of Ontario, Toronto.
Ontario Ministry of Tourism and Recreation (1985b) *Better Planning for Better Recreation*. Government of Ontario, Toronto.
O'Riordan, T. (1971) *Perspectives on Resource Management*. Pion, London.
O'Riordan, T. (1976) *Environmentalism*. Pion, London.
O'Riordan, T. and Cameron, J. (1994) The history and contemporary significance of the precautionary principle. In: O'Riordan, T. and Cameron, J. (eds) *Interpreting the Precautionary Principle*. Earthscan, London, pp. 12–30.
Orr, D.W. (1992) *Ecological Literacy: Education and the Transition to a Postmodern World*. State University of New York Press, New York.
Ortolano, L. (1984) *Environmental Planning and Decision Making*. John Wiley & Sons, New York.
Otto, J.E. and Ritchie, J.R.B. (1996) The service orientation in tourism. *Tourism Management* 17(3), 165–174.
[ORRRC] Outdoor Recreation Resources Review Commission (1962) *Outdoor recreation for America*. Government Printing Office, Washington, DC.
Ozark Ecotours (2001) Ozark Ecotours Reservation Form. Site accessed on 10 March 2001 http://www.ozarkecotours.com/reservation.html
Paaby, P., Clark, D.B. and Gonzalez, H. (1991) Training rural residents as naturalist guides: Evaluation of a pilot project in Costa Rica. *Conservation Biology* 18, 542–546.
Page, S.J. and Dowling, R.K. (2002) *Ecotourism*. Pearson Education, London.
Page, S.J., Clift, S. and Clark, N. (1994) Tourist health: the precautions, behaviour and health problems of British tourists in Malta. In: Seaton, A.V. (ed.) *Tourism: the State of the Art*. John Wiley & Sons, New York, pp. 799–817.
Parkinson, R.S. & Associates (1982) *Managing Health Promotion in the Workplace: Guidelines for Implementation and Evaluation*. Mayfield, Palo Alto, California.
Patterson, C. (1997) *The Business of Ecotourism*. Explorer's Guide Publishing, Rhinelander, Wisconsin.
Patterson, F.C. (1991) *A Systems Approach to Recreation Programming*. Waveland Press, Prospect Heights, Illinois.
Paul, S. (1987) *Community Participation in Development Projects: the World Bank Experience*. World Bank, Washington, DC.
Pavesic, D.V. (1989) Psychological aspects of menu pricing. *International Journal of Hospitality Management* 8(1), 43–49.
Payne, R.J. and Graham, R. (1993) Visitor planning and management in parks and protected areas. In: Dearden, P. and Rollins, R. (eds) *Parks and Protected Areas in Canada*. Oxford University Press, Toronto, pp. 185–210.
Pearce, D.G. (1979) Towards a geography of tourism. *Annals of Tourism Research* 6(3), 245–272.
Pearce, D.G. (1985) Tourism and environmental research: a review. *International Journal of Environmental Studies* 25, 247–255.

Pearce, D.G. (1989) *Tourist Development.* Longman, London.

Peattie, K. (1995) *Environmental Marketing Management: Meeting the Green Challenge.* Pitman, London.

Perrings, C., Turner, R.K. and Folke, C. (1995) *Ecological Economics: the Study of Interdependent Economic and Ecological Systems.* Beijer Discussion Paper Series No. 55. Beijer Institute, Stockholm.

Peterson, G. (1996) Four corners of human ecology: different paradigms of human relationships with earth. In: Driver, B.L., Dustin, D., Baltic, T., Elsner, G. and Peterson, G. (eds) *Nature and the Human Spirit: Toward an Expanded Land Management Ethic.* Venture Publishing, State College, Pennsylvania, pp. 25–40.

Phipps, M. and Swiderski, M. (1990) The 'soft' skills of outdoor leadership. In: Miles, J.C. and Priest, S. (eds) *Adventure Education.* Venture Publishing, State College, Pennsylvania, pp. 221–232.

PLAE, Inc. (1993) *Universal Access to Outdoor Recreation: a Design Guide.* PLAE Inc., Berkeley, California.

Plog, S.C. (1972) Why destination areas rise and fall in popularity. Paper presented to the Travel Research Association, Southern California Chapter, Los Angeles.

Plog, S.C. (1987) Understanding psychographics in tourism research. In: Ritchie, J.R.B. and Goeldner, C.R. (eds) *Travel, Tourism and Hospitality Research: a Handbook for Managers and Researchers.* John Wiley & Sons, New York, pp. 203–213.

Plummer, J. (1989) Changing values: the new emphasis on self-actualization. *The Futurist.*

Porter, P.W. (1978) Geography as human ecology. *American Behavioural Scientist* 22(1), 15–39.

Priest, S. (1990) The semantics of adventure education. In: Miles, J.C. and Priest, S. (eds) *Adventure Education.* Venture Publishing, State College, Pennsylvania, pp. 113–118.

Priest, S. and Gass, M. (1997) *Effective Leadership in Adventure Programming.* Human Kinetics, Champaign, Illinois.

Priest, C. and Gass, M. (1999) Six generations of facilitation skills. In: Miles, C.J. and Priest, S. (eds) *Adventure Programming.* Venture Publishing, State College, Pennsylvania, pp. 215–218.

Province of British Columbia (1981) *Outdoor Safety and Survival.* Ministry of Environment, Victoria, British Columbia.

Raven, P.H., Evert, R.F. and Eichhorn, S.E. (1986) *Biology of Plants.* Worth Publishers, New York.

Rees, W. (1996) Revisiting carrying capacity: area-based indicators of sustainability. *Population and Environment: A Journal of Interdisciplinary Studies* 17(3), accessed at http://www. dieoff. org/ page110.htm.

Rees, W. (1988) A role for environmental assessment in achieving sustainable development. *Environmental Impact Assessment Review* 8, 273–291.

Rees, W. and Wackernagel, M. (1996) *Our Ecological Footprint: Reducing Human Impact on the Earth.* New Society Publishers, Gabriola Island, British Columbia.

Reichel, A., Lowengart, O. and Milman, A. (2000) Rural tourism in Israel: service quality and orientation. *Tourism Management* 21, 451–459.

Rest, J.R. (1975) Recent research on an objective test moral judgement: how the important issues of a moral dilemma are defined. In: DePalma, D.J. and Foley, J.M. (eds) *Moral Development: Current Theory and Research.* John Wiley & Sons, New York, pp. 75–94.

Reynolds, P.C. and Braithwaite, D. (2001) Towards a conceptual framework for wildlife tourism. *Tourism Management* 22, 31–42.

Robinson, D. and Twynam, D. (1995) Alternative tourism, indigenous peoples, and environment: the case of Sagarmatha (Everest) National Park, Nepal. *Environments* 23(3), 13–35.

Robinson, J., Francis, G., Legge, R. and Lerner, S. (1990) Defining a sustainable society: values, principles, and definitions. *Alternatives* 17, 36–46.

Roe, E. (1998) *Taking Complexity Seriously: Policy Analysis, Triangulation and Sustainable Development.* Kluwer Academic Publishers, Boston, Massachusetts.

Rollins, R. (1993) Managing the national parks. In: Dearden, P. and Rollins, R. (eds) *Parks and Protected Areas in Canada.* Oxford University Press, Toronto, pp. 75–96.

Rolston III, H. (1991) Creation and recreation: environmental benefits and human leisure. In: Driver, B.L., Brown, P.J. and Peterson, G.L. (eds) *Benefits of Leisure.* Venture, State College, Pennsylvania, pp. 393–403.

Romero, S. (2000) Amazon seeks to green economy. *The Toronto Star* (Canada) 19 June, E6.

Roseland, M. (1992) *Toward Sustainable Communities.* National Round Table on the Environment and the Economy, Ottawa.

Ross, S.A. and Sanchez, J.L. (1990) Recreational sun exposure in Puerto Rico: trends and cancer risk awareness. *Journal of the American Academy of Dermatology* 23(6), 1090–1092.

Russell, R.V. (1982) *Planning Programs in Recreation*. Mosby, St Louis, Missouri.
Sanders, I.T. (1958) Theories of community development. *Rural Sociology* 23(1), 1–12.
San Diego Chapter of the Sierra Club (1993) *Wilderness Basics: The Complete Handbook for Hikers and Backpackers*. The Mountaineers, Seattle, Washington.
Saskatchewan Tourism Education Council (1997) *Tourism Small Business Operator*. STEC, Saskatoon, Saskatchewan.
Saveriades, A. (2000) Establishing the social tourism carrying capacity for the tourist resorts of the east coast of the Republic of Cyprus. *Tourism Management* 21, 147–156.
Schermerhorn, J.R. (1996) *Management*, 5th edn. John Wiley & Sons, New York.
Scheyvens, R. (1999) Ecotourism and the employment of local communities. *Tourism Management* 20, 245–249.
Schilit, W.K. (1987) How to write a winning business plan. *Business Horizons* September–October, 1–24.
Schultz, J.H., McAvoy, L.H. and Dustin, D.L. (1988) What are we in business for? *Parks and Recreation* 23(1), 51–53.
Scoones, I. (1999) New ecology and the social sciences: what prospects for a fruitful engagement? *Annual Reviews Anthropology* 28, 479–507.
Searle, M.S. and Brayley, R.E. (1993) *Leisure Services in Canada*. Venture Publishing, State College, Pennsylvania.
Sessoms, H.D. and Stevenson, J.L. (1981) *Leadership and Group Dynamics in Recreation Services*. Allyn and Bacon, Boston, Massachusetts.
Shackleford, P. (1985) The World Tourism Organization – 30 Years of commitment to environmental protection. *International Journal of Environmental Studies* 25, 257–263.
Shafer, C.L. (1990) *Nature Reserves: Island Theory and Conservation Practice*. Smithsonian Institution Press, Washington, DC.
Shelby, B., Vaske, J.J. and Donnelly, M.P. (1996) Norms, standards, and natural resources. *Leisure Sciences* 18, 103–123.
Short, J.R. (1991) *Imagined Country: Society, Culture and Environment*. Routledge, London.
Shultz, J.H., McAvoy, L.H. and Dustin, D.L. (1988) *Parks and Recreation* 23, 51–53.
Siegel, L.N., Attkisson, C.C. and Carson, L.G. (1987) Need identification and program planning in the community context. In: Cox, F.M., Erlich, J.L., Rothman, J. and Tropman, J.E. (eds) *Strategies of Community Development*, 4th edn. F.E. Peacock, Itasca, Illinois.
Simberloff, D.S. and Abele, L.G. (1976) Island biogeography theory and conservation practice. *Science* 191, 285–286.
Simer, P. and Sullivan, J. (1985) *The National Outdoor Leadership School's Wilderness Guide*. Fireside Books, New York.
Simmons, I.G. (1990) *Changing the Face of the Earth: Culture, Environment, History*. Basil Blackwell, Oxford.
Simpson, K. (2001) Strategic planning and community involvement as contributors to sustainable tourism development. *Current Issues in Tourism* 4(1), 3–41.
Skaros, S. (1998) Managing medical emergencies in the outdoors. In: Dougherty, N.J. IV (ed.) *Outdoor Recreation Safety*. Human Kinetics, Champaign, Illinois, pp. 25–42.
Smith, G. (1998) Data requested for survey of international standards and certification programs. TRINET Communication, 9 November.
Smith, J.W., Carlson, R.E., Donaldson, G.W. and Masters, H.B. (1972) *Outdoor Education*, 2nd edn. Prentice-Hall, Englewood Cliffs, New Jersey.
Smith, S.L.J. (1998) Tourism as an industry: debates and concepts. In: Ioannides, D. and Debbage, K.G. (eds) *The Economic Geography of the Tourist Industry*. Routledge, London.
Smith, S.L.J. (1990a) *Dictionary of Concepts in Recreation and Leisure Studies*. Greenwood Press, New York.
Smith, S.L.J. (1990b) *Tourism Analysis*. John Wiley & Sons, New York.
Snepenger, D.J. and Bowyer, R.T. (1990) Differences among nonresident tourists making consumptive and nonconsumptive uses of Alaskan wildlife. *Arctic* 43(3), 262–266.
Snyder, N.H., Dowd J.J. Jr and Morse Houghton, D. (1994) *Vision, Values, and Courage: Leadership for Quality Management*. Maxwell Macmillan, Toronto.
Spears, L. (1995) *Reflections on Leadership: How Robert K. Greenleaf's Theory of Servant Leadership Influenced Today's Top Management Thinkers*. John Wiley & Sons, Toronto.
St John Ambulance (1995) *First on the Scene: the Complete Guide to First Aid and CPR*. St John Ambulance, Ottawa, Canada.

Stapp, W.B., Wals, A.E.J. and Stankorb, S.L. (1996) *Environmental Education for Empowerment: Action Research and Community Problem Solving*. Kendall/Hunt, Dubuque, Iowa.

Stankey, G.H. and McCool, S.F. (1984) Carrying capacity in recreational settings: evolution, appraisal, and application. *Leisure Sciences* 6(4), 453–473.

Steele, P. (1993) The economics of eco-tourism. *In Focus* 9, 7–9.

Sterrer, W. (1982) People using resources. In: Hayward, S.J., Gomez, V.H. and Sterrer, W. (eds) *Bermuda's Delicate Balance: People and the Environment*. The Bermuda National Trust, Hamilton, Bermuda, pp. 3–9.

Svendon, G.L.H. and Svendson, G.T. (2000) Measuring social capital: the Danish co-operative dairy movement. *European Society for Rural Sociology* 40(1), 72–86.

Swarbrooke, J. (1999) *Sustainable Tourism Management*. CAB International, Wallingford, UK.

Taylor, P. (1975) *Principles of Ethics: an Introduction*. Dickenson, Encino, California.

Taylor, S., Simpson, J. and Howie, H. (1998) Financing small businesses. In: Thomas, R. (ed.) *The Management of Small Tourism and Hospitality Firms*. Cassell, London, pp. 58–77.

Thomas, R., Friel, M., Jameson, S. and Parsons, D. (1997) *The National Survey of Small Tourism and Hospitality Firms Annual Report 1996–97*. Centre for the Study of Small Tourism and Hospitality Firms, Leeds Metropolitan University, Leeds.

Tillman, A. (1973) *The Program Book for Recreation Professionals*. Mayfield Publishing, Palo Alto, California.

Timmons, J. (1994) *New Venture Creation*. Irwin, Boston, Massachusetts.

Tims, D. (1996) The perspective of outfitters and guides. In: Driver, B.L., Dustin, D., Baltic, T., Elsner, G. and Peterson, G. (eds) *Nature and the Human Spirit: Toward an Expanded Land Management Ethic*. Venture, State College, Pennsylvania, pp. 177–184.

Tisdell, C. (1996) Ecotourism, economics, and the environment: observations from China. *Journal of Travel Research* 34(4), 11–19.

Trefethen, J.B. (1975) *An American Crusade for Wildlife*. Winchester Press, New York.

Tribe, J., Font, X., Griffiths, N., Vickery, R. and Yale, K. (2000) *Environmental Management for Rural Tourism and Recreation*. Cassell, London.

Tucker, K. and Sundberg, M. (1988) *International Trade in Services*. Routledge, London.

Turner, L. (1976) The international division of leisure: tourism and the Third World. *Annals of Tourism Research* 4(1), 12–24.

Turner, L. and Ashe, J. (1975) *The Golden Hordes*. Constable, London.

UNESCO–UNEP (1988) International Strategy for Action in the Field of Environmental Education and Training for the 1990s. *Congress on Environmental Education and Training*, Moscow, Paris, and Nairobi.

United Nations (1959) *Public Administration Aspects of Community Development Programs*. UN Technical Assistance Program, New York.

US National Parks Net (1999) Death Valley National Park. http://www.nps.gov/deva/. Site accessed 21 June 1999.

van der Smissen, B. (1998) Legal responsibility for recreational safety. In: Dougherty, N.J. IV (ed.) *Outdoor Recreation Safety*. Human Kinetics, Champaign, Illinois, pp. 11–24.

van der Merwe, C. (1996) How it all began: the man who 'coined' ecotourism tells us what it means. *African Wildlife* 50(3), 7–8.

van der Wagen, L. and Davies, C. (1998) *Supervision and Leadership in Tourism and Hospitality*. Hospitality Press, Melbourne.

van Matre, S. (1979) *Sunship Earth: An Acclimatization Program for Outdoor Learning*. American Camping Association, Martinsville, Indiana.

Var, T. (1992) Travel and tourism statistics. *Annals of Tourism Research* 19, 589–592.

Veal, A.J. (1992) *Research Methods for Leisure and Tourism: a Practical Guide*. Longman Group, Essex, UK.

Vinton, D.A. and Farley, E.M. (1979) *Camp Staff Training Series: Knowing the Campers: a Competency-Based Manual for Camp Staff Training*. Project REACH, Lexington, Kentucky.

Vitousek, P.M., Ehrlich, P.R., Ehrlich, A.H. and Matson, P.A. (1986) Human appropriation of the products of photosynthesis. *BioScience* 36, 368–373.

Vogt, C.A. and Fesenmaier, D.R. (1995) Tourists and retailers' perceptions of service. *Annals of Tourism Research* 22(4), 763–780.

Voth, D.E. and Brewster, M. (1989) An overview of international community development. In: Christenson, J.A. and Robinson, J.R. (eds) *Community Development in Perspective*. Iowa State University Press, Ames, Iowa, pp. 280–306.

Wagar, J.A. (1964) The carrying capacity of wildlands for recreation. *Society of American Foresters, Forest Service Monograph* 7, 23.

Wallace, G.N. and Smith, M.D. (1997) A comparison of motivations, preferred management actions, and setting preferences among Costa Rican, North American and European visitors to five protected areas in Costa Rica. *Journal of Park and Recreation Administration* 15, 59–82.

Wallace, R.A., King, J.L. and Sanders, G.P. (1981) *Biology: the Science of Life*. Scott, Foresman, Dallas, Texas.

Walle, A.H. (1995) Business ethics and tourism: from micro to macro perspectives. *Tourism Management* 16(4), 263–268.

Warner, M. (1999) Social capital construction and the role of the local state. *Rural Sociology* 64(3), 373–393.

Warren, N.M. (1998) *Developing Recreation Trails in Nova Scotia*. Nova Scotia Trails Federation, Nova Scotia.

Watson, R. (1996) Risk management: a plan for safer activities. *CAHPERD Journal* (Canadian Association of Health, Physical Education, Recreation and Dance) Spring, 13–17.

Wearing, S. (1995) Professionalisation and accreditation of ecotourism. *Leisure and Recreation* 37(4), 31–36.

Wearing, S. and Neil, J. (1999) *Ecotourism: Impacts, Potential and Possibilities*. Butterworth Heinemann, Oxford.

Weaver, D.B. (2001) Ecotourism in the context of other tourism types. In: Weaver, D.B. (ed.) *The Encyclopedia of Ecotourism*. CAB International, Wallingford, UK, pp. 73–83.

Weaver, D.B. (2002) Hard-core ecotourists in Lamington National Park, Australia. *Journal of Ecotourism* 1(1), 19–35.

Weaver, D.B. and Fennell, D.A. (1997) Rural tourism in Canada: the Saskatchewan vacation farm operator as entrepreneur. In: Page, S.J. and Getz, D. (eds) *The Business of Rural Tourism: International Perspectives*. International Thomson Business Press, London, pp. 77–92.

Weaver, D.B. and Oppermann, M. (1999) *Tourism Management*. John Wiley & Sons, New York.

Weiler, B. (1993) Nature-based tour operators: are they environmentally friendly or are they faking it? *Tourism Recreation Research* 18(1), 55–60.

Weiler, B. (1999) Assessing the interpretation competencies of ecotour guides. *Journal of Interpretation Research* 4(1), 80–83.

Weiler, B. and Ham, S. (2001a) Perspectives and thoughts on tour guiding. In: Lockwood, A. and Medlki, S. (eds) *Tourism and Hospitality in the 21st Century*. Butterworth Heinnemann, Oxford, pp. 255–263.

Weiler, B. and Ham, S. (2001b) Tour guides and interpretations. In: Weaver, D.B. (ed.) *The Encyclopedia of Ecotourism*, CAB International, Wallingford, UK, pp. 549–563.

Wellman, J.D. (1987) *Wildland Recreation Policy: an Introduction*. John Wiley & Sons, New York.

Wheeler, M. (1994) The emergence of ethics in tourism and hospitality. *Progress in Tourism, Recreation, and Hospitality Management* 6, 46–56.

Wheeller, B. (1994) Egotourism, sustainable tourism and the environment – a symbiotic, symbolic or shambolic relationship. In: Seaton, A.V. (ed.) *Tourism: the State of the Art*. John Wiley & Sons, Chichester, UK, pp. 647–654.

White, L. (1967) The historical roots of our ecological crisis. *Science* 155, 1203–1207.

Wiard, B. (1996) How to clean yourself without soiling the backwoods environment. *Backpacker Magazine* May, 136–140.

Wight, P. (1993) Sustainable ecotourism: balancing economic, environmental and social goals within an ethical framework. *Journal of Tourism Studies* 4(2), 54–66.

Wilderness Inquiry (1999) *Wilderness Inquiry Trail Staff Policy Manual*. Wilderness Inquiry, Minneapolis, Minnesota.

Wilks, J. and Atherton, T. (1994) Health and safety in Australian marine tourism: a social, medical, and legal appraisal. *The Journal of Tourism Studies* 5(2), 2–16.

Williams, H. and Watts, C. (2002) *Steps to Success: Global Good Practices in Tourism Human Resources*. Prentice Hall, Toronto.

Williams, P. (1992) A local framework for ecotourism development. *Western Wildlands* 18(3), 14–19.

Windsor, R., Baranowski, T., Clark, N. and Cutter, G. (1994) *Evaluation of Health Promotion, Health Education, and Disease Prevention Programs*. Mayfeld, Mountain View, California.
Wiseman, J. (1986) *The SAS Survival Handbook*. HarperCollins, London.
Wolfe, R.I. (1964) Perspectives on outdoor recreation: a bibliographical survey. *Geographical Review* 54, 203–238.
Wolfe, R.I. (1966) Recreational travel: the new migration. *Canadian Geographer* 10(1), 1–14.
Woodley, S. (1993) Tourism and sustainable development in parks and protected areas. In: Nelson, J.G., Bulter, R.W. and Wall, G. (eds) *Tourism and Sustainable Development: Monitoring, Planning, Managing*. Department of Geography Publication Series Number 37. University of Waterloo, Ontario, pp. 83–96.
World Commission on Environment and Development (1987) *Our Common Future*. Oxford University Press, Oxford.
World Resources Institute (1992) *Global Biodiversity Strategy*. WRI, Washington, DC.
Worthen, B.R., Sanders, J.R. and Fitzpatrick, J.L. (1997) *Program Evaluation: Alternative Approaches and Practical Guidelines*. Longman, New York.
Wyatt, S. (1997) Dialogue, reflection, and community. *The Journal of Experiential Education* 20(2), 80–85.
Young, B. (1983) Touristization of a traditional Maltese fishing-farming village: a general model. *Tourism Management* 4(1), 35–41.
Young, O.R. (1995) The problem of scale in human/environment relationships. In: Keohane, R.O. and Ostrom, E. (eds) *Local Commons and Global Interdependence*. Sage, London, pp. 27–45.
Yukl, G.A. (1989) *Leadership in Organizations*, 2nd edn. Prentice Hall, Englewood Cliffs, New Jersey.
Zimbardo, P.G. (1979) *Psychology and Life*, 10th edn. Scott, Foresman, Dallas, Texas.
Zimmerer, K.S. (1994) Human geography and the 'new ecology': the prospect and promise of integration. *Annals of the Association of American Geographers* 84 (1), 108–125.
Zimmermann, E.W. (1933/1955) *World Resources and Industries*. Harper and Brothers, New York.
Zuckerman, M. (1979) *Sensation Seeking: Beyond the Optimal Level of Arousal*. LEA, Hillsdale, New Jersey.
Zurick, D.N. (1992) Adventure travel and sustainable tourism in the peripheral economy of Nepal. *Annals of the Association of Geographers* 82(4), 608–628.

Appendix 1

Older Adult Orienteering Questionnaire*

The following questions are designed to uncover some basic attitudes and feelings towards the orienteering. Please either circle the appropriate numbers, or [✓] tick the appropriate box(es) for each of the following questions.

1. Please indicate the extent to which you agree or disagree with each reason as it relates to ***your potential*** **for participation in orienteering**, according to the following 5 categories: **1 = strongly disagree; 2 = moderately disagree; 3 = neutral; 4 = moderately agree; 5 = strongly agree**. Please circle only 1 number for each item.

a. I would not have enough time to participate in orienteering . . 1 2 3 4 5
b. I may have to travel too far to participate in orienteering . . . 1 2 3 4 5
c. I would not feel safe while participating in orienteering . . . 1 2 3 4 5
d. Because of my health, I would be unable to participate . . . 1 2 3 4 5
e. I don't know anyone with whom I could participate . . . 1 2 3 4 5
f. I wouldn't have a way to get to the facilities to participate . . . 1 2 3 4 5
g. My family or household wouldn't like me to participate . . . 1 2 3 4 5
h. I worry that I wouldn't be able to afford to participate . . . 1 2 3 4 5

- Would your participation in orienteering be limited because of ... (Tick all that apply)

[ ] **Mobility** problems [ ] **Agility** problems [ ] **Eyesight** problems
[ ] **Hearing** problems [ ] **Speaking** problems [ ] Not applicable

3. Are you interested in learning or improving your skills in orienteering?

[ ] Yes [ ] No

*If no, please briefly state why? (Then proceed to question 11.)

____

____

*If yes, please continue through the following questions.

4. In general, why are you interested in orienteering? (Tick one only)

[ ] Fun/recreation [ ] Competition [ ] Both [ ] Neither

5. With whom would you most probably choose to participate in orienteering? (Tick one only)

[ ] Friends [ ] Family [ ] Individually [ ] Doesn't matter

*Reduced in size for the purposes of this book.

6. How far would you be willing to walk and/or jog during an orienteering event?

[ ] Up to a $\frac{1}{4}$ mile [ ] Up to a $\frac{1}{2}$ mile [ ] Up to 1 mile
[ ] Up to $1\frac{1}{2}$ miles [ ] Up to 2 miles [ ] More than 2 miles

7. What would be the longest amount of time that you would want to be involved in jogging/walking through an orienteering course?

[ ] Up to a half hour [ ] Up to an hour [ ] Up to 90 minutes
[ ] Up to 2 hours [ ] As long as it takes

8. Would you be willing to devote between 2 and 3 hours per week (for 3 or 4 weeks) to learn about orienteering?

[ ] Yes [ ] No

9. The cost for a 12-hour seminar on orienteering will be $25.00. Based on this fee, would you be interested in participating?

[ ] Yes [ ] No

10. When is the best time of the week for you to attend such a seminar? (Please tick one only.)

[ ] Week**end** [ ] Week**day** [ ] Both

11. What is your gender? [ ] Male [ ] Female

12. What is your age? ____ years

13. Marital status? [ ] Never married [ ] Married
[ ] Divorced/separated [ ] Widowed

14. What is your retirement status? (Please circle one)

Fully retired Partially retired
Fully employed Not applicable

Thank-you for participating in this study.

Appendix 2
Medical Form

Personal information

Name: ______________________ Age: ______________

Social insurance no.: ______________ Health no.: ____________

Blood type: ______________________

In case of emergency, contact (name, address, phone no.): ______________

Medical information

Family doctor (name, address, phone no.): ______________________

____

Do you have allergies? Yes ❑ No ❑

If yes, please list what these allergies are: ______________________

____

Are you currently on medication? Yes ❑ No ❑

If yes, please identify and the reason why: ______________________

____

Current health status. Please indicate if you have any physical conditions that would interfere with your participation in this trip. If necessary, consult your physician.

Hearing/vision problems (apart from glasses)	Yes ❑ No ❑
Respiratory problems	Yes ❑ No ❑
Back problems	Yes ❑ No ❑
Joint problems (ankle, hips, knees, etc.)	Yes ❑ No ❑
Any recent serious illness	Yes ❑ No ❑
Any recent hospitalizations	Yes ❑ No ❑
Frequent muscle cramps	Yes ❑ No ❑
High or low blood sugar	Yes ❑ No ❑
Seizure disorders	Yes ❑ No ❑
Other:	Yes ❑ No ❑

Appendix 3
Packing Information*

Clothing

Leggings	Shorts may be worn in most places; however, long cotton or nylon field trousers are advisable, especially when in some of the forested areas and trails. Make sure to bring a bathing suit. Consider long underwear in colder areas.
Shirts	Cotton or nylon, short-sleeved or T-shirts, at least one long-sleeved shirt to keep sun and insects off. Consider long underwear in colder areas.
Hat	A wide-brimmed or peaked hat for sun protection.
Sweater	Light-weight sweater or fleece jacket for cooler evenings.
Raingear	Waterproof gear is a must.
Shoes and socks	Trainers or walking boots with a good solid grip and cotton or man-made socks. Bring two or three pair of shoes depending on conditions.

Accessories

Daypack	Bring a small backpack to carry water bottles, extra socks, moist travel wipes, field guides, film and personal items.
Sunscreen	We will be spending most of the daylight hours in the field, so make sure you have sunglasses, a good sunblock (at least 15), and lip balm.
Torch (flashlight)	Bring a strong torch (and extra batteries) in case there are electrical shortages or there is a night walk.
Water bottle	Keep a water bottle in your pack while travelling and participating in activities.
Binoculars	For nature tours, binoculars are very important.
Insect repellent	Very important especially in areas that are moist, and during the evening.
Miscellaneous	A spare sink stopper, a wash cloth/J-cloth, beach towel, soap, clothes-line, a utility knife, moist travel wipes, spare eyeglasses.
Personal health	Bring sufficient quantities of any medications required, as well as: antibiotic cream, antacid and anti-diarrhoea tablets, cold and flu medications, aspirin or equivalent, adhesive bandages or moleskin, calamine lotion or after-bite cream.
Equipment	If taking expensive camera or video equipment, make a list of these items along with description and serial numbers. It is also good practice to register new equipment with Customs, or carry a copy of the original receipt to avoid problems when bringing it back.

*Clothes will vary depending on the trip and region. Adapted from Quest Nature Tours brochure.

Appendix 4
The First Aid Kit*

Recommended items	Quantity	Use
Adhesive bandages, 1 or 2 cm ($\frac{1}{2}$″ or $\frac{3}{4}$″)	20–30	For cuts and abrasions
Cloth tape, 2.5 cm (1″) roll	1	To secure splints
Gauze bandage, 5 cm (2″) roll and 10 cm (4″) roll	1 each	To secure dressings
Gauze pads, 10 cm (4″) size	6–10	To cover abrasions
Moleskin or molefoam, 15 cm (6″) square	1	To prevent and treat blisters
Triangular bandage, 1 m (36″)	1	Multiple uses
Elastic bandage, 10 cm (4″)	1	To support sprains and secure dressings
Aspirin, 325 mg tablets	30–50	For pain and fever
Antibiotic ointment, 15 mg tube	1	To prevent skin infections
Antihistamine, 8 mg chlorphenamine or 25 mg acrivastine (Benadryl) tablets	6–10	To counter allergic reactions
Decongestant with antihistamine tablets	6–10	For nasal congestion
Ibuprofen 200 mg or paracetamol (Tylenol) with codeine (32 mg)	6–10	To treat pain
Iodine solution, 2% tincture or 10% Betadine	1 oz	For wound disinfection
Thermometer	1	To measure body temperature
Lip balm with sunscreen	1 tube	To prevent sun damage
Insect repellent	1 oz tube	To discourage insect bites
Wire-mesh splint	1	For a cervical collar or splints
Sawyer extractor	1	Removal of venom from insect/snake bites

In addition, see van der Smissen (1998). The following are suggested medicines taken from a medical journal article *Post Graduate Medicine* (1985) 78(2). Travel Medicine. Each of the first three items requires a prescription. Proprietary names are in parentheses.

Suggested items	Quantity	Use
Cavit (obtain from a dentist)	7 g tube	For tooth fillings
Acetazolamide (Diamox), 250 mg tablets	14	For acute mountain sickness
Promethazine (Phenergan), 25 mg suppos.	4	For nausea and vomiting
Loperamide (Imodium), 2 mg tablets	12/person/week	For diarrhoea
Phenylephrine (Neo-Synephrine)	15 g tube	Nasal congestion, sinusitis, nasal bleeding
Bacitracin-polymyxin ointment	30 g tube	For burns, abrasions, blisters
Pseudoephedrine (Sudafed) tablets	12–24	For nasal congestion
Corticosteroid cream	15 g tube	For rashes, swelling and itching from plant contact or insect bites

*Source: Murdock and Trygstad (1993).

Appendix 5a
Injury Checklist

Casualty's full name ______________________________ Age __________

Address ______________________________ Medical no. __________

Family physician ____________________ Closest relative ______________

Airway OK? __________ Breathing OK? __________ Pulse present? __________

Neck spine OK? __________ Excessive bleeding stopped? ______________

What happened? ______________________________

____

What signs do you see? ______________________________

What does the casualty feel? ______________________________

Head-to-Toe Examination

Watch his face as you examine him. Note all tenderness, bruising, deformity, bleeding, swelling, loss of use (ANYTHING UNUSUAL). Work down the list.

Head ____________________

Eyes ____________________

Ears ____________________

Nose ____________________

Mouth ____________________

Face ____________________

Face colour ____________________

Breath odour ____________________

Sweating ____________________

Neck ____________________

Neck veins ____________________

Collarbones ____________________

Ribs ____________________

Chest rise – both sides ____________________

Belly ____________________

Tightness (where?) ____________________

Bloating ____________________

Pelvis ____________________

Messed pants ____________________

Legs ____________________

Joints normal ____________________

Foot colour ____________________

Foot temperature (compare) ____________________

Feeling in feet ____________________

Equal strength in feet ____________________

Arms/hands ____________________

Joints normal ____________________

Hand colour ____________________

Hand temperature (compare) ____________________

Feeling in hands ____________________

Equal grip ____________________

Medic alert ____________________

Notes on above ______________________________

____

____

Vital Signs

Time												
Pulse												
Breath rate												
Consciousness												
Pupils												
Skin												
Temperature												

Current medical conditions ______________ Current medications ____________________

Significant past history __________________ Allergies ____________________________

What do you think is wrong? ____

____

____

Treatment Given – Casualty Condition Record

Describe exact treatment and the casualty's condition at the time. Keep a complete record.

Time/date: ____

____

____

Source: Merry, 1994.

Appendix 5b
Illness Checklist

Casualty's full name ______________________ Age __________
Address ______________________ Medical no. __________
Family physician ______________ Closest relative ______________
What is the main complaint? ______________________________
____
Describe pain (e.g. sharp, dull, crushing, etc.)______________________
When did it start?______________ Did it start gradually/suddenly?__________
Timing of pain (e.g. continuous, comes & goes, how often, how long between) __________
____
Does it move outward?______________ (Where to where?) ______________
When did it start?______________ (Sudden or gradual?) ______________
Has this happened before? ______________________________
What makes it feel better/worse? ______________________________
Other complaints? ______________________________
Is a doctor being seen now?______________ For what? ______________
Taking any medicines?______________ What kind? ______________
____
How much, how often? ______________________________
Allergies? ______________ To what?______________

Head-to-toe Examination

Head ______________ Lungs fill equally ______________
Eyes ______________ Messed pants ______________
Ears ______________ Unusual discharge______________
Nose ______________ Hands/arms ______________
Mouth ______________ Colour/temp. ______________
Face colour ______________ Equal strength ______________
Breath odour ______________ Feeling in hands/arms ______________
Sweating ______________ Legs/feet ______________
Neck ______________ Colour/temp. ______________

Neck veins ______________________ Equal strength ______________________

Belly – tenderness ______________________ Feeling in legs/feet ______________________

Belly – tightness ______________________ Medic alert ______________________

Bloating ______________________

Notes on above ______________________

Other things observed ______________________

____

____

Vital Signs

Time												
Pulse												
Breath rate												
Consciousness												
Pupils												
Skin												
Temperature												

What do you think is wrong?______________________

____

Treatment Given – Casualty Condition Record

Describe exact treatment and the casualty's condition at the time. Keep a complete record.

Time/date: ______________________

____

____

Source: Merry, 1994.

Appendix 5c
Consciousness Record

Casualty's full name ______________________________ Age __________

Address ______________________________ Medical no. __________

Family physician ____________________ Closest relative ____________________

Check the casualty's condition at least once every hour. Record sudden changes immediately. Send or take this record and the injury or illness checklist with the casualty to medical help.

Time															
Glasgow Coma Scale															
Eyes opening	Spontaneous	4													
	To voice	3													
	To pain	2													
	None	1													
Motor response	Obeys command	6													
	Localized pain	5													
	Withdraw pain	4													
	Flexion (pain)	3													
	Extension (pain)	2													
	None	1													
Verbal response	Oriented	5													
	Confused	4													
	Inappropriate words	3													
	Incomprehensible sounds	2													
	None	1													
Glasgow Coma Scale total															

Time												
Pulse rate (per minute)												
Breathing Rate (per minute)												
Skin colour												
Nausea, vomiting												
Sees clearly (yes/no)												
Personality change												
Unequal pupil size or reaction												
Twitching or seizures												

Source: Merry, 1994.

Appendix 6
Survival Kit

Suggested survival kit. The recommended survival kit is packed in a light-weight watertight metal container which could also be used as an emergency cooking pot. Because of its size, about 10 cm by 13 cm, it can easily be attached to your belt. Should you fall into a river or through ice, you may have to discard your backpack, but you'll still have your survival kit, which consists of:

1. Medium compress bandage
2. Small compress bandage
3. Several large adhesive bandages
4. Small adhesive bandages (5)
5. Dental floss
6. Curved needle
7. Sting stop (2)
8. Painkiller (e.g. aspirin 375 mg + codeine 8 mg) (12)
9. Aluminium foil
10. Salt (several)
11. Coffee (2)
12. Dextrose – plain
13. Dextrose – chocolate
14. Knife (tape protecting edge)
15. Spoon
16. Soups – chicken and beef (23)
17. Water purification tablets (e.g. Halazone) (12)
18. Tea (three varieties) (5)
19. Plastic bags (2 large garbage or tube size and 2 small)
20. Signalling mirror
21. Emergency saw
22. Wire for suspending cooking container
23. 6 m nylon shroud cord
24. 15 m fishing line
25. Small hooks (20)
26. Medium hooks (8)
27. Compass
28. Whistle and chain (under freezing conditions – plastic)
29. Space blanket
30. Single-edge razor blade
31. Snare wire – 8 m wrapped around knife
32. Solid fuel, barbecue starter (fire lighting)
33. Flint and steel
34. Magnifying glass
35. Waterproof container with cotton wool
36. Tin package for kit and cooking container – sealed with vinyl tape
38. Soap
39. Ground-to-air signal card
40. Spare prescription eye glasses (if applicable)

Source: Province of British Columbia, 1981.

Appendix 7

Checklist for Effective Leadership of Hiking and Backpacking Daytrip

Hiking and Backpacking

Daytripping

Experience – Has at least 10 days personal and/or leadership hiking and/or backpacking experience over the last 5 years. [ ]

Fitness – The level of cardiovascular and muscular endurance required will vary with the duration and intensity of the hike planned, but they must be well over and above that required to complete the trip. The leader must have sufficient mental and physical energy reserves to deal with any emergencies occurring at or near the end of the day. [ ]

Navigation – Has travelled the route previously and/or studied topographical map and route report and talked to reliable others who have been there within the preceding year. While pre-travel of routes is desirable, this is not always feasible or necessary. [ ]

- Based on previous experience in the area and/or map reading, can select a safe and appropriate route for the group and the time available. Is aware that maps may be out of date or otherwise in error and is prepared to make route readjustments in the field. [ ]
- Must have strong map reading and compass skills if going off-trail and/or in unfamiliar terrain (see Navigation and Guidance, p. 23). [ ]

Environmental factors – Is aware of any potentially hazardous spots along the route (e.g. ledge walks, creek crossings, etc.) and is prepared to deal with these. (e.g. avoidance, making hand rails or bridges, giving cautions, etc.) [ ]

Has a strong environmental ethic and knows how to lead a minimal impact hike (e.g. avoiding multiple trailing, not unnecessarily disturbing wildlife or vegetation, dealing with human and other wastes properly, etc.). [ ]

Emergency training for:

Physical injury – Knows ABC's of basic life support; can deal with interruptions in airway, breathing and circulation. Cardiopulmonary resuscitation (CPR) training is highly desirable. [ ]

- Can prevent, recognize and treat common hiking related injuries (e.g. blisters) and conditions (e.g. dehydration, sunburn, hypothermia, etc.). [ ]
- Knows how to deal with conditions specific to group members (i.e. knows how to deal with an epileptic seizure, diabetic reaction or allergies if participants with these conditions are to be present). [ ]
- Knows how to deal with hazards unique to the area (e.g. poisonous snakes, insects, plants, etc.) [ ]
- Knows evacuation routes and location of nearest vehicles, telephone, residents and support services. Has phone numbers of support services. [ ]

Lost participant – Has an understanding of basic search procedures, demarcation of search areas and allocation of priorities. Can assume a leadership role in organizing available people toward finding a lost member without endangering them also. [ ]

Group lost or stranded – If the possibility exists of becoming lost or otherwise delayed so as to be caught out overnight, the leader must be prepared to employ his available resources to shelter the group, keep them warm and set up a distress signal if necessary. [ ]

Source: Hanna, 1986.

Appendix 8
Sample Waiver Form

This is a sample of the type of waiver form often used by outfitters in western Canada. **Do not use verbatim for your organization without consulting your lawyer.**

RELEASE OF ALL CLAIMS AND WAIVER OF LIABILITY

BY SIGNING THIS YOU GIVE UP CERTAIN RIGHTS, INCLUDING THE RIGHT TO SUE.
(*This is plain language, makes signer take notice!*)

TO: DEATH MOUNTAIN OUTDOOR EPICS LTD and HER MAJESTY THE QUEEN IN RIGHT OF THE PROVINCE OF …

(*The province is mentioned so that they cannot be sued on the basis of such ridiculous, but legally possible arguments as, 'they are the landowners where it happened' or 'they were allowing the operator to operate.'*)

In conideration of my decision (*'since I have decided'*) to attend a course or outdoor excursion led or instructed by DEATH MOUNTAIN OUTDOOR EPICS LTD., I AGREE to this Release of Claims, Waiver of Liability and Assumption of Risks (collectively 'this AGREEMENT').

I WAIVE ANY AND ALL CLAIMS I may have against, and RELEASE FROM ALL LIABILITY, and AGREE NOT TO SUE DEATH MOUNTAIN OUTDOOR EPICS LTD AND THE PROVINCE AND THEIR DIRECTORS, OFFICERS, EMPLOYEES, GUIDES, AGENTS AND REPRESENTATIVES, (henceforth 'The Releasees') for ANY PERSONAL INJURY, DEATH, PROPERTY DAMAGE OR LOSS sustained by me as a result of my participation in any kind of activity, course or trip with 'The Releasees', DUE TO ANY CAUSE WHATSOEVER, INCLUDING WITHOUT LIMITATION, NEGLIGENCE ON THE PART OF THE RELEASEES.

INITIALS ....

(*Initial to indicate the signer has read it and understands there is no one he can sue, even if people are negligent.*)

I AM AWARE that outdoor recreational activities have certain inherent RISKS associated with being in the natural environment. These include, but ARE NOT LIMITED TO, rapid and extreme weather changes, terrain hazards and obstacles, wild animals, other people, fast water, changes in river beds and rapids, falling trees, avalanches, floods, use of stoves and possible adverse effects of food. I also understand that the outdoor environment is continually changing and that Death Mountain Outdoor Epics' staff and the Releasees may fail to predict natural hazards, including, but not limited to, unstable snow, avalanches, rockfalls and water hazards. I am also aware that ACTIVITIES MAY TAKE PLACE IN LOCATIONS WHERE MEDICAL TREATMENT OR RESCUE MAY NOT BE AVAILABLE due to such factors as remoteness, communication difficulties, poor travelling conditions, bad weather, etc.

(*Puts some onus on the participant, and also prevents them bringing forward stupid, but potentially legally acceptable arguments, such as 'I didn't know there were wild animals in the park.' A waiver for a specific activity might go into a lot more detail of the hazards of that activity.*)

I FREELY ACCEPT ALL THE RISKS associated with the course or trip in which I have agreed to participate.

INITIALS ....

(*Initial to indicate they have read this paragraph. 'Freely accept' is a key phrase.*)

I ACKNOWLEDGE that the inherent risks of many outdoor activities contribute to the enjoyment of such activities.

I am freely entering into this Agreement, and in doing so I am NOT RELYING ON ANY ORAL OR WRITTEN STATEMENTS made by the Releasees, including those in brochures or letters of correspondence, to induce me to participate in a course or excursion organized, led or instructed by 'the Releasees'.

(*'Freely entering' is a key phrase – if people are forced into signing a waiver, it may be less valid. Many outfitters now send out waivers with booking information, and may require them to be signed at that time. If you sign at the outfitter's location after booking and at the start of your trip, there could be deemed to be some coercion.*)

I confirm that I am nineteen or more years of age and that I have READ AND UNDERSTAND THIS AGREEMENT PRIOR TO SIGNING IT and that this Agreement shall be BINDING UPON MY HEIRS, NEXT OF KIN, EXECUTORS AND SUCCESSORS.

Signed this .... day of ...., 19.......

In the presence of Witness (Parents if younger than 19 years of age):

Signature: .... Witness: ....

Please PRINT full name and mailing addresses:

....

....

....

(*The witness is important, and a prudent operator will be very careful to check that all the signing is done completely and correctly. Some people seem to try to hand in incomplete waivers and operators must watch for this!*)

Source: Morton, 1997.

Appendix 9
Ozark Ecotours Reservation Form

Full day Ecotours start at $50 and include lunch in a dramatic outdoor setting. All Ecotours meet at the Tourist Information Center just off the Jasper square. Check schedule for departure times and prices. You can reserve a tour in one of several ways.

1. Pick up the telephone and call 1–877–622–5901. Our staff can book your tour with major credit card. This is the very best way, due to the limited number of participants on each tour.

2. You can print out the form below on your printer and mail in with payment. Conformations will be made by phone to your daytime number.

- - - - - - - - - - - - - - -snip -

Mail to: Ozark Ecotours, Post Office Box 513, Jasper, AR 72641-0513, Phone 877-622-5901, Fax 870-446-2701

Reservation Form

Name: ____________________

Address: ____________________

City: ____________________

State: ____________________

Zip: ____________________

Daytime phone no. ____________________

Ecotour no. ____________________

Tour Name ____________________

Tour Date ____________________

Debit/Credit Card ____________________

Expiry Date ____________________

Card type ____________________

Card number ____________________

***Policy:** If you cancel 10 days or more in advance of your Ecotour you have the option of signing up for one of this seasons remaining tours or receiving a refund minus a $10 Reservation fee. If you cancel 9 days or less before your Ecotour you may sign up for one of this seasons remaining tours, no money will be refunded. Tour and/or tour guides may be substituted due to events beyond our control.*

Source: Ozark Ecotours http://www.ozarkecotours.com/reservation.html

Appendix 10
Birds of Trinidad and Tobago – Itinerary

Day 1: Arrive in Trinidad

On our evening arrival at Piarco Airport at Port-of-Spain; we'll clear customs and drive to Asa Wright Nature Centre and Lodge where a welcome fresh juice or rum punch will be awaiting us. We'll gather together in the Main House to get acquainted and discuss the week's programme. Note that flight schedules vary. We may adjust the programme to match our designated group flight. Overnight at Asa Wright Nature Centre and Lodge.

Day 2: Asa Wright Nature Centre

We'll awaken this morning to the raucous calls of crested oropendulas and a host of other exotic sounds, and the wonderful aroma of fresh-brewed coffee. Drinking morning coffee on Asa Wright's veranda is a memorable experience. A first-time visitor may see 20–30 new birds before breakfast. On our introductory walk, we'll most likely see, among others, such bird species as violaceous trogon, channel-billed toucan, chestnut woodpecker, white-bearded manakin, bearded bellbird, rufous-browed peppershrike, turquoise and bay-headed tanager.

The afternoon will be free of organized activities so we can relax, enjoy the trails and adjust to the tropical sun and heat. It is most likely that we'll end up on the veranda of the Main House where the feeders and excellent views into the surrounding trees provide one of the world's most pleasant and exciting ornithological experiences. We'll be on the lookout for such regular visitors as ruby-topaz hummingbird, tufted coquette, whitechested emerald, barred antshrike, and green honeycreeper. In the evening there may even be time for a night walk to listen for spectacled owl and semi-collared nighthawk, and tree frogs. Overnight at Asa Wright Nature Centre and Lodge.

Day 3: Aripo Savannah/Arena Forest

This all-day journey to the lowland Aripo Savannah and Arena Forest could provide yet another batch of new species. In the savannah, we'll look for plumbeous kite, greenrumped parrotlet, red-breasted blackbird, savannah hawk, pearl kite, ruby-topaz hummingbird, masked yellowthroat, yellow rumped cacique, blue-black grassquit and ruddy seedeater. The Aripo Savannah is an area of extremely acidic soil with poor drainage, and is a remnant of what was a major habitat type of lower Trinidad. This soil supports sundew and other interesting plants. These remnant savannahs are now surrounded by extensively altered landscapes where much sugar cane is grown, as well as small-scale agriculture and housing development sprawl. Eventually, we will reach the Arena Forest where we may see short-tailed swift, white-bellied antbird and Trinidad euphonia. Overnight at Asa Wright Nature Centre and Lodge

Day 4: Caroni Swamp

This morning we will continue our exploration of the trails at Asa Wright. We will be hoping for ornate hawk-eagle, white hawk, blue-headed and orange-winged parrots and black-faced antthrush. After lunch, we'll depart for the famous Caroni Swamp and its spectacular avian highlights. Neotropical cormorant, anhinga, striated heron, green-throated mango, pied water-tyrant, bicolored conebill, straight-billed woodcreeper, yellow-headed caracara and red-capped cardinal are among the many new species we'll hope to see here. The Caroni is a very specialized mangrove forest that contains several genera and species of mangroves and shows classic examples of plant adaptation in a brackish water community. We might also see golden tree boa and silky anteaters resting in the branches. As we continue through the swamp in our outboard motor-powered boat, our day will end at dusk as we sit and watch the spectacular flights of scarlet ibis as they return to their mangrove-island roosts. Truly, one of the most dramatic and spine-tingling natural moments one could experience. As we return to the dock, we'll use our spotlight to search the mangrove-lined channels for boat-billed herons and the mysterious sounding common potoo. Overnight at Asa Wright Nature Centre and Lodge.

Day 5: Blanchisseuse

We'll travel north today up the valley to the main ridge of Trinidad's Northern Range, where the elevation reaches about 6000 m. We'll be on the lookout for species that are not as likely to be found at Asa Wright: bat falcon, blue-headed parrot, collared trogon, rufous-breasted wren, speckled tanager and many others. From the top we will make our way to the coast, where we may enjoy a swim in the ocean and observe caracara, pygmy kingfisher or dray kingbird. King vultures have been seen here regularly in recent years. Overnight at Asa Wright Nature Centre and Lodge.

Day 6: Asa Wright Nature Centre

Before breakfast, we'll again watch the many species arriving at the feeders overlooking the valley. Then we'll hike the trails to Dunstan Cave, a beautiful riparian grotto located on the grounds of the sanctuary, to view a breeding colony of the fascinating, nocturnal oilbird.

After our visit to the caves, our day will continue with a tour of the several trails that traverse this rich and diverse wildlife sanctuary. Photographers may wish to photograph the hummingbirds and tanagers as they feed at the many feeders spotted around the grounds. We'll walk the trails to the more enclosed forest sections, where we will look for the leks of white-bearded and golden-headed manakins and we'll scan the canopy and mid-level vegetation to look for channel-billed toucans and the bizarrely adorned bearded bellbird. A natural rock pool lined with ferns and bamboo may tempt you for a refreshing dip in the heat of the day. Overnight at Asa Wright Nature Centre and Lodge.

Day 7: Tobago/Grafton Sanctuary/Buccoo Reef

Today we will take a short morning flight to Tobago. We will make our way to Bon Accord Lagoon, on the Bon Accord estate, home to crabs, waterfowl and waders. This small but important wetland is the habitat of white-cheeked pintail, copper-rumped and ruby topaz hummingbirds, bare-eyed thrush, scrub greenlet and black-faced grassquit. After a picnic lunch we will visit the Grafton Caledonia Wildlife Sanctuary. Following the devastation of

Hurricane Flora in 1963, this former cocoa plantation has been allowed to return to its natural state and is now a wildlife sanctuary. As we walk the forest trails, we will be looking for rufous-vented chacalaca, blue-crowned motmot, blue-backed manakin, white-fringed antwren and white-tailed sabrewing, while keeping an eye open for green iguana. We will end our day at Blue Waters Inn, pleasantly located on the edge of a sand beach. Overnight at Asa Wright Nature Centre and Lodge.

Day 8: Gilpin Trace/Roxborough Road

Gilpin Trace is a trail in what may be the oldest protected rainforest reserve in the western hemisphere; the Bloody Bay Rain Forest Reserve. This reserve was created by the island's governors in 1764 to 'preserve the rains', and is a sanctuary for an amazing variety of plant and wildlife species. The trail is at least 1000 years old and descends steeply toward the Caribbean. As we walk in the reserve, we will be looking for rufous tailed jacamar, blue-backed manakin, orange-winged parrots, collared trogon, olivaceous wood-creeper and gray-throated leafscraper. Overnight at Asa Wright Nature Centre and Lodge.

Day 9: Little Tobago Island

In the morning, we will visit Little Tobago Island, one of the most important seabird sanctuaries in the Caribbean. This island lies about 3.5 km offshore and is the nesting ground of the lovely red-billed tropicbird, as well as substantial colonies of brown boobies, Audubon shearwaters, bridled and sooty terns, magnificent frigatebird and numerous other species. We'll travel there by glass-bottomed boat and admire the spectacular fish, coral and sponges in the shallow and clear water. There is good hiking on the island and we'll climb to see the spectacular view of the Caribbean and Atlantic convergence, where red-billed tropicbirds swoop over the cliffs. After lunch, we'll have time to enjoy swimming and snorkelling off the beach at the Inn, or walk up the hill behind the Inn. Overnight at Asa Wright Nature Centre and Lodge.

Day 10: Departure Day

After an early breakfast, we'll transfer to the airport at Crown Point for the short flight back to Port of Spain where we will connect with our flights home. We may adjust the programme on the last day to match changing flight schedules.

Index

Page references to tables, charts and other illustrations are given in *italic*, with the exception of references to 'programming tips', which are indicated as such in the entry.